5th EDITION

Essential Mathematics with Applications

Vernon C. Barker
Palomar College

Richard N. Aufmann
Palomar College

Joanne S. Lockwood
Plymouth State College

HOUGHTON MIFFLIN COMPANY

Boston   New York

Senior Sponsoring Editor: Maureen O'Connor
Senior Associate Editor: Dawn M. Nuttall
Project Editor: Christina Lillios
Senior Production/Design Coordinator: Carol Merrigan
Senior Manufacturing Coordinator: Marie Barnes
Editorial Assistant: Lauren Gagliardi

Cover designer: Harold Burch Design, NYC

Photo Credits

Page 1, Matthew Borkoski/Stock, Boston, Inc.; page 61, Will & Deni McIntyre/Photo Researchers, Inc.; page 121, Gary Landsman/The Stock Market; page 167, Oliver LeClerc/Gamma Liaison; page 195, Ben Osborne/TSI; page 227, Jeff Greenberg/Photo Researchers, Inc.

Art Credits

Page 39, (table sources) USA TODAY, September 15, 1993 and Craig Carter, *Complete Baseball Record Book*—1994 (The Sporting News); page 58, (figure) copyright © 1993, USA TODAY. Reprinted with permission.; page 90, (top figure) reprinted by permission of Wall Street Journal, © 1997 Dow Jones & Company, Inc. All rights reserved; page 127, (figure) reprinted courtesy of West Shore Acres Display Garden; page 130, (top figure) copyright © 1993, USA TODAY. Reprinted with permission.

Printed in the U.S.A.

ISBN Number: 0-395-90710-1

3456789-WC-02 01 00 99

Contents

3 Decimals   121

4 Ratio and Proportion 167

5 Percents 195

Preface

The fifth edition of *Essential Mathematics with Applications* provides mathematically sound and comprehensive coverage of the topics considered essential in a basic college mathematics course. The text has been designed not only to meet the needs of the traditional college student but also to serve the needs of returning students whose mathematical proficiency may have declined during years away from formal education.

In this new edition of *Essential Mathematics with Applications*, we have continued to integrate some of the approaches suggested by AMATYC. Each chapter begins with a mathematical vignette in which there may be a historical note, application, or curiosity related to mathematics. At the end of each section there are "Applying the Concepts" exercises that include writing, synthesis, critical thinking, and challenge problems. At the end of each chapter there is a "Focus on Problem Solving" that introduces students to various problem-solving strategies. This is followed by "Projects and Group Activities" that can be used for cooperative learning activities.

One of the main challenges for students is the ability to translate verbal phrases into mathematical expressions. One reason for this difficulty is that students are not exposed to verbal phrases until later in most texts. In *Essential Mathematics with Applications*, we introduce verbal phrases for operations as we introduce the operation. For instance, after addition concepts have been presented, we provide exercises which say "Find the sum of ...." or "What is 6 more than 7?" In this way, students are constantly confronted with verbal phrases and must make a mathematical connection between the phrase and a mathematical operation.

INSTRUCTIONAL FEATURES

Interactive Approach

Essential Mathematics with Applications uses an interactive style that provides a student with an opportunity to try a skill as it is presented. Each section is divided into objectives, and every objective contains one or more sets of matched-pair examples. The first example in each set is worked out; the second example, called "You Try It," is for the student to work. By solving this problem, the student practices concepts as they are presented in the text. There are complete worked-out solutions to these examples in an appendix at the end of the book. By comparing their solution to the solution in the appendix, students are able to obtain immediate feedback on and reinforcement of the concept.

Emphasis on Problem-Solving Strategies

Essential Mathematics with Applications features a carefully developed approach to problem solving that emphasizes developing strategies to solve problems. Students are encouraged to develop their own strategies, to draw diagrams, and to write strategies as part of their solution to a problem. In each case, model strategies are presented as guides for students to follow as they attempt the "You Try It" problem. Having students provide strategies is a natural way to incorporate writing into the math curriculum.

Emphasis on Applications

The traditional approach to teaching algebra covers only the straightforward manipulation of numbers and variables and thereby fails to teach students the practical value of algebra. By contrast, *Essential Mathematics with Applications* contains an extensive collection of contemporary application problems. Wherever appropriate, the last objective of a section presents applications that require the student to use the skills covered in that section to solve practical problems. This carefully integrated applied approach generates student awareness of the value of algebra as a real-life tool.

Completely Integrated Learning System Organized by Objectives

Each chapter begins with a list of the learning objectives included within that chapter. Each of the objectives is then restated in the chapter to remind the student of the current topic of discussion. The same objectives that organize the text are also used as the structure for exercises, testing programs, and the Computer Tutor. For each objective in the text, there is a corresponding computer tutorial and a corresponding set of test questions.

AN INTERACTIVE APPROACH

Instructors have long realized the need for a text that requires students to use a skill as it is being taught. *Essential Mathematics with Applications* uses an interactive technique that meets this need. Every objective, including the one shown on the next page, contains at least one pair of examples. One of the examples is worked. The second example in the pair (You Try It) is not worked so that students may "interact" with the text by solving it. To provide immediate feedback, a complete worked-out solution to this example is provided in the Solutions Section at the end of the book. The benefit of this interactive style is that students can immediately determine whether a new skill has been learned before attempting a homework assignment or moving on to the next skill.

An explanatory passage begins each objective.

Paired examples follow the explanatory passage.

The interactive key is the You Try It in each pair. It has not been worked so that the student may practice the skill, referring to the worked example at the left if necessary.

Reference to the Solutions Section allows the student to check full solutions immediately.

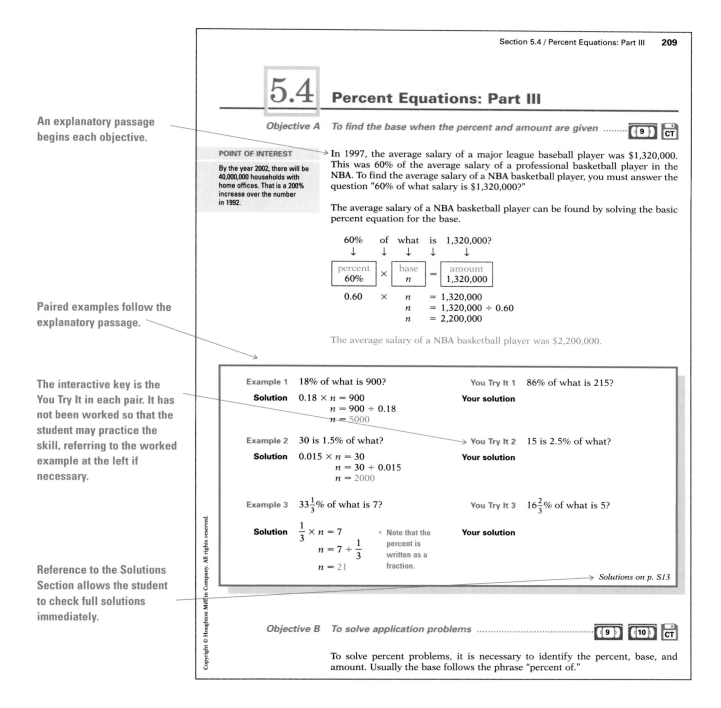

5.4 Percent Equations: Part III

Objective A *To find the base when the percent and amount are given* .... 9 CT

POINT OF INTEREST

By the year 2002, there will be 40,000,000 households with home offices. That is a 200% increase over the number in 1992.

In 1997, the average salary of a major league baseball player was $1,320,000. This was 60% of the average salary of a professional basketball player in the NBA. To find the average salary of a NBA basketball player, you must answer the question "60% of what salary is $1,320,000?"

The average salary of a NBA basketball player can be found by solving the basic percent equation for the base.

60% of what is 1,320,000?

$$\boxed{\begin{array}{c}\text{percent}\\60\%\end{array}} \times \boxed{\begin{array}{c}\text{base}\\n\end{array}} = \boxed{\begin{array}{c}\text{amount}\\1,320,000\end{array}}$$

$$0.60 \times n = 1,320,000$$
$$n = 1,320,000 \div 0.60$$
$$n = 2,200,000$$

The average salary of a NBA basketball player was $2,200,000.

Example 1   18% of what is 900?

Solution $0.18 \times n = 900$
$n = 900 \div 0.18$
$n = 5000$

You Try It 1   86% of what is 215?

Your solution

Example 2   30 is 1.5% of what?

Solution $0.015 \times n = 30$
$n = 30 \div 0.015$
$n = 2000$

You Try It 2   15 is 2.5% of what?

Your solution

Example 3 $33\frac{1}{3}\%$ of what is 7?

Solution $\frac{1}{3} \times n = 7$
$n = 7 \div \frac{1}{3}$
$n = 21$

• Note that the percent is written as a fraction.

You Try It 3 $16\frac{2}{3}\%$ of what is 5?

Your solution

Solutions on p. S13

Objective B *To solve application problems* .... 9 10 CT

To solve percent problems, it is necessary to identify the percent, base, and amount. Usually the base follows the phrase "percent of."

AN EMPHASIS ON APPLICATIONS

The traditional teaching approach neglects the difficulties that students have in making the transition from arithmetic to algebra. One of the most troublesome and uncomfortable transitions for the student is from concrete arithmetic to symbolic algebra. *Essential Mathematics with Applications* recognizes the formidable task the student faces by introducing variables in a very natural way—through applications of mathematics. A secondary benefit of this approach is that the student becomes aware of the value of algebra as a real-life tool.

The solution of an application problem in *Essential Mathematics with Applications* is always accompanied by two parts: **Strategy** and **Solution**. The strategy is a written description of the steps that are necessary to solve the problem; the solution is the implementation of the strategy. This format provides students with a structure for problem solving. It also encourages students to write strategies for solving problems which, in turn, fosters organizing problem-solving strategies in a logical way.

---

206   Chapter 5 / Percents

Objective B   **To solve application problems** .... 〈 9 〉 〈 10 〉 CT

To solve percent problems, remember that it is necessary to identify the percent, base, and amount. Usually the base follows the phrase "percent of."

Example 4
The monthly house payment for the Kaminski family is $787.50. What percent of the Kaminskis' monthly income of $3750 is the house payment?

Strategy
To find what percent of the income the house payment is, write and solve the basic percent equation, using n to represent the percent. The base is $3750 and the amount is $787.50.

Solution
$n \times \$3750 = \$787.50$
$ n = \$787.50 \div \$3750$
$ n = 0.21 = 21\%$

The house payment is 21% of the monthly income.

Example 5
On one Thursday night, 33.4 million of the approximately 64.5 million people watching television on the four major networks were not watching *Seinfeld*. What percent of these viewers were watching *Seinfeld*? Round the answer to the nearest percent.

Strategy
To find the percent of viewers watching *Seinfeld*:

- Subtract to find the number of people who were watching *Seinfeld* (64.5 million − 33.4 million).
- Write and solve the basic percent equation, using n to represent the percent. The base is 64.5 and the amount is the number of people watching *Seinfeld*.

Solution
64.5 million − 33.4 million = 31.1 million people were watching *Seinfeld*.

$n \times 64.5 = 31.1$
$ n = 31.1 \div 64.5$
$ n \approx 0.482$

Approximately 48% of the viewers were watching *Seinfeld*.

You Try It 4
Tomo Nagata had an income of $33,500 and paid $5025 in income tax. What percent of the income is the income tax?

Your strategy

Your solution

You Try It 5
Of the approximately 1,300,000 enlisted women and men in the U.S. military, 416,000 are over the age of 30. What percent of the enlisted people are under the age of 30?

Your strategy

Your solution

Solutions on pp. S12–S13

A strategy that the student may use in solving an application problem is stated.

The strategy is used in the solution of the worked example.

Students are encouraged to write a strategy for the application problem they solve.

OBJECTIVE-SPECIFIC APPROACH

Many mathematics texts are not organized in a manner that facilitates management of learning. Typically, students are left to wander through a maze of apparently unrelated lessons, exercise sets, and tests. *Essential Mathematics with Applications* solves this problem by organizing all lessons, exercise sets, computer tutorials, and tests around a carefully constructed hierarchy of objectives. The advantage of this objective-by-objective organization is that it enables the student who is uncertain at any step in the learning process to refer easily to the original presentation and review that material.

The Objective-Specific Approach also gives the instructor greater control over the management of student progress. The Computerized Test Generator and the printed Test Bank are organized by the same objectives as the text. These references are provided with the answers to the test items, thereby allowing the instructor to quickly determine those objectives on which a student may need additional instruction.

The Computer Tutor is also organized around the objectives of the text. As a result, supplemental instruction is available on any objectives that are troublesome for a student.

A numbered objective statement names the topic of each lesson.

Section 5.4 / Percent Equations: Part III   **209**

5.4   Percent Equations: Part III

Objective A   *To find the base when the percent and amount are given* .... ⟨ 9 ⟩ ⎡CT⎤

The exercise sets correspond to the objectives in the text.

Section 5.4 / Percent Equations: Part III   **211**

5.4 Exercises

Objective A

Solve. Round to the nearest hundredth.

1. 12% of what is 9?   **2.** 38% of what is 171?

The answers to the odd-numbered exercises are provided in the Answer Section.

SECTION 5.4

1. 75   **3.** 50   **5.** 100   **7.** 85   **9.** 1200   **11.** 19.2   **13.** 7.5   **15.** 32   **17.** 200   **19.** 80   **21.** 9
23. 504   **25.** 108   **27.** 7122.15   **29.** The average size of a house in 1977 was 1680 square feet.   **31.** The selling

---

The answers to the Chapter Review Exercises, Chapter Test, and the Cumulative Review Exercises show the objective to study if the student incorrectly answers the exercise.

CHAPTER REVIEW

1. 60 [5.2A]   **2.** 20% [5.3A]   **3.** 175% [5.1B]   **4.** 75 [5.4A]   **5.** $\frac{3}{25}$ [5.1A]   **6.** 19.36 [5.2A]

7. 150% [5.3A]   **8.** 504 [5.4A]   **9.** 0.42 [5.1A]   **10.** 5.4 [5.2A]   **11.** 157.5 [5.4A]   **12.** 0.076 [5.1A]

CHAPTER TEST

1. 0.973 [5.1A]   **2.** $\frac{5}{6}$ [5.1A]   **3.** 30% [5.1B]   **4.** 163% [5.1B]   **5.** 150% [5.1B]   **6.** $66\frac{2}{3}$% [5.1B]

7. 50.05 [5.2A]   **8.** 61.36 [5.2A]   **9.** 76% of 13 [5.2A]   **10.** 212% of 12 [5.2A]   **11.** The company spends

---

CUMULATIVE REVIEW

1. 4 [1.6B]   **2.** 240 [2.1A]   **3.** $10\frac{11}{24}$ [2.4C]   **4.** $12\frac{41}{48}$ [2.5C]   **5.** $12\frac{4}{7}$ [2.6B]   **6.** $\frac{7}{24}$ [2.7B]

7. $\frac{1}{3}$ [2.8B]   **8.** $\frac{13}{36}$ [2.8C]   **9.** 3.08 [3.1B]   **10.** 1.1196 [3.3A]   **11.** 34.2813 [3.5A]   **12.** 3.625 [3.6A]

ADDITIONAL LEARNING AIDS

Chapter Opener

The Chapter Opener relates a historical, contemporary, or interesting note about mathematics or its application.

Focus on Problem Solving

At the end of each chapter there is a Focus on Problem Solving, the purpose of which is to introduce the student to various successful problem-solving strategies. Each Focus consists of a problem and an appropriate strategy to solve the problem. Strategies such as guessing, trying to solve a simpler but similar problem, drawing a diagram, and looking for patterns are some of the techniques that are demonstrated.

Projects and Group Activities

The Projects and Group Activities feature can be used as extra credit or cooperative learning activities. The projects cover various aspects of mathematics including the use of calculators, extended applications, additional problem-solving strategies, and other topics related to mathematics.

Chapter Summaries

At the end of each chapter there is a Chapter Summary that includes Key Words and Essential Rules that were covered in the chapter. These chapter summaries provide a single point of reference as the student prepares for a test.

Study Skills

The To the Student Preface on page xv provides suggestions for using this text and approaches to creating good study habits. Students are referred to this Preface at appropriate places in the text.

Computer Tutor

This state-of-the-art Tutor is a networkable, interactive, algorithmically-driven software package. Features include full-color graphics, a glossary, extensive hints, animated solution steps, and a comprehensive class management system. Written by Dick Aufmann, the tutorial and the text are in the same voice.

Glossary

A Glossary at the end of the book includes definitions of terms used in the text.

Margin Notes

There are three types of margin notes in the student text. *Point of Interest* notes interesting sidelights of the topic being discussed. The *Take Note* feature warns students that a procedure may be particularly involved or reminds students that there are certain checks of their work that should be performed. *Calculator Notes* provide suggestions for using a calculator in certain situations. In addition, there are *Instructor Notes* that are printed only in the Instructor's Annotated Edition. These notes provide suggestions for presenting the material or related material that can be used in class.

Index of Applications

The Index of Applications illustrates the power and scope of mathematics and its application. This may help some students see the benefits of mathematics as a tool that is used in everyday experiences.

EXERCISES

End-of-Section Exercises

Essential Mathematics with Applications contains more than 3000 exercises. At the end of each section there are exercise sets that are keyed to the corresponding learning objectives. The exercises are carefully developed to ensure that students can apply the concepts in the section to a variety of problem situations. Data Analysis exercises are identified by ⬤. Calculator exercises are identified by ▦.

Applying the Concepts Exercises

The End-of-Section Exercises are followed by Applying the Concepts Exercises. These sections contain a variety of exercise types, including:

- challenge problems
- problems that require that the student determine if a statement is always true, sometimes true, or never true
- problems that ask students to determine incorrect procedures

Writing Exercises

Within the "Applying the Concepts Exercises," there are Writing Exercises denoted by ∥. These exercises ask students to write about a topic in the section or to research and report on a related topic.

Chapter Review Exercises

Review Exercises are found at the end of each chapter. These exercises are selected to help the student integrate all of the topics presented in the chapter. The answers to all review exercises are given in the answer section at the end of the book. Along with the answer, there is a reference to the objective that pertains to each exercise.

Chapter Test Exercises

The Chapter Test Exercises are designed to simulate a possible test of the material in the chapter. The answers to all Chapter Test Exercises are given in the answer section at the end of the book. Along with the answer, there is a reference to the objective that pertains to each exercise.

Cumulative Review Exercises

Cumulative Review Exercises, which appear at the end of each chapter (beginning with Chapter 2), help students maintain skills learned in previous chapters. The answers to all Cumulative Review Exercises are given in the answer section. Along with the answer, there is a reference to the objective that pertains to each exercise.

NEW TO THIS EDITION

The material in Chapter 6, *Applications for Business and Consumers,* has been updated to reflect current interest rates and prices.

We have added some problems that have too much data, thereby requiring the student to select the information needed to solve the problem.

For some exercises, not enough information is given to reach a single answer. Thus, there is more than one answer that satisfies the conditions of the problem.

The skill development exercises were thoroughly reviewed to ensure that there was an adequate representation of various problem types. As a result of this review, we have changed or replaced some drill exercises to include problem types that were missing.

Approximately one-third of all the application problems were changed to reflect current data and trends. New application problems were added to demonstrate to students the variety of problems that require mathematical analysis.

Career notes were added to the chapter opener pages to illustrate to students the diverse ways mathematics is used in the workplace.

We have more than doubled the number of projects and group activities. Some of these projects have suggested internet sites so that the student may continue to explore a topic.

In response to suggestions by users, the Chapter Review Exercises are no longer categorized by section. Thus there are no organizational clues to students as to the type of skill needed to solve an exercise. The answers to all Chapter Review Exercises are in the answer appendix. Along with the answer, there is a reference to the objective that pertains to each exercise.

SUPPLEMENTS FOR THE INSTRUCTOR

Instructor's Answer Booklet

The Instructor's Answer Booklet contains the answers to all end-of-section exercise sets, Chapter Reviews, Chapter Tests, and Cumulative Reviews.

ADDITIONAL SUPPLEMENTS

The following are available to accompany *Essential Mathematics with Applications*. Please ask your Houghton Mifflin sales representative for details.

Instructor's Annotated Edition
Instructor's Resource Manual with Chapter Tests
Computerized Test Generator
Printed Test Bank
Solutions Manual
Student Solutions Manual
Computer Tutor
Videotapes

ACKNOWLEDGMENTS

We sincerely wish to thank the following reviewers, who reviewed the manuscript in various stages of development, for their valuable contributions.

Lawrence Chernoff, Miami-Dade Community College, FL
Jeanne Marie Draper
Rose Ann Haw, Mesa Community College, AZ
Rosalie K. Hojegian, Passaic County Community College, NJ
Susan Howey, Harford Community College, MD
Ann Johnson, Community College of Denver, CO
Dr. Barbara Kistler, Lehigh Carbon Community College, PA
Mort E. Mattson, Lansing Community College, MI
Debi McCandrew, Florence-Darlington Technical College, SC
Donna L. McCart, Southern Vermont College
Fadi Nasr, Mt. San Jacinto College, CA
Sister Elizabeth Ogilvie, Horry Georgetown Technical College, SC
Thomas J. Pandolfini, Jr., Johnson & Wales University, RI
Dennis Reissig, Suffolk Community College, Selden, NY
Deana J. Richmond
Jennifer J. Sanders, Bryant & Stratton Business Institute, NY
Alice Stauffer, Slippery Rock University, PA
Desmond Tynan, Holyoke Community College, MA
George Welch, Laredo Community College, TX
Special thanks to Dan Clegg of Palomar College for some of the new application problems.

To the Student

Take an active role in the learning process.

Many students feel that they will never understand math, while others appear to do very well with little effort. Oftentimes what makes the difference is that successful students take an active role in the learning process.

Do the homework.

Attend class regularly.

Participate in class.

Learning mathematics requires your *active* participation. Although doing homework is one way you can actively participate, it is not the only way. First, you must attend class regularly and become an active participant. Second, you must become actively involved with the textbook.

Essential Mathematics with Applications was written and designed with you in mind as a participant. Here are some suggestions on how to use the features of this textbook.

Use the features of the text.

    Read the objective statement.

    Read the objective material.

    Study the in-text examples.

There are 6 chapters in this text. Each chapter is divided into sections, and each section is subdivided into learning objectives. Each learning objective is labeled with a letter from A to D.

First, read each objective statement carefully so you will understand the learning goal that is being presented. Next, read the objective material carefully, being sure to note each bold word. These words indicate important concepts that you should familiarize yourself with. Study carefully each in-text example (denoted by an orange arrow), noting the techniques and strategies used to solve the example.

Use the boxed examples.

1. Study the example on the left.

2. Solve the You Try It example.

3. Check your work against the solution in the back of the book.

You will then come to the key learning feature of this text, the *boxed examples*. These examples have been designed to assist you in a very specific way. Notice that in each example box, the example on the left is completely worked out and the "You Try It" example on the right is not. *You* are expected to work the right-hand example (in the space provided) in order to immediately test your understanding of the material you have just studied.

You should study the worked-out example carefully by working through each step presented. This allows you to focus on each step and reinforces the technique for solving that type of problem. You can then use the worked-out example as a model for solving similar problems.

Next, try to solve the "You Try It" example using the problem-solving techniques that you have just studied. When you have completed your solution, check your work by turning to the page in the Appendix where the complete solution can be found. The page number on which the solution appears is printed at the bottom of the example box in the right-hand corner. By checking your solution, you will know immediately whether or not you fully understand the skill you just studied.

Do the exercises.

Check your answers to the odd-numbered exercises.

When you have completed studying an objective, do the exercises in the exercise set that correspond to that objective. The exercises are labeled with the same letter as the objective. Math is a subject that needs to be learned in small sections and practiced continually in order to be mastered. Doing all of the exercises in each exercise set will help you to master the problem-solving techniques necessary for success. As you work through the exercises for an objective, check your answers to the odd-numbered exercises with those in the back of the book.

Read the Chapter Summary.

Do the Chapter Review exercises.

After completing a chapter, read the Chapter Summary. This summary highlights the important topics covered in the chapter. Following the Chapter Summary are a Chapter Review, a Chapter Test, and a Cumulative Review (beginning with Chapter 2). Doing the review exercises is an important way of testing your understanding of the chapter. The answer to each review exercise is given at the back of the book. Each answer to the reviews and test is followed by a reference that

Check your answers.

Restudy objectives
you missed.

tells that objective that exercise was taken from. For example, (4.2B) means Section 4.2, Objective B. After checking your answers, restudy any objective that you missed. It may be very helpful to retry some of the exercises for that objective to reinforce your problem-solving techniques.

Do the Chapter Test.

The Chapter Test should be used to prepare for an exam. We suggest that you try the Chapter Test a few days before your actual exam. Take the test in a quiet place and try to complete the test in the same amount of time you will be allowed for your exam. When taking the Chapter Test, practice the strategies of successful test takers: 1) scan the entire test to get a feel for the questions; 2) read the directions carefully; 3) work the problems that are easiest for you first; and perhaps most importantly, 4) try to stay calm.

Check your answers.

Restudy objectives
you missed.

When you have completed the Chapter Test, check your answers. If you missed a question, review the material in that objective and rework some of the exercises from that objective. This will strengthen your ability to perform the skills in that objective.

The Cumulative Review allows you to refresh the skills you have learned in previous chapters. This is very important in mathematics. By consistently reviewing previous material, you will retain the skills already learned as you build new ones.

Remember, to be successful: attend class regularly; read the textbook carefully; actively participate in class; work with your textbook using the "You Try It" examples for immediate feedback and reinforcement of each skill; do all the homework assignments; review constantly; and work carefully.

Index of Applications

C H A P T E R

1

Whole Numbers

Medical technicians use a petri dish, named after Julius Petri (1852–1921), to count the number of bacteria in a culture. By using whole numbers to count the bacteria at various times, medical technicians help researchers determine the growth rate of the bacteria. By the end of an experiment, there may be millions of bacteria in the dish.

Objectives

Section 1.1
To identify the order relation between two numbers
To write whole numbers in words and in standard form
To write whole numbers in expanded form
To round a whole number to a given place value

Section 1.2
To add whole numbers
To solve application problems

Section 1.3
To subtract whole numbers without borrowing
To subtract whole numbers with borrowing
To solve application problems

Section 1.4
To multiply a number by a single digit
To multiply larger whole numbers
To solve application problems

Section 1.5
To divide by a single digit with no remainder in the quotient
To divide by a single digit with a remainder in the quotient
To divide by larger whole numbers
To solve application problems

Section 1.6
To simplify expressions that contain exponents
To use the Order of Operations Agreement to simplify expressions

Section 1.7
To factor numbers
To find the prime factorization of a number

Family Tree for Numbers

Our number system is called the Hindu-Arabic system because it has its ancestry in India and was refined by the Arabs. But despite the influence of these cultures on our system, there is some evidence that our system may have originated in China around 1400 B.C. That is 34 centuries ago.

The family tree shown here illustrates the most widely believed account of the history of our number system. In the 16th century, with Gutenberg's invention of the printing press, symbols for our numbers started to become standardized.

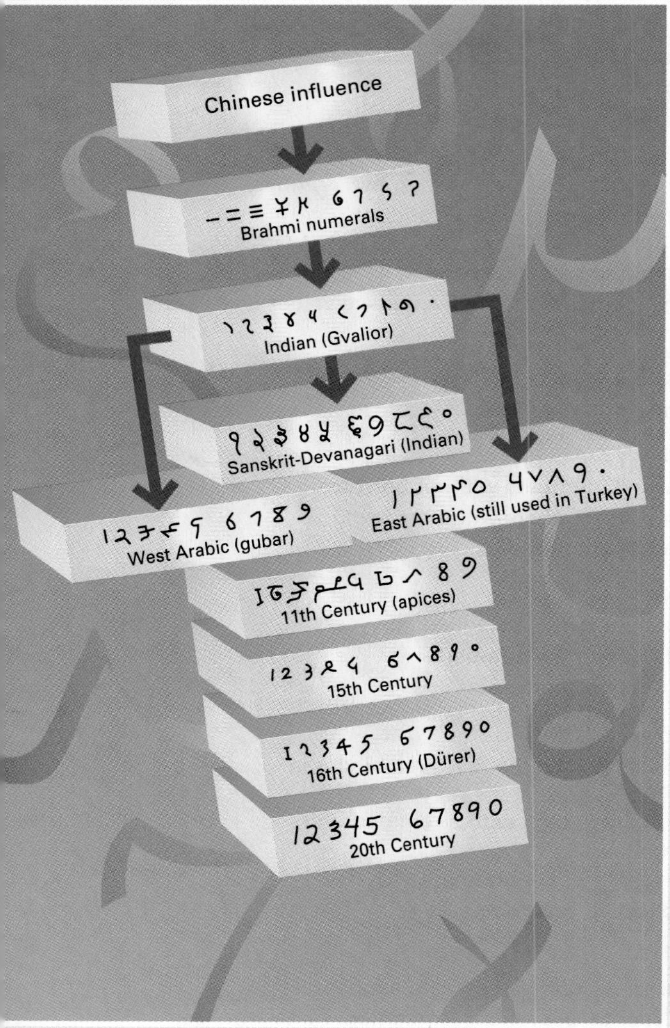

Chinese influence

Brahmi numerals

Indian (Gvalior)

Sanskrit-Devanagari (Indian)

West Arabic (gubar)

East Arabic (still used in Turkey)

11th Century (apices)

15th Century

16th Century (Dürer)

20th Century

1.1 Introduction to Whole Numbers

Objective A   *To identify the order relation between two numbers* ....

The **whole numbers** are 0, 1, 2, 3, 4, 5, 6, 7, 8, 9, 10, 11, 12, 13, 14, ...

The three dots mean that the list continues on and on and that there is no largest whole number.

Just as distances are associated with the markings on the edge of a ruler, the whole numbers can be associated with points on a line. This line is called the **number line.** The arrow on the number line indicates that there is no largest whole number.

The Number Line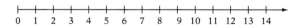

The **graph of a whole number** is shown by placing a heavy dot on the number line directly above the number. Here is the graph of 7 on the number line:

The number line can be used to show the order of whole numbers. A number that appears to the left of a given number is **less than** the given number. The symbol for "is less than" is <. A number that appears to the right of a given number is **greater than** the given number. The symbol for "is greater than" is >.

Four is less than seven.
4 < 7

Twelve is greater than seven.
12 > 7

Example 1	Graph 11 on the number line.	**You Try It 1**	Graph 9 on the number line.
Solution	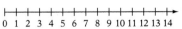	**Your solution**	
Example 2	Place the correct symbol, < or >, between the two numbers. **a.** 39 24 **b.** 0 51	**You Try It 2**	Place the correct symbol, < or >, between the two numbers. **a.** 45 29 **b.** 27 0
Solution	**a.** 39 > 24 **b.** 0 < 51	**Your solution**	**a.** **b.**

Solutions on p. S1

Objective B   *To write whole numbers in words and in standard form* ··········

When a whole number is written using the digits 0, 1, 2, 3, 4, 5, 6, 7, 8, and 9, it is said to be in **standard form.** The position of each digit in the number determines the digit's **place value.** The diagram below shows a **place-value chart** naming the first twelve place values. The number 37,462 is in standard form and has been entered in the chart.

In the number 37,462, the position of the digit 3 determines that its place value is ten-thousands.

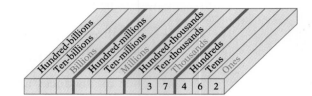

When a number is written in standard form, each group of digits separated by a comma is called a **period.** The number 3,786,451,294 has four periods. The period names are shown in red in the place-value chart above.

To write a number in words, start from the left. Name the number in each period. Then write the period name in place of the comma.

3,786,451,294 is read "three billion seven hundred eighty-six million four hundred fifty-one thousand two hundred ninety-four."

To write a whole number in standard form, write the number named in each period, and replace each period name with a comma.

Four million sixty-two thousand five hundred eighty-four is written 4,062,584. The zero is used as a place holder for the hundred-thousands' place.

Example 3   Write 25,478,083 in words.

Solution   Twenty-five million four hundred seventy-eight thousand eighty-three

You Try It 3   Write 36,462,075 in words.

Your solution

Example 4   Write three hundred three thousand three in standard form.

Solution   303,003

You Try It 4   Write four hundred fifty-two thousand seven in standard form.

Your solution

Solutions on p. S1

Objective C   *To write whole numbers in expanded form* ························

The whole number 26,429 can be written in **expanded form** as

20,000 + 6000 + 400 + 20 + 9

The place-value chart can be used to find the expanded form of a number.

2 Ten-thousands	+	6 Thousands	+	4 Hundreds	+	2 Tens	+	9 Ones
20,000	+	6000	+	400	+	20	+	9

The number 420,806 is written in expanded form below.

Note the effect of having zeros in the number.

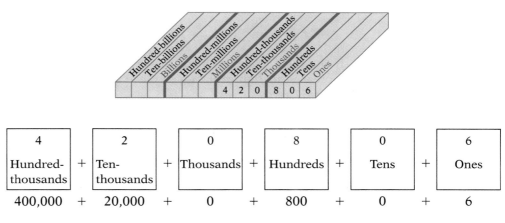

4 Hundred-thousands	+	2 Ten-thousands	+	0 Thousands	+	8 Hundreds	+	0 Tens	+	6 Ones
400,000	+	20,000	+	0	+	800	+	0	+	6

or simply 400,000 + 20,000 + 800 + 6

Example 5   Write 23,859 in expanded form.

Solution   20,000 + 3000 + 800 + 50 + 9

Example 6   Write 709,542 in expanded form.

Solution   700,000 + 9000 + 500 + 40 + 2

You Try It 5   Write 68,281 in expanded form.

Your solution

You Try It 6   Write 109,207 in expanded form.

Your solution

Solutions on p. S1

Objective D   *To round a whole number to a given place value* ....

When the distance to the moon is given as 240,000 miles, the number represents an approximation to the true distance. Giving an approximate value for an exact number is called **rounding.** A number is always rounded to a given place value.

37 is closer to 40 than it is to 30. 37 rounded to the nearest ten is 40.

3673 rounded to the nearest ten is 3670. 3673 rounded to the nearest hundred is 3700.

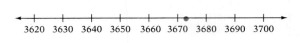

A whole number is rounded to a given place value without using the number line by looking at the first digit to the right of the given place value.

If the digit to the right of the given place value is less than 5, that digit and all digits to the right are replaced by zeros.

➡ Round 13,834 to the nearest hundred.

```
            ┌──── Given place value
    13,834
         └──── 3 < 5
```

13,834 rounded to the nearest hundred is 13,800.

If the digit to the right of the given place value is greater than or equal to 5, increase the digit in the given place value by 1, and replace all other digits to the right by zeros.

➡ Round 386,217 to the nearest ten-thousand.

```
            ┌──── Given place value
    386,217
          └──── 6 > 5
```

386,217 rounded to the nearest ten-thousand is 390,000.

Example 7   Round 525,453 to the nearest ten-thousand.

Solution
```
            ┌──── Given place value
    525,453
          └──── 5 = 5
```

525,453 rounded to the nearest ten-thousand is 530,000.

Example 8   Round 1972 to the nearest hundred.

Solution
```
          ┌──── Given place value
    1972
        └──── 7 > 5
```

1972 rounded to the nearest hundred is 2000.

You Try It 7   Round 368,492 to the nearest ten-thousand.

Your solution

You Try It 8   Round 3962 to the nearest hundred.

Your solution

Solutions on p. S1

1.1 Exercises

TAKE NOTE

To the Student in the front of the book discusses the exercise sets in this textbook.

Objective A

Graph the number on the number line.

1. 3

2. 5

3. 9

4. 0

Place the correct symbol, < or >, between the two numbers.

5. 37   49

6. 58   21

7. 101   87

8. 16   5

9. 245   158

10. 2701   2071

11. 0   45

12. 107   0

13. 815   928

Objective B

Write the number in words.

14. 2675

15. 3790

16. 42,928

17. 58,473

18. 356,943

19. 498,512

20. 3,697,483

21. 6,842,715

Write the number in standard form.

22. Eighty-five

23. Three hundred fifty-seven

24. Three thousand four hundred fifty-six

25. Sixty-three thousand seven hundred eighty

26. Six hundred nine thousand nine hundred forty-eight

27. Seven million twenty-four thousand seven hundred nine

Objective C

Write the number in expanded form.

28. 5287   **29.** 6295   **30.** 58,943   **31.** 453,921

32. 200,583   **33.** 301,809   **34.** 403,705   **35.** 3,000,642

Objective D

Round the number to the given place value.

36. 926 Tens   **37.** 845 Tens

38. 1439 Hundreds   **39.** 3973 Hundreds

40. 43,607 Thousands   **41.** 52,715 Thousands

42. 647,989 Ten-thousands   **43.** 253,678 Ten-thousands

APPLYING THE CONCEPTS

Answer true or false for Exercise 44a and 44b. If the answer is false, give an example to show that it is false.

44. **a.** If you are given two distinct whole numbers, then one of the numbers is always greater than the other number.
   b. A rounded-off number is always less than its exact value.

45. What is the largest three-digit whole number? What is the smallest five-digit whole number?

46. In the Roman numeral system, IV = 4 and VI = 6. Does the position of the I in this system change the value of the number it represents? Determine the value of IX and XI.

47. If 3846 is rounded off to the nearest ten and then that number is rounded to the nearest hundred, is the result the same as what you get when you round 3846 to the nearest hundred? If not, which of the two methods is correct for rounding to the nearest hundred?

1.2 Addition of Whole Numbers

Objective A   *To add whole numbers* ....

Addition is the process of finding the total of two or more numbers.

By counting, we see that the total of $3 and $4 is $7.

$3   + $4   = $7

Addend   **Addend**   **Sum**

Addition can be illustrated on the number line by using arrows to represent the addends. The size, or magnitude, of a number can be represented on the number line by an arrow.

The number 3 can be represented anywhere on the number line by an arrow that is 3 units in length.

To add on the number line, place the arrows representing the addends head to tail, with the first arrow starting at zero. The sum is represented by an arrow starting at zero and stopping at the tip of the last arrow.

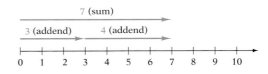

$$3 + 4 = 7$$

More than two numbers can be added on the number line.

$$3 + 2 + 4 = 9$$

Some special properties of addition that are used frequently are given below.

Addition Property of Zero

Zero added to a number does not change the number.

$$4 + 0 = 4$$
$$0 + 7 = 7$$

Commutative Property of Addition

Two numbers can be added in either order; the sum will be the same.

$$4 + 8 = 8 + 4$$
$$12 = 12$$

Associative Property of Addition

Grouping the addition in any order gives the same result. The parentheses are grouping symbols and have the meaning "do the operations inside the parentheses first."

$$(4 + 2) + 3 = 4 + (2 + 3)$$
$$6 \quad + 3 = 4 + \quad 5$$
$$9 = 9$$

POINT OF INTEREST

The first use of the plus sign appeared in 1489 in *Mercantile Arithmetic*. It was used to indicate a surplus and not as the symbol for addition. That use did not appear until around 1515.

The number line is not useful for adding large numbers. The basic addition facts for adding one digit to one digit should be memorized. Addition of larger numbers requires the repeated use of the basic addition facts.

To add large numbers, begin by arranging the numbers vertically, keeping the digits of the same place value in the same column.

➡ Add: 321 + 6472

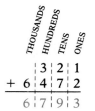

$$\begin{array}{r} 3\ 2\ 1 \\ +\ 6\ 4\ 7\ 2 \\ \hline 6\ 7\ 9\ 3 \end{array}$$

• Add the digits in each column.

There are several words or phrases in English that indicate the operation of addition. Here are some examples.

added to	3 added to 5	5 + 3
more than	7 more than 5	5 + 7
the sum of	the sum of 3 and 9	3 + 9
increased by	4 increased by 6	4 + 6
the total of	the total of 8 and 3	8 + 3
plus	5 plus 10	5 + 10

CALCULATOR NOTE

A scientific calculator is a useful tool in mathematical computation. To add 24 + 71 with your calculator, enter the following:

24 + 71 =

➡ What is the sum of 24 and 71?

The phrase *the sum of* means to add.

$$\begin{array}{r} 24 \\ +\ 71 \\ \hline 95 \end{array}$$

The sum of 24 and 71 is 95.

When the sum of the digits in a column exceeds 9, the addition will involve "carrying."

➡ Add: 487 + 369

$$\begin{array}{r} 1 \\ 4\ 8\ 7 \\ +\ 3\ 6\ 9 \\ \hline 6 \end{array}$$

• Add the ones' column.
7 + 9 = 16 (1 ten + 6 ones).
Write the 6 in the ones' column and carry the 1 ten to the tens' column.

$$\begin{array}{r} 1\ 1 \\ 4\ 8\ 7 \\ +\ 3\ 6\ 9 \\ \hline 5\ 6 \end{array}$$

• Add the tens' column.
1 + 8 + 6 = 15 (1 hundred + 5 tens).
Write the 5 in the tens' column and carry the 1 hundred to the hundreds' column.

$$\begin{array}{r} 1\ 1 \\ 4\ 8\ 7 \\ +\ 3\ 6\ 9 \\ \hline 8\ 5\ 6 \end{array}$$

• Add the hundreds' column.
1 + 4 + 3 = 8 (8 hundreds).
Write the 8 in the hundreds' column.

Example 1 Find the total of 17, 103, and 8.

Solution

     ₁

    17
  103
+   8
───
  128

You Try It 1 What is 347 increased by 12,453?

Your solution

Example 2 Add: 89 + 36 + 98

Solution

   ₂
 89
 36
+ 98
───
223

You Try It 2 Add: 95 + 88 + 67

Your solution

Example 3 Add: 41,395
            4,327
       497,625
   + 32,991

Solution

  _{1 1 2 2 1}
  41,395
   4,327
497,625
+ 32,991
─────
576,338

You Try It 3 Add: 392
      4,079
   89,035
  + 4,992

Your solution

Solutions on p. S1

ESTIMATION

Estimation and Calculators

At some places in the text, you will be asked to use your calculator. Effective use of a calculator requires that you estimate the answer to the problem. This helps ensure that you have entered the numbers correctly and pressed the correct keys.

For example, if you use your calculator to find 22,347 + 5896 and the answer in the calculator's display is 131,757,912, you should realize that you have entered some part of the calculation incorrectly. In this case, you pressed ☒ instead of ⊞. By estimating the answer to a problem, you can help ensure the accuracy of your calculations. The symbol ≈ is used to denote **approximately equal.**

For example, to estimate the answer to 22,347 + 5896, round each number to the same place value. In this case, we will round to the nearest thousand. Then add.

$$
\begin{array}{r}
22{,}347 \approx 22{,}000 \\
+\ \ 5{,}896 \approx +\ \ 6{,}000 \\
\hline
28{,}000
\end{array}
$$

The sum 22,347 + 5896 is approximately 28,000. Knowing this, you would know that 131,757,912 is much too large and is therefore incorrect.

To estimate the sum of two numbers, first round each whole number to the same place value and then add. Compare this answer with the calculator's answer.

Objective B   To solve application problems ....

To solve an application problem, first read the problem carefully. The **Strategy** involves identifying the quantity to be found and planning the steps that are necessary to find that quantity. The **Solution** involves performing each operation stated in the Strategy and writing the answer.

The table below displays the number of snowmobiles registered in four states in 1996.

This information can also be displayed using a bar graph.

State	Number of Snowmobiles
Michigan	270,266
Minnesota	254,510
Wisconsin	193,184
New York	89,617

Source: American Council of Snowmobile Associations

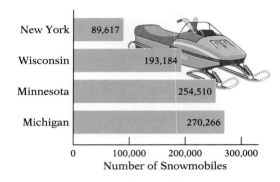

New York   89,617
Wisconsin   193,184
Minnesota   254,510
Michigan   270,266

0   100,000   200,000   300,000
Number of Snowmobiles

➡ Find the total number of snowmobiles registered in the four states as shown in the table and graph above.

Strategy   To find the total number of snowmobiles registered in the four states, read the table to find the number of snowmobiles registered in each state. Then add the numbers.

Solution
$$
\begin{array}{r}
270{,}266 \\
254{,}510 \\
193{,}184 \\
+\ \ \ 89{,}617 \\
\hline
807{,}577
\end{array}
$$

There were 807,577 snowmobiles registered in the four states in 1996.

Example 4
Your paycheck shows deductions of $225 for savings, $98 for taxes, and $27 for insurance. Find the total of the three deductions.

Strategy
To find the total of the deductions, add the three amounts ($225, $98, and $27).

Solution
$$
\begin{array}{r}
\$225 \\
98 \\
+\ \ \ 27 \\
\hline
\$350
\end{array}
$$

The total of the three deductions is $350.

You Try It 4
Anna Barrera has a monthly budget of $475 for food, $275 for car expenses, and $120 for entertainment. Find the total amount budgeted for the three items each month.

Your strategy

Your solution

Solution on p. S1

1.2 Exercises

Objective A

Add.

1. 17
    + 11

2. 25
    + 63

3. 83
    + 42

4. 63
    + 94

5. 77
    + 25

6. 63
    + 49

7. 56
    + 98

8. 86
    + 68

9. 658
    + 831

10. 842
    + 936

11. 735
    + 93

12. 189
    + 50

13. 859
    + 725

14. 637
    + 829

15. 470
    + 749

16. 427
    + 690

17. 36,925
    + 65,392

18. 56,772
    + 51,239

19. 50,873
    + 28,453

20. 34,872
    + 46,079

21. 878
    737
    + 189

22. 768
    461
    + 669

23. 319
    348
    + 912

24. 292
    579
    + 315

25. 9409
    3253
    + 7078

26. 8188
    8020
    + 7104

27. 2038
    2243
    + 3139

28. 4252
    6882
    + 5235

29. 67,428
    32,171
    + 20,971

30. 52,801
    11,664
    + 89,638

31. 76,290
    43,761
    + 87,402

32. 43,901
    98,301
    + 67,943

Add.

33. 20,958 + 3218 + 42

34. 80,973 + 5168 + 29

35. 392 + 37 + 10,924 + 621

36. 694 + 62 + 70,129 + 217

37. 294 + 1029 + 7935 + 65

38. 692 + 2107 + 3196 + 92

39. 97 + 7234 + 69,532 + 276

40. 87 + 1698 + 27,317 + 727

41. What is 9874 plus 4509?

42. What is 7988 plus 5678?

43. What is 3487 increased by 5986?

44. What is 99,567 added to 126,863?

45. What is 23,569 more than 9678?

46. What is 7894 more than 45,872?

47. What is 479 added to 4579?

48. What is 23,902 added to 23,885?

49. Find the total of 659, 55, and 1278.

50. Find the total of 4561, 56, and 2309.

51. Find the sum of 34, 329, 8, and 67,892.

52. Find the sum of 45, 1289, 7, and 32,876.

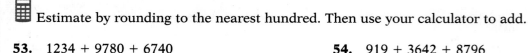 Estimate by rounding to the nearest hundred. Then use your calculator to add.

53. 1234 + 9780 + 6740

54. 919 + 3642 + 8796

55. 241 + 569 + 390 + 1672

56. 107 + 984 + 1035 + 2904

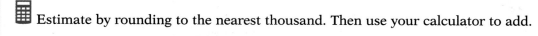 Estimate by rounding to the nearest thousand. Then use your calculator to add.

57.
$$\begin{array}{r} 32{,}461 \\ 9{,}844 \\ + 59{,}407 \\ \hline \end{array}$$

58.
$$\begin{array}{r} 29{,}036 \\ 22{,}904 \\ + 7{,}903 \\ \hline \end{array}$$

59.
$$\begin{array}{r} 25{,}432 \\ 62{,}941 \\ + 70{,}390 \\ \hline \end{array}$$

60.
$$\begin{array}{r} 66{,}541 \\ 29{,}365 \\ + 98{,}742 \\ \hline \end{array}$$

Estimate by rounding to the nearest ten-thousand. Then use your calculator to add.

61.
```
   67,421
   82,984
   66,361
   10,792
 + 34,037
```

62.
```
   21,896
    4,235
   62,544
   21,892
 +  1,334
```

63.
```
  281,421
    9,874
   34,394
  526,398
 + 94,631
```

64.
```
  542,698
   97,327
    7,235
   73,667
 + 173,201
```

Estimate by rounding to the nearest million. Then use your calculator to add.

65.
```
  28,627,052
     983,073
 +  3,081,496
```

66.
```
   1,792,085
  29,919,301
 +  3,406,882
```

67.
```
  12,377,491
   3,409,723
   7,928,026
 + 10,705,682
```

68.
```
  46,751,070
   6,095,832
     280,011
 +  1,563,897
```

Objective B   *Application Problems*

69. In 1996, the United States had $290 billion in total trade with Canada, $183 billion with Japan, and $130 billion with Mexico. Find the total amount of these countries' trade with the United States.

70. The attendance at a Friday night San Diego Padres game was 16,542. The attendance at the Saturday night game was 20,763. Find the total attendance for the two games.

71. Dan Marino threw 5 passes for 42 yards in the first quarter, 7 passes for 117 yards in the second quarter, 9 passes for 66 yards in the third quarter, and 8 passes for 82 yards in the fourth quarter. Find the total number of yards Dan Marino gained by passing.

The graph at the right shows the box-office income of several selected Disney productions. Use this information for Exercises 72 to 75.

72. Estimate the total income from the five Disney productions.

73. Find the total income from the five Disney productions.

74. Find the total income from the two productions with the lowest box-office incomes.

75. Does the total income from the two productions with the lowest box-office incomes exceed the income from the production of *Aladdin*?

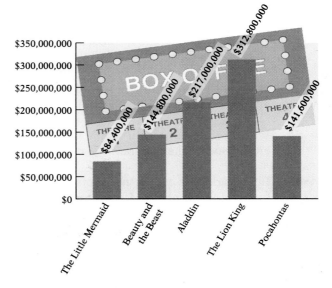

Source: Exhibitor Relations Co. Inc.

76. A student has $2135 in a checking account to be used for the fall semester. During the summer the student makes deposits of $518, $678, and $468.
   a. Find the total amount deposited.
   b. Find the new checking account balance, assuming that there are no withdrawals.

77. The odometer on a moving van reads 68,692. The driver plans to drive 515 miles the first day, 492 miles the second day, and 278 miles the third day.
   a. How many miles will be driven during the three days?
   b. What will the odometer reading be at the end of the trip?

The accompanying table shows the average amount of money all Americans have invested in selected assets and the amounts invested for Americans between the ages of 16 and 34.

78. What is the total average amount for all Americans in checking accounts, savings accounts, and U.S. Savings Bonds?

79. What is the total average amount for Americans ages 16 to 34 in checking accounts, savings accounts, and U.S. Savings Bonds?

80. What is the total average amount for Americans ages 16 to 34 in all categories except home equity and retirement?

81. Is the sum of the average amounts invested in home equity and retirement for all Americans greater than or less than that same sum for Americans between the ages of 16 and 34?

	All Americans	Ages 16 to 34
Checking accounts	$487	$375
Savings accounts	3,494	1,155
U.S. Savings Bonds	546	266
Money market	10,911	4,427
Stocks/mutual funds	4,510	1,615
Home equity	43,070	17,184
Retirement	9,016	4,298

APPLYING THE CONCEPTS

82. How many two-digit numbers are there? How many three-digit numbers are there?

83. If you roll two ordinary six-sided dice and add the two numbers that appear on top, how many different sums are possible?

84. If you add two *different* whole numbers, is the sum always greater than either one of the numbers? If not, give an example.

85. If you add two whole numbers, is the sum always greater than either one of the numbers? If not, give an example. (Compare this with the previous exercise.)

86. Make up a word problem for which the answer is the sum of 34 and 28.

87. Call a number "lucky" if it ends in a 7. How many lucky numbers are less than 100?

1.3 Subtraction of Whole Numbers

Objective A   *To subtract whole numbers without borrowing* ....

Subtraction is the process of finding the difference between two numbers.

By counting, we see that the difference between $8 and $5 is $3.

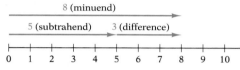

$$\$8 \quad - \quad \$5 \quad = \quad \$3$$

Minuend   Subtrahend   Difference

The difference $8 - 5$ can be shown on the number line.

Note from the number line that addition and subtraction are related.

$$
\begin{array}{rr}
\text{Subtrahend} & 5 \\
+ \text{ Difference} & +\,3 \\
\hline
= \text{Minuend} & 8
\end{array}
$$

The fact that the sum of the subtrahend and the difference equals the minuend can be used to check subtraction.

To subtract large numbers, begin by arranging the numbers vertically, keeping the digits that have the same place value in the same column. Then subtract the digits in each column.

➡ Subtract $8955 - 2432$ and check.

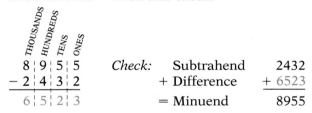

Check:

$$
\begin{array}{lr}
\text{Subtrahend} & 2432 \\
+ \text{ Difference} & +\,6523 \\
\hline
= \text{Minuend} & 8955
\end{array}
$$

Example 1   Subtract $6594 - 3271$ and check.

Solution

$$
\begin{array}{r}
6594 \\
-\,3271 \\
\hline
3323
\end{array}
$$

Check:

$$
\begin{array}{r}
3271 \\
+\,3323 \\
\hline
6594
\end{array}
$$

You Try It 1   Subtract $8925 - 6413$ and check.

Your solution

Example 2   Subtract $15,762 - 7541$ and check.

Solution

$$
\begin{array}{r}
15,762 \\
-\,7,541 \\
\hline
8,221
\end{array}
$$

Check:

$$
\begin{array}{r}
7,541 \\
+\,8,221 \\
\hline
15,762
\end{array}
$$

You Try It 2   Subtract $17,504 - 9302$ and check.

Your solution

Solutions on p. S1

Objective B *To subtract whole numbers with borrowing* ....

In all the subtraction problems in the previous objective, for each place value the lower digit was not larger than the upper digit. When the lower digit is larger than the upper digit, subtraction will involve "borrowing."

➡ Subtract: 692 − 378

| Because 8 > 2, borrowing is necessary. 9 tens = 8 tens + 1 ten. | Borrow 1 ten from the tens' column and write 10 in the ones' column. | Add the borrowed 10 to 2. | Subtract the digits in each column. |

The phrases below are used to indicate the operation of subtraction. An example is shown at the right of each phrase.

minus	8 minus 5	8 − 5
less	9 less 3	9 − 3
less than	2 less than 7	7 − 2
the difference between	the difference between 8 and 2	8 − 2
decreased by	5 decreased by 1	5 − 1

➡ Find the difference between 1234 and 485 and check.

From the phrases that indicate subtraction, the difference between 1234 and 485 is 1234 − 485.

$$
\begin{array}{ccc}
& \overset{2}{\cancel{3}} \; \overset{14}{\cancel{4}} \\
1 \;\; 2 \;\; \cancel{3} \;\; \cancel{4} \\
- \quad 4 \;\; 8 \;\; 5 \\
\hline
\qquad\qquad 9
\end{array}
\qquad
\begin{array}{c}
1 \quad 12 \quad 14 \\
1 \;\; \cancel{2} \;\; \cancel{3} \;\; \cancel{4} \\
- \quad 4 \;\; 8 \;\; 5 \\
\hline
4 \;\; 9
\end{array}
\qquad
\begin{array}{c}
0 \quad 11 \quad 12 \quad 14 \\
\cancel{1} \;\; \cancel{2} \;\; \cancel{3} \;\; \cancel{4} \\
- \quad 4 \;\; 8 \;\; 5 \\
\hline
7 \;\; 4 \;\; 9
\end{array}
\qquad
\textit{Check:}
\qquad
\begin{array}{c}
1 \; 1 \\
485 \\
+ \; 749 \\
\hline
1234
\end{array}
$$

Subtraction with a zero in the minuend involves repeated borrowing.

➡ Subtract: 3904 − 1775

$$
\begin{array}{c}
\quad 8 \quad 10 \\
3 \;\; \cancel{9} \;\; \cancel{0} \;\; 4 \\
- 1 \;\; 7 \;\; 7 \;\; 5 \\
\hline
\end{array}
\qquad
\begin{array}{c}
\qquad 9 \\
\quad 8 \quad \cancel{10} \quad 14 \\
3 \;\; \cancel{9} \;\; \cancel{0} \;\; \cancel{4} \\
- 1 \;\; 7 \;\; 7 \;\; 5 \\
\hline
\end{array}
\qquad
\begin{array}{c}
\qquad 9 \\
\quad 8 \quad \cancel{10} \quad 14 \\
3 \;\; \cancel{9} \;\; \cancel{0} \;\; \cancel{4} \\
- 1 \;\; 7 \;\; 7 \;\; 5 \\
\hline
2 \;\; 1 \;\; 2 \;\; 9
\end{array}
$$

| 5 > 4 There is a 0 in the tens' column. Borrow 1 hundred (= 10 tens) from the hundreds' column and write 10 in the tens' column. | Borrow 1 ten from the tens' column and add 10 to the 4 in the ones' column. | Subtract the digits in each column. |

Example 3   Subtract 4392 − 678 and check.

Solution

$$
\begin{array}{r}
\overset{3}{\cancel{4}}\ \overset{13}{\cancel{3}}\ \overset{8}{\cancel{9}}\ \overset{12}{\cancel{2}} \\
-\ \ \ 6\ \ 7\ \ 8 \\
\hline
3\ \ 7\ \ 1\ \ 4
\end{array}
$$

Check:
$$
\begin{array}{r}
678 \\
+\ 3714 \\
\hline
4392
\end{array}
$$

You Try It 3   Subtract 3481 − 865 and check.

Your solution

Example 4   Find 23,954 less than 63,221 and check.

Solution

$$
\begin{array}{r}
\overset{5}{\cancel{6}}\ \overset{12}{\cancel{3}}, \overset{11}{\cancel{2}}\ \overset{11}{\cancel{2}}\ \overset{11}{\cancel{1}} \\
-\ 2\ 3, 9\ 5\ 4 \\
\hline
3\ 9, 2\ 6\ 7
\end{array}
$$

Check:
$$
\begin{array}{r}
23,954 \\
+39,267 \\
\hline
63,221
\end{array}
$$

You Try It 4   Find 54,562 decreased by 14,485 and check.

Your solution

Example 5   Subtract 46,005 − 32,167 and check.

Solution

$$
\begin{array}{r}
\overset{5}{\cancel{4}} \overset{10}{\cancel{6}},\cancel{0}05 \\
-\ 32,167
\end{array}
$$

 • There are two zeros in the minuend. Borrow 1 thousand from the thousands' column and write 10 in the hundreds' column.

$$
\begin{array}{r}
\overset{5}{\cancel{4}} \overset{9}{}\overset{10}{}\overset{10}{\cancel{6}},\cancel{0}\cancel{0}5 \\
-\ 32,167
\end{array}
$$

 • Borrow 1 hundred from the hundreds' column and write 10 in the tens' column.

$$
\begin{array}{r}
\overset{5}{\cancel{4}} \overset{9}{}\overset{9}{}\overset{10}{}\overset{10}{}\overset{15}{\cancel{6}},\cancel{0}\cancel{0}\cancel{5} \\
-\ 32,167 \\
\hline
13,838
\end{array}
$$

 • Borrow 1 ten from the tens' column and add 10 to the 5 in the ones' column.

Check:
$$
\begin{array}{r}
32,167 \\
+\ 13,838 \\
\hline
46,005
\end{array}
$$

You Try It 5   Subtract 64,003 − 54,936 and check.

Your solution

Solutions on p. S1

ESTIMATION

Estimating the Difference Between Two Whole Numbers

Estimate and then use your calculator to find 323,502 − 28,912.

To estimate the difference between two numbers, round each number to the same place value. In this case we will round to the nearest ten-thousand. Then subtract. The estimated answer is 290,000.

$$
\begin{array}{r}
323,502 \approx\ \ 320,000 \\
-\ 28,912 \approx\ -\ 30,000 \\
\hline
290,000
\end{array}
$$

Now use your calculator to find the exact result. The exact answer is 294,590.

$$323502\ \boxed{-}\ 28912\ \boxed{=}\ 294590$$

Objective C   *To solve application problems* ··································

 The table at the right shows that west-
ern migration has slowed. The table
shows the number of U.S. residents who
have migrated to the states listed during the
years 1995 and 1996. Use the table for
Example 6 and You Try It 6.

State	1995	1996
Arizona	100,170	72,465
Colorado	45,546	30,049
Idaho	18,350	11,039
Montana	10,064	5250
New Mexico	11,963	4692
Oregon	33,438	33,386

Source: Analysis of U.S. Census data by Paul Overberg, USA
Today, March 21, 1997

Example 6
Find the difference between the numbers of
residents who migrated to Arizona in 1995
and 1996.

Strategy
To find the difference, subtract the number
of residents who migrated in 1996 (72,465)
from the number of residents who migrated
in 1995 (100,170).

Solution
$$\begin{array}{r} 100{,}170 \\ -\ 72{,}465 \\ \hline 27{,}705 \end{array}$$

27,705 fewer residents migrated to Arizona
in 1996 than in 1995.

You Try It 6
How many more residents migrated to
Colorado than to Montana in 1996?

Your strategy

Your solution

Example 7
You had a balance of $815 in your checking
account. You then wrote checks in the
amount of $112 for taxes, $57 for food, and
$39 for shoes. What is your new checking
account balance?

Strategy
To find your new checking account balance:
- Add to find the total of the three checks
  ($112 + $57 + $39).
- Subtract the total of the three checks from
  the old balance ($815).

Solution

$$\begin{array}{r} 112 \\ 57 \\ +\ 39 \\ \hline 208 \end{array} \text{ total of checks} \qquad \begin{array}{r} 815 \\ -\ 208 \\ \hline 607 \end{array}$$

Your new checking account balance is $607.

You Try It 7
Your total salary is $638. Deductions of
$127 for taxes, $18 for insurance, and $35
for savings are taken from your pay. Find
your take-home pay.

Your strategy

Your solution

Solutions on p. S1

1.3 Exercises

· ·

Objective A

Subtract.

1. $\begin{array}{r} 9 \\ -\ 5 \\ \hline \end{array}$
2. $\begin{array}{r} 8 \\ -\ 7 \\ \hline \end{array}$
3. $\begin{array}{r} 8 \\ -\ 4 \\ \hline \end{array}$
4. $\begin{array}{r} 7 \\ -\ 3 \\ \hline \end{array}$
5. $\begin{array}{r} 10 \\ -\ 0 \\ \hline \end{array}$

6. $\begin{array}{r} 11 \\ -\ 4 \\ \hline \end{array}$
7. $\begin{array}{r} 12 \\ -\ 8 \\ \hline \end{array}$
8. $\begin{array}{r} 19 \\ -\ 8 \\ \hline \end{array}$
9. $\begin{array}{r} 15 \\ -\ 6 \\ \hline \end{array}$
10. $\begin{array}{r} 16 \\ -\ 7 \\ \hline \end{array}$

11. $\begin{array}{r} 25 \\ -\ 3 \\ \hline \end{array}$
12. $\begin{array}{r} 55 \\ -\ 4 \\ \hline \end{array}$
13. $\begin{array}{r} 68 \\ -\ 8 \\ \hline \end{array}$
14. $\begin{array}{r} 77 \\ -\ 3 \\ \hline \end{array}$
15. $\begin{array}{r} 89 \\ -\ 23 \\ \hline \end{array}$

16. $\begin{array}{r} 54 \\ -\ 21 \\ \hline \end{array}$
17. $\begin{array}{r} 88 \\ -\ 57 \\ \hline \end{array}$
18. $\begin{array}{r} 1202 \\ -\ 701 \\ \hline \end{array}$
19. $\begin{array}{r} 1305 \\ -\ 404 \\ \hline \end{array}$
20. $\begin{array}{r} 1763 \\ -\ 801 \\ \hline \end{array}$

21. $\begin{array}{r} 1497 \\ -\ 706 \\ \hline \end{array}$
22. $\begin{array}{r} 8974 \\ -3972 \\ \hline \end{array}$
23. $\begin{array}{r} 2836 \\ -1711 \\ \hline \end{array}$
24. $\begin{array}{r} 8976 \\ -7463 \\ \hline \end{array}$
25. $\begin{array}{r} 9273 \\ -6142 \\ \hline \end{array}$

26. $77 - 36$
27. $129 - 82$
28. $132 - 61$
29. $969 - 44$
30. $1347 - 103$

31. $4865 - 304$
32. $1525 - 702$
33. $9999 - 6794$
34. $7806 - 3405$
35. $8843 - 7621$

36. What is 3795 minus 1092?

37. What is 9071 minus 6050?

38. Find the difference between 9763 and 541.

39. Find the difference between 6094 and 3072.

40. What is 3701 less than 6932?

41. What is 2031 less than 5071?

42. Find 6509 decreased by 3102.

43. Find 7994 decreased by 7782.

44. Find 23,907 less 12,705.

45. Find 65,986 less 5741.

Objective B

Subtract.

46.   71
     − 18

47.   93
     − 28

48.   47
     − 18

49.   44
     − 27

50.   37
     − 29

51.   50
     − 27

52.   70
     − 33

53.   993
     − 537

54.   250
     − 192

55.   840
     − 783

56.   768
     − 194

57.   770
     − 395

58.   674 − 337

59.   3526 − 387

60.   1712 − 289

61.   4350 − 729

62.   1702 − 948

63.   1607 − 869

64.   5933 − 3754

65.   7293 − 3748

66.   9407 − 2918

67.   3706 − 2957

68.   8605 − 7716

69.   8052 − 2709

70.   80,305 − 9176

71.   70,702 − 4239

72.   10,004 − 9306

73.   80,009 − 63,419

74.   70,618 − 41,213

75.   80,053 − 27,649

76.   70,700 − 21,076

77.   80,800 − 42,023

78.   2600
     − 1972

79.   8400
     − 3762

80.   9003
     − 2471

81.   6004
     − 2392

82.   8202
     − 3916

83.   7050
     − 4137

84.   7015
     − 2973

85.   4207
     − 1624

86.   7005
     − 1796

87.   8003
     − 2735

88.   20,005
     − 9,627

89.   80,004
     − 8,237

90. Find 10,051 less 9027.

91. Find 17,031 less 5792.

92. Find the difference between 1003 and 447.

93. What is 29,874 minus 21,392?

94. What is 29,797 less than 68,005?

95. What is 69,379 less than 70,004?

96. What is 25,432 decreased by 7994?

97. What is 86,701 decreased by 9976?

Estimate by rounding to the nearest ten-thousand. Then use your calculator to subtract.

98.   80,032
      − 19,605

99.   90,765
      − 60,928

100.   32,574
       − 10,961

101.   96,430
      − 59,762

102.   567,423
       − 208,444

103.   300,712
       − 198,714

Objective C   *Application Problems*

104. You have $304 in your checking account. If you write a check for $139, how much is left in your checking account?

105. Sam Akyol buys a Geo Tracker for $15,392. How much did Sam save from a sticker price of $17,871?

106. The tennis coach at a high school purchased a video camera that costs $1079 and made a down payment of $180. Find the amount that remains to be paid.

107. Rod Guerra, an engineer, purchased a used car that cost $5225 and made a down payment of $450. Find the amount that remains to be paid.

108. Merger-related job cuts decreased from 72,083 in 1995 to 42,603 in 1996. What was the decrease in merger-related cuts from 1995 to 1996?

109. In May 1997, the population of the United States was 267,103,163 and the national debt was approximately $5,348,000,000,000. In May 1996, the debt was approximately $5,128,000,000,000. How much did the national debt increase in one year?

110. At the end of a vacation trip, the odometer of your car read 63,459 miles. If the odometer reading at the start of the trip was 62,963, what was the length of the trip?

111. Florida had 20,188,506 acres of wetlands 200 years ago. Today Florida has 10,901,793 acres of wetlands. How many acres of wetlands has Florida lost over the last 200 years?

The table shows the cost of a Ford Escort and the cost of a Chevrolet Cavalier. The table also shows four popular options that are available on these models. Use the table for Exercises 112 and 113.

Source: Consumer Guide, 1997 cars
Signet Reference AE 9241, pages 284, 314

	Ford	Chevrolet
4-door notchback	$11,015	$11,180
Automatic transmission	795	815
CD player	515	652
Cruise control package	495	456
Alloy wheels	265	295

112. What is the difference in cost between the two notchbacks without any additional options?

113. Which of these cars, equipped with all the options listed, costs more?

The graph at the right shows the number of months required to recover all interest and principal from the Social Security fund after a person retires. Use this graph for Exercises 114 and 115.

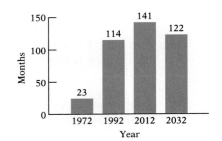

114. How much longer did it take to withdraw principal and interest in 1992 than it did in 1972?

115. How much longer will it take to withdraw principal and interest in 2012 than it did in 1992?

APPLYING THE CONCEPTS

116. Answer true or false.
a. The phrases "the difference between 9 and 5" and "5 less than 9" mean the same thing.
b. $9 - (5 - 3) = (9 - 5) - 3$.
c. Subtraction is an associative operation. *Hint*: See part b of this exercise.

117. Explain how you can check the answer to a subtraction problem.

118. Make up a word problem for which the difference between 15 and 8 is the answer.

1.4 Multiplication of Whole Numbers

Objective A   *To multiply a number by a single digit* ....

Six boxes of toasters are ordered. Each box contains eight toasters. How many toasters are ordered?

This problem can be worked by adding 6 eights.

$8 + 8 + 8 + 8 + 8 + 8 = 48$

This problem involves repeated addition of the same number and can be worked by a shorter process called **multiplication.** Multiplication is the repeated addition of the same number.

$8 + 8 + 8 + 8 + 8 + 8 = 48$

The numbers that are multiplied are called **factors.** The answer is called the **product.**

or

$$6 \quad \times \quad 8 \qquad = 48$$
Factor   Factor   Product

The product of 6×8 can be represented on the number line. The arrow representing the whole number 8 is repeated 6 times. The result is the arrow representing 48.

The times sign "$\times$" is one symbol that is used to mean multiplication. Another common symbol used is a dot placed between the numbers.

$$7 \times 8 = 56 \qquad 7 \cdot 8 = 56$$

As with addition, there are some useful properties of multiplication.

Multiplication Property of Zero

The product of a number and zero is zero.

$0 \times 4 = 0$
$7 \times 0 = 0$

Multiplication Property of One

The product of a number and one is the number.

$1 \times 6 = 6$
$8 \times 1 = 8$

Commutative Property of Multiplication

Two numbers can be multiplied in either order. The product will be the same.

$4 \times 3 = 3 \times 4$
$12 = 12$

Associative Property of Multiplication

Grouping the numbers to be multiplied in any order gives the same result. Do the multiplication inside the parentheses first.

$(4 \times 2) \times 3 = 4 \times (2 \times 3)$
$8 \quad \times 3 = 4 \times \quad 6$
$24 = 24$

The basic facts for multiplying one-digit numbers should be memorized. Multiplication of larger numbers requires the repeated use of the basic multiplication facts.

➡ Multiply: 37 × 4

$$
\begin{array}{r}
\overset{2}{} \\
3\;7 \\
\times4 \\
\hline
8
\end{array}
$$

• $4 \times 7 = 28$ (2 tens + 8 ones).
Write the 8 in the ones' column and carry the 2 to the tens' column.

$$
\begin{array}{r}
\overset{2}{} \\
3\;7 \\
\times4 \\
\hline
14\;\;8
\end{array}
$$

• The 3 in 37 is 3 tens.
$$4 \times 3 \text{ tens} = 12 \text{ tens}$$
Add the carry digit. $\quad + 2 \text{ tens}$
$$\overline{14 \text{ tens}}$$

• Write the 14.

The phrases below are used to indicate the operation of multiplication. An example is shown at the right of each phrase.

times	7 times 3	$7 \cdot 3$
the product of	the product of 6 and 9	$6 \cdot 9$
multiplied by	8 multiplied by 2	$2 \cdot 8$

Example 1 Multiply: 735×9

Solution
$$
\begin{array}{r}
\overset{3\;4}{} \\
735 \\
\times9 \\
\hline
6615
\end{array}
$$

You Try It 1 Multiply: 648×7

Your solution

Solution on p. S2

Objective B   To multiply larger whole numbers ....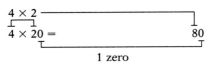

Note the pattern when the following numbers are multiplied.

Multiply the nonzero part of the factors.

Now attach the same number of zeros to the product as the total number of zeros in the factors.

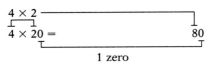

4 × 2
4 × 20 =   80
1 zero

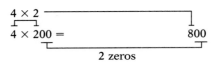

4 × 2
4 × 200 =   800
2 zeros

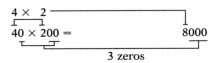

4 × 2
40 × 200 =   8000
3 zeros

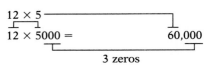

12 × 5
12 × 5000 =   60,000
3 zeros

➡ Find the product of 47 and 23.

Multiply by the ones' digit.	Multiply by the tens' digit.	Add.
47 × 23 141 (= 47 × 3)	47 × 23 141 940 (= 47 × 20) Writing the 0 is optional.	47 × 23 141 940 1081

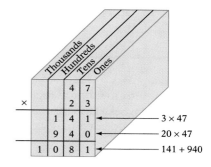

The place-value chart illustrates the placement of the products.

Note the placement of the products when we are multiplying by a factor that contains a zero.

➡ Multiply: 439 × 206

$$439$$
$$\times\ 206$$
$$2634$$
$$000 \quad 0 \times 439$$
$$878$$
$$90{,}434$$

When working the problem, we usually write only one zero. Writing this zero ensures the proper placement of the products.

$$439$$
$$\times\ 206$$
$$2634$$
$$8780$$
$$90{,}434$$

Example 2   Find 829 multiplied by 603.

Solution
$$829$$
$$\times\ 603$$
$$2487$$
$$49740$$
$$499{,}887$$

You Try It 2   Multiply: 756 × 305

Your solution

Solution on p. S2

ESTIMATION

Estimating the Product of Two Whole Numbers

Estimate and then use your calculator to find 3267 × 389.

To estimate a product, round each number so that all the digits are zero except the first digit. Then multiply. The estimated answer is 1,200,000.

$$3267 \approx \quad 3000$$
$$\times\ 389 \approx \quad \times\ 400$$
$$\overline{\qquad\qquad 1{,}200{,}000}$$

Now use your calculator to find the exact answer. The exact answer is 1,270,863.

3267 ⌧× 389 ⌧= 1270863

Objective C   To solve application problems ···

Example 3
An auto mechanic receives a salary of $525 each week. How much does the auto mechanic earn in 4 weeks?

Strategy
To find the mechanic's earnings for 4 weeks, multiply the weekly salary ($525) by the number of weeks (4).

Solution
```
     $525
×       4
   $2100
```
The mechanic earns $2100 in 4 weeks.

You Try It 3
A new-car dealer receives a shipment of 37 cars each month. Find the number of cars the dealer will receive in 12 months.

Your strategy

Your solution

Example 4
A press operator earns $320 for working a 40-hour week. This week the press operator also worked 7 hours of overtime at $13 an hour. Find the press operator's total pay for the week.

Strategy
To find the press operator's total pay for the week:

- Find the overtime pay by multiplying the hours of overtime (7) by the overtime rate of pay ($13).
- Add the weekly salary ($320) to the overtime pay.

Solution
```
    $13              $320
×     7            +   91
  $91 overtime pay   $411
```
The press operator earned $411 this week.

You Try It 4
The buyer for Ross Department Store can buy 80 men's suits for $4800. Each sports jacket will cost the store $23. The manager orders 80 men's suits and 25 sports jackets. What is the total cost of the order?

Your strategy

Your solution

Solutions on p. S2

1.4 Exercises

Objective A

Multiply.

1. 3 × 4	**2.** 2 × 8	**3.** 5 × 7	**4.** 6 × 4	**5.** 5 × 5
6. 7 × 7	**7.** 0 × 7	**8.** 8 × 0	**9.** 8 × 9	**10.** 7 × 6
11. 66 × 3	**12.** 70 × 4	**13.** 67 × 5	**14.** 127 × 9	**15.** 623 × 4
16. 802 × 5	**17.** 607 × 9	**18.** 300 × 5	**19.** 600 × 7	**20.** 906 × 8
21. 703 × 9	**22.** 127 × 5	**23.** 632 × 3	**24.** 559 × 4	**25.** 632 × 8
26. 524 × 4	**27.** 337 × 5	**28.** 841 × 6	**29.** 6709 × 7	**30.** 3608 × 5
31. 8568 × 7	**32.** 5495 × 4	**33.** 4780 × 4	**34.** 3690 × 5	**35.** 9895 × 2

36. Find the product of 5, 7, and 4.

37. Find the product of 6, 2, and 9.

38. Find the product of 5304 and 9.

39. Find the product of 458 and 8.

40. What is 3208 multiplied by 7?

41. What is 5009 multiplied by 4?

42. What is 3105 times 6?

43. What is 8957 times 8?

Objective B

Multiply.

44. 16
× 21

45. 18
× 24

46. 35
× 26

47. 27
× 72

48. 693
× 91

49. 581
× 72

50. 419
× 80

51. 727
× 60

52. 8279
× 46

53. 9577
× 35

54. 6938
× 78

55. 8875
× 67

56. 7035
× 57

57. 6702
× 48

58. 3009
× 35

59. 6003
× 57

60. 809
× 530

61. 607
× 460

62. 800
× 325

63. 700
× 274

64. 987
× 349

65. 688
× 674

66. 312
× 134

67. 423
× 427

68. 379
× 500

69. 684
× 700

70. 985
× 408

71. 758
× 209

72. 3407
× 309

73. 5207
× 902

74. 4258
× 986

75. 6327
× 876

76. What is 5763 times 45?

77. What is 7349 times 7?

78. Find the product of 2, 19, and 34.

79. Find the product of 6, 73, and 43.

80. What is 376 multiplied by 402?

81. What is 842 multiplied by 309?

82. Find the product of 233,489 and 3005.

83. Find the product of 34,985 and 9007.

Estimate and then use your calculator to multiply.

84. 8745
         × 63

85. 4732
         × 93

86. 39,246
         × 29

87. 64,409
         × 67

88. 2937
         × 206

89. 8941
         × 726

90. 3097
         × 1025

91. 6379
         × 2936

92. 32,508
         × 591

93. 62,504
         × 923

94. 81,405
         × 902

95. 66,735
         × 844

Objective C *Application Problems*

96. Rob Hill owns a compact car that averages 43 miles on 1 gallon of gas. How many miles could the car travel on 12 gallons of gas?

97. A plane flying from Los Angeles to Boston uses 865 gallons of jet fuel each hour. How many gallons of jet fuel were used on a 6-hour flight?

98. Anthony Davis purchased 225 shares of Public Service of Colorado for $40 per share. The stock has a yearly dividend of $2 a share, paid in 4 quarterly payments. Find the yearly income from the 225 shares of stock.

99. Assume that a machine at the Coca Cola Bottling Company can fill and cap 4200 bottles of Coke in 1 hour. How many bottles of Coke can the machine fill and cap in 40 hours?

100. A computer graphics screen has 640 rows of pixels, and there are 480 pixels per row. Find the total number of pixels on the screen.

101. Jack Murphy Stadium in San Diego has a capacity of 60,836. For the first 11 home games, the average attendance was 30,013. For the first 4 road games, the average attendance was 16,754. Estimate the total attendance for the 11 home games. Find the total attendance for the first 15 games.

A lighting consultant to a bank suggests that the bank lobby contain 43 can lights, 15 high-intensity lights, 20 fire safety lights, and one chandelier. The table at the right gives the costs for each type of light from two companies. Use this table for Exercises 102 and 103.

	Company A	Company B
Can lights	$2 each	$3 each
High-intensity	$6 each	$4 each
Fire safety	$12 each	$11 each
Chandelier	$998 each	$1089 each

102. Which company offers the lights for the lower total price?

103. How much can the lighting designer save by purchasing the lights from the company that offers the lower total price?

The table at the right shows the hourly wages of four different job classifications at a small construction company. Use this table for Exercises 104 to 106.

Type of Work	Wage per Hour
Electrician	$17
Plumber	$15
Clerk	$8
Bookkeeper	$10

104. The owner of this company wants to provide the electrical installation for a new house. Based on the architectural plans for the house, it is estimated that it will require 3 electricians each working 50 hours to complete the job. What is the estimated cost for the electricians' labor?

105. Carlos Vasquez, a plumbing contractor, hires 4 plumbers from this company at the hourly wage given in the table. If each plumber works 23 hours, what are the total wages paid by Carlos?

106. The owner of this company estimates that remodeling a kitchen will require 1 electrician working 30 hours and 1 plumber working 33 hours. This project also requires 3 hours of clerical work and 4 hours of bookkeeping. What is the total cost for these four components of this remodeling?

APPLYING THE CONCEPTS

107. Determine whether each of the following statements is always true, sometimes true, or never true.
   a. A whole number times zero is zero.
   b. A whole number times one is the whole number.
   c. The product of two whole numbers is greater than either one of the whole numbers.

108. According to the National Safety Council, in a recent year a death resulting from an accident occurred at the rate of 1 every 5 minutes. At this rate, how many accidental deaths occurred each hour? Each day? Throughout the year? Explain how you arrived at your answers.

109. In Brazil, 713 acres are deforested each hour. At this rate, how many acres of deforestation occur each day? In a 30-day month? Throughout the year? Explain how you arrived at your answers.

110. Pick your favorite number between 1 and 9. Multiply the number by 3. Multiply that product by 37,037. How is the product related to your favorite number? Explain why this works. (Suggestion: Multiply 3 and 37,037 first.)

111. There are quite a few tricks based on whole numbers. Here's one about birthdays. Write down the month in which you were born. Multiply by 5. Add 7. Multiply by 20. Subtract 100. Add the day of the month on which you were born. Multiply by 4. Subtract 100. Multiply by 25. Add the year you were born. Subtract 3400. The answer is the month/day/year of your birthday.

1.5   Division of Whole Numbers

Objective A   *To divide by a single digit with no remainder in the quotient* ....

Division is used to separate objects into equal groups.

A store manager wants to display 24 new objects equally on 4 shelves. From the diagram, we see that the manager would place 6 objects on each shelf.

The manager's division problem can be written as follows:

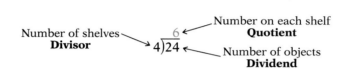

Number of shelves — **Divisor**

Number on each shelf — **Quotient**

Number of objects — **Dividend**

$$4)\overline{24}$$

Note that the quotient multiplied by the divisor equals the dividend.

$$4)\overline{24}^{\,6} \quad \text{because} \quad \boxed{\underset{\text{Quotient}}{6}} \times \boxed{\underset{\text{Divisor}}{4}} = \boxed{\underset{\text{Dividend}}{24}}$$

$$9)\overline{54}^{\,6} \quad \text{because} \quad 6 \quad \times \quad 9 \quad = \quad 54$$

$$8)\overline{40}^{\,5} \quad \text{because} \quad 5 \quad \times \quad 8 \quad = \quad 40$$

Here are some important quotients and the properties of zero in division:

Important Quotients

Any whole number, except zero, divided by itself is 1.

$$8)\overline{8}^{\,1} \qquad 14)\overline{14}^{\,1} \qquad 10)\overline{10}^{\,1}$$

Any whole number divided by 1 is the whole number.

$$1)\overline{9}^{\,9} \qquad 1)\overline{27}^{\,27} \qquad 1)\overline{10}^{\,10}$$

Properties of Zero in Division

Zero divided by any other whole number is zero.

$$7)\overline{0}^{\,0} \qquad 13)\overline{0}^{\,0} \qquad 10)\overline{0}^{\,0}$$

Division by zero is not allowed.

$$0)\overline{8}^{\,?}$$

There is no number whose product with 0 is 8.

When the dividend is a larger whole number, the digits in the quotient are found in steps.

➡ Divide $4\overline{)3192}$ and check.

$$
\begin{array}{r}
7 \\
4\overline{)3192} \\
-28 \\
\hline
39
\end{array}
$$

- Think $4\overline{)31}$.
- Subtract 7×4.
- Bring down the 9.

$$
\begin{array}{r}
79 \\
4\overline{)3192} \\
-28 \\
\hline
39 \\
-36 \\
\hline
32
\end{array}
$$

- Think $4\overline{)39}$.
- Subtract 9×4.
- Bring down the 2.

$$
\begin{array}{r}
798 \\
4\overline{)3192} \\
-28 \\
\hline
39 \\
-36 \\
\hline
32 \\
-32 \\
\hline
0
\end{array}
$$

- Think $4\overline{)32}$.
- Subtract 8×4.

Check:
$$
\begin{array}{r}
798 \\
\times 4 \\
\hline
3192
\end{array}
$$

The place-value chart can be used to show why this method works.

$$
\begin{array}{r}
\text{HUNDREDS}\ \text{TENS}\ \text{ONES} \\
7\ \ 9\ \ 8 \\
4\overline{)3\ \ 1\ \ 9\ \ 2} \\
-2\ \ 8\ \ 0\ \ 0 \quad \text{7 hundreds} \times 4 \\
\hline
3\ \ 9\ \ 2 \\
-3\ \ 6\ \ 0 \quad \text{9 tens} \times 4 \\
\hline
3\ \ 2 \\
-3\ \ 2 \quad \text{8 ones} \times 4 \\
\hline
0
\end{array}
$$

There are other ways of expressing division.

54 divided by 9 equals 6.

54 ÷ 9 equals 6.

$\dfrac{54}{9}$ equals 6.

Example 1   Divide 7)56 and check.

Solution
$$
\begin{array}{r}
8 \\
7\overline{)56}
\end{array}
$$

Check: $8 \times 7 = 56$

You Try It 1   Divide 9)63 and check.

Your solution

Example 2   Divide $2808 \div 8$ and check.

Solution
$$
\begin{array}{r}
351 \\
8\overline{)2808} \\
-24 \\
\hline
40 \\
-40 \\
\hline
08 \\
-8 \\
\hline
0
\end{array}
$$

Check: $351 \times 8 = 2808$

You Try It 2   Divide $4077 \div 9$ and check.

Your solution

Example 3   Divide 7)2856 and check.

Solution
$$
\begin{array}{r}
408 \\
7\overline{)2856} \\
-28 \\
\hline
05 \\
-0 \\
\hline
56 \\
-56 \\
\hline
0
\end{array}
$$

• Think 7)5. Place 0 in quotient.
  Subtract 0 × 7.
  Bring down the 6.

Check: $408 \times 7 = 2856$

You Try It 3   Divide 9)6345 and check.

Your solution

Solutions on p. S2

Objective B   To divide by a single digit with a remainder in the quotient ....

Sometimes it is not possible to separate objects into a whole number of equal groups.

A baker has 14 muffins to pack into 3 boxes. Each box holds 4 muffins. From the diagram, we see that after the baker places 4 muffins in each box, there are 2 left over. The 2 is called the **remainder.**

The clerk's division problem could be written

$$
\begin{array}{r}
4 \\
3{\overline{\smash{\big)}\,14}} \\
-12 \\
\hline
2
\end{array}
$$

4 ← Quotient (Number in each box)

Divisor (Number of boxes) → 3) 14 ← Dividend (Total number of objects)

2 ← Remainder (Number left over)

The answer to a division problem with a remainder is frequently written

$$
3{\overline{\smash{\big)}\,14}}^{\,4\ \text{r2}}
$$

Note that

4 × 3 Quotient Divisor	+	2 Remainder	=	14 Dividend

Example 4 Divide $4{\overline{\smash{\big)}\,2522}}$ and check.

Solution

$$
\begin{array}{r}
630 \ \text{r2} \\
4{\overline{\smash{\big)}\,2522}} \\
-24 \\
\hline
12 \\
-12 \\
\hline
02 \\
-\ 0 \\
\hline
2
\end{array}
$$

● Think $4{\overline{\smash{\big)}\,2}}$. Place 0 in quotient.

Subtract 0 × 4.

Check: (630 × 4) + 2 =
2520 + 2 = 2522

You Try It 4 Divide $6{\overline{\smash{\big)}\,5225}}$ and check.

Your solution

Example 5 Divide $9{\overline{\smash{\big)}\,27{,}438}}$ and check.

Solution

$$
\begin{array}{r}
3{,}048 \ \text{r6} \\
9{\overline{\smash{\big)}\,27{,}438}} \\
-27 \\
\hline
0\ 4 \\
-\ 0 \\
\hline
43 \\
-36 \\
\hline
78 \\
-72 \\
\hline
6
\end{array}
$$

● Think $9{\overline{\smash{\big)}\,4}}$.
● Subtract 0 × 9.

Check: (3048 × 9) + 6 =
27,432 + 6 = 27,438

You Try It 5 Divide $7{\overline{\smash{\big)}\,21{,}409}}$ and check.

Your solution

Solutions on p. S2

Objective C   **To divide by larger whole numbers** ....

When the divisor has more than one digit, estimate at each step by using the first digit of the divisor. If that product is too large, lower the guess by 1 and try again.

➡ Divide $34\overline{)1598}$ and check.

$$
\begin{array}{r}
5 \\
34\overline{)\ 1598} \\
-170 \\
\hline
\end{array}
$$

- Think $3\overline{)15}$.
- Subtract 5×34.

170 is too large. Lower the guess by 1 and try again.

$$
\begin{array}{r}
4 \\
34\overline{)\ 1598} \\
-136 \\
\hline
238
\end{array}
$$

- Subtract 4×34.

$$
\begin{array}{r}
47 \\
34\overline{)\ 1598} \\
-136 \\
\hline
238 \\
-238 \\
\hline
0
\end{array}
$$

- Think $3\overline{)23}$.
- Subtract 7×34.

Check:
$$
\begin{array}{r}
47 \\
\times 34 \\
\hline
188 \\
141 \\
\hline
1598
\end{array}
$$

The phrases below are used to indicate the operation of division. An example is shown at the right of each phrase.

| the quotient of | the quotient of 9 and 3 | $9 \div 3$ |
| divided by | 6 divided by 2 | $6 \div 2$ |

Example 6   Find 7077 divided by 34 and check.

You Try It 6   Divide $4578 \div 42$ and check.

Solution

$$
\begin{array}{r}
208\ \text{r}5 \\
34\overline{)\ 7077} \\
-68 \\
\hline
27 \\
-\ 0 \\
\hline
277 \\
-272 \\
\hline
5
\end{array}
$$

- Think $34\overline{)27}$.
- Place 0 in the quotient.
- Subtract 0×34.

Check: $(208 \times 34) + 5 =$
$7072\ \ \ + 5 = 7077$

Your solution

Solution on p. S2

Example 7   Find the quotient of 21,312 and 56 and check.

Solution

$$
\begin{array}{r}
380 \text{ r}32 \\
56\overline{)21{,}312} \\
-16\ 8 \\
\hline
4\ 51 \\
-4\ 48 \\
\hline
32 \\
-\ 0 \\
\hline
32
\end{array}
$$

• Think $5\overline{)21}$.

  4×56 is too large. Try 3.

Check: $(380 \times 56) + 32 =$
  $21{,}280\ \ + 32 = 21{,}312$

You Try It 7   Divide $18{,}359 \div 39$ and check.

Your solution

Example 8   Divide $427\overline{)24{,}782}$ and check.

Solution

$$
\begin{array}{r}
58 \text{ r}16 \\
427\overline{)24{,}782} \\
-21\ 35 \\
\hline
3\ 432 \\
-3\ 416 \\
\hline
16
\end{array}
$$

Check: $(58 \times 427) + 16 =$
  $24{,}766\ \ + 16 = 24{,}782$

You Try It 8   Divide $534\overline{)33{,}219}$ and check.

Your solution

Example 9   Divide $386\overline{)206{,}149}$ and check.

Solution

$$
\begin{array}{r}
534 \text{ r}25 \\
386\overline{)206{,}149} \\
-193\ 0 \\
\hline
13\ 14 \\
-11\ 58 \\
\hline
1\ 569 \\
-1\ 544 \\
\hline
25
\end{array}
$$

Check: $(534 \times 386) + 25 =$
  $206{,}124\ \ + 25 = 206{,}149$

You Try It 9   Divide $515\overline{)216{,}848}$ and check.

Your solution

Solutions on p. S3

ESTIMATION

Estimating the Quotient of Two Whole Numbers

Estimate and then use your calculator to find 36,936 ÷ 54.

To estimate a quotient, round each number so that all the digits are zero except the first digit. Then divide.

$$36{,}936 \div 54 \approx$$
$$40{,}000 \div 50 = 800$$

The estimated answer is 800.

Now use your calculator to find the exact answer.

36936 [÷] 54 [=] 684

The exact answer is 684.

Objective D   ***To solve application problems*** ....

The *average* of several numbers is the sum of all the numbers divided by the number of numbers.

$$\text{Average test score} = \frac{498}{6} = 83$$

⇒ The table at the right shows the number of home dates and the total home attendance of the baseball teams in the National League. Find the average home attendance for the Chicago team. Round to the nearest whole number.

Team	Home Dates	Home Att.
Atlanta	69	3,296,967
Chicago	74	2,446,912
Cincinnati	69	2,189,722
Colorado	67	3,872,511
Florida	67	2,629,117
Houston	74	1,913,944
Los Angeles	71	2,796,973
Montreal	72	1,374,708
New York	70	1,708,606
Philadelphia	74	2,869,808
Pittsburgh	70	1,514,206
St. Louis	72	2,578,292
San Diego	73	1,277,268
San Francisco	73	2,362,524

Strategy

Determine from the table the total attendance and the number of home dates for the Chicago team. To find the average home attendance, divide the total attendance (2,446,912) by the number of home dates (74).

Solution

```
              33,066
      74) 2,446,912
          -2 22
            226
           -222
             49
            - 0
             491
            -444
             472
            -444
              28
```

● When rounding to the nearest whole number, compare twice the remainder to the divisor. If twice the remainder is less than the divisor, drop the remainder. If twice the remainder is greater than or equal to the divisor, add 1 to the units digit of the quotient.

● Twice the remainder is 2 × 28 = 56. Because 56 < 74, drop the remainder.

The average attendance is 33,066.

Example 10
Ngan Hui, a freight supervisor, shipped 192,600 bushels of wheat in 9 railroad cars. Find the amount of wheat shipped in each car.

Strategy
To find the amount of wheat shipped in each car, divide the number of bushels (192,600) by the number of cars (9).

Solution

$$
\begin{array}{r}
21,400 \\
9\overline{)\,192,600} \\
-18 \\
\hline
12 \\
-9 \\
\hline
36 \\
-36 \\
\hline
0
\end{array}
$$

Each car carried 21,400 bushels of wheat.

Example 11
The car you are buying costs $11,216. A down payment of $2000 is required. The remaining balance is paid in 48 equal monthly payments. What is the monthly payment?

Strategy
To find the monthly payment:

- Find the remaining balance by subtracting the down payment ($2000) from the total cost of the car ($11,216).
- Divide the remaining balance by the number of equal monthly payments (48).

Solution

$$
\begin{array}{r}
11,216 \\
-\ 2,000 \\
\hline
9,216
\end{array}
$$
remaining balance

$$
\begin{array}{r}
192 \\
48\overline{)\,9216} \\
-48 \\
\hline
441 \\
-432 \\
\hline
96 \\
-96 \\
\hline
0
\end{array}
$$

The monthly payment is $192.

You Try It 10
Suppose a Firestone retail outlet can store 270 tires on 15 shelves. How many tires can be stored on each shelf?

Strategy

Your solution

You Try It 11
A soft-drink manufacturer produces 12,600 cans of soft drink each hour. Cans are packed 24 to a case. How many cases of soft drink are produced in 8 hours?

Your strategy

Your solution

Solutions on p. S3

1.5 Exercises

· ·

Objective A

Divide.

1. $4\overline{)8}$ **2.** $3\overline{)9}$ **3.** $6\overline{)36}$ **4.** $9\overline{)81}$

5. $7\overline{)49}$ **6.** $5\overline{)80}$ **7.** $6\overline{)96}$ **8.** $6\overline{)480}$

9. $4\overline{)840}$ **10.** $3\overline{)690}$ **11.** $7\overline{)308}$ **12.** $7\overline{)203}$

13. $9\overline{)6327}$ **14.** $4\overline{)2120}$ **15.** $8\overline{)7280}$ **16.** $9\overline{)8118}$

17. $3\overline{)64,680}$ **18.** $4\overline{)50,760}$ **19.** $6\overline{)21,480}$ **20.** $5\overline{)18,050}$

21. Find the quotient of 1446 and 3.   **22.** Find the quotient of 4123 and 7.

23. What is 7525 divided by 7?   **24.** What is 32,364 divided by 4?

Objective B

Divide.

25. $4\overline{)9}$ **26.** $2\overline{)7}$ **27.** $5\overline{)27}$ **28.** $9\overline{)88}$ **29.** $3\overline{)40}$

30. $6\overline{)97}$ **31.** $8\overline{)83}$ **32.** $5\overline{)54}$ **33.** $7\overline{)632}$ **34.** $4\overline{)363}$

35. $4\overline{)921}$ **36.** $7\overline{)845}$ **37.** $8\overline{)1635}$ **38.** $5\overline{)1548}$ **39.** $7\overline{)9432}$

40. $7\overline{)8124}$ **41.** $3\overline{)5162}$ **42.** $5\overline{)3542}$ **43.** $8\overline{)3274}$

44. $4\overline{)15,300}$ **45.** $7\overline{)43,500}$ **46.** $8\overline{)72,354}$ **47.** $5\overline{)43,542}$

48. Find the quotient of 3107 and 8.

49. Find the quotient of 8642 and 8.

50. What is 45,738 divided by 4? Round to the nearest ten.

51. What is 37,896 divided by 9? Round to the nearest hundred.

52. What is 3572 divided by 7? Round to the nearest ten.

53. What is 78,345 divided by 4? Round to the nearest hundred.

Objective C

Divide.

54. $27\overline{)96}$ **55.** $44\overline{)82}$ **56.** $42\overline{)87}$ **57.** $67\overline{)93}$

58. $41\overline{)897}$ **59.** $32\overline{)693}$ **60.** $23\overline{)784}$ **61.** $25\overline{)772}$

62. $74\overline{)600}$ **63.** $92\overline{)500}$ **64.** $70\overline{)329}$ **65.** $50\overline{)467}$

66. $36\overline{)7225}$ **67.** $44\overline{)8821}$ **68.** $19\overline{)3859}$ **69.** $32\overline{)9697}$

70. $88\overline{)3127}$ **71.** $92\overline{)6177}$ **72.** $33\overline{)8943}$ **73.** $27\overline{)4765}$

74. $22\overline{)98,654}$ **75.** $77\overline{)83,629}$ **76.** $64\overline{)38,912}$ **77.** $78\overline{)31,434}$

78. $206\overline{)3097}$ **79.** $504\overline{)6504}$ **80.** $654\overline{)1217}$ **81.** $546\overline{)2344}$

82. Find the quotient of 5432 and 21.

83. Find the quotient of 8507 and 53.

84. What is 37,294 divided by 72?

85. What is 76,788 divided by 46?

86. Find 23,457 divided by 43. Round to the nearest hundred.

87. Find 341,781 divided by 43. Round to the nearest ten.

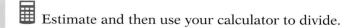 Estimate and then use your calculator to divide.

88. $76\overline{)389{,}804}$

89. $53\overline{)117{,}925}$

90. $29\overline{)637{,}072}$

91. $67\overline{)738{,}072}$

92. $38\overline{)934{,}648}$

93. $34\overline{)906{,}304}$

94. $309\overline{)876{,}324}$

95. $642\overline{)323{,}568}$

96. $209\overline{)632{,}016}$

97. $614\overline{)332{,}174}$

98. $179\overline{)5{,}734{,}444}$

99. $374\overline{)7{,}712{,}254}$

Objective D   *Application Problems*

100. Assume that the American Red Cross, the Boys Club, the Girls Club, and the North County Crisis Center collected $548,000 to promote and provide community services. What amount did each organization receive if the money was divided evenly?

101. An insurance agent drives a car with a 16-gallon gas tank. The agent used 128 gallons of gas in traveling 3456 miles. Find the average number of miles traveled on each gallon of gas.

The five highest paid baseball players for the year 1997 are listed at the right. This is the contract salary and does not include endorsements. The following exercises do not include everything the players are responsible for. However, it is interesting to see the income for each event for the different players.

Name	Annual Income
Albert Belle	$10,000,000
Cecil Fielder	$9,237,000
Barry Bonds	$8,513,000
Roger Clemens	$8,250,000
Jeff Bagwell	$8,015,000

Source: USA TODAY research, USA TODAY, March 4, 1997

102. Assume that Roger Clemens pitched in 40 games. Find the average income per game that Clemens pitched.

103. Albert Belle was signed because of his home run hitting ability. Assume that Belle breaks the record and hits 64 home runs in a season. Find the income per home run that Belle receives.

104. Assume that Cecil Fielder plays in 150 games. Find the average salary that Fielder receives per game.

105. A computer can store 2,211,840 bytes of information on 6 disks. How many bytes of information can be stored on 1 disk?

106. Ken Martinez, a computer analyst, received $5376 for working 168 hours on a computer consulting project. Find the hourly rate that Ken charged.

107. Marie Sarro, a design consultant, made a down payment of $1560 on a car costing $10,536.
   a. What is the remaining balance to be paid?
   b. The balance is to be paid in 48 equal monthly payments. Find the monthly payment.

The table at the right shows the highest paid athletes in four professional sports for the year 1996. Michael Jordan plays professional basketball, Albert Belle plays professional baseball, Mario Lemieux plays professional hockey, and Troy Aikman plays professional football. Use this list for Exercises 108 to 111.

Name	Annual Income
Jordan	$30,140,000
Belle	$10,000,000
Lemieux	$11,320,000
Aikman	$5,370,000

Source: USA TODAY research, USA TODAY, April 3, 1997

108. There are 82 games in a professional basketball season. Assuming that Michael Jordan played in all 82 games, find his average salary per game. Round to the nearest whole number.

109. The average worker in the United States works approximately 2000 hours per year. Assume that Mario Lemieux worked 2000 hours. Find his average hourly wage.

110. Approximately how many times Troy Aikman's annual income was Michael Jordan's annual income? Round to the nearest whole number.

111. There are 162 games in a professional baseball season and 16 games in a professional football season. Find Albert Belle's and Troy Aikman's average salary per game. Approximately how many times Albert Belle's average salary per game was Troy Aikman's average salary per game? Round to the nearest whole number.

APPLYING THE CONCEPTS

112. A palindromic number is a whole number that remains unchanged when its digits are written in reverse order. For instance, 292 is a palindromic number. Find the smallest three-digit palindromic number that is divisible by 4.

113. Find the smallest four-digit palindromic number (see Exercise 112) that is divisible by 8. (Helpful fact: A whole number whose last three digits are divisible by 8 is itself divisible by 8.)

114. The number 10,981 is not divisible by 4. By rearranging the digits, find the largest possible number that is divisible by 4.

115. Determine whether each of the following statements is true or false.

   a. Any whole number divided by zero is zero.   **b.** $\frac{0}{0} = 1$

   c. Zero divided by any whole number, except zero, is zero.

116. Explain how the answer to a division problem can be checked.

1.6 Exponential Notation and the Order of Operations Agreement

Objective A   *To simplify expressions that contain exponents* ....

Repeated multiplication of the same factor can be written in two ways:

$$3 \cdot 3 \cdot 3 \cdot 3 \cdot 3 \quad \text{or} \quad 3^5 \leftarrow \textbf{exponent}$$

The exponent indicates how many times the factor occurs in the multiplication. The expression 3^5 is in **exponential notation.**

It is important to be able to read numbers written in exponential notation.

$6 = 6^1$ is read "six to the first power" or just "six." Usually the exponent 1 is not written.

$6 \cdot 6 = 6^2$ is read "six squared" or "six to the second power."

$6 \cdot 6 \cdot 6 = 6^3$ is read "six cubed" or "six to the third power."

$6 \cdot 6 \cdot 6 \cdot 6 = 6^4$ is read "six to the fourth power."

$6 \cdot 6 \cdot 6 \cdot 6 \cdot 6 = 6^5$ is read "six to the fifth power."

Each place value in the place-value chart can be expressed as a power of 10.

Ten =	10	=	10	$= 10^1$
Hundred =	100	=	$10 \cdot 10$	$= 10^2$
Thousand =	1000	=	$10 \cdot 10 \cdot 10$	$= 10^3$
Ten-Thousand =	10,000	=	$10 \cdot 10 \cdot 10 \cdot 10$	$= 10^4$
Hundred-Thousand =	100,000	=	$10 \cdot 10 \cdot 10 \cdot 10 \cdot 10$	$= 10^5$
Million =	1,000,000	=	$10 \cdot 10 \cdot 10 \cdot 10 \cdot 10 \cdot 10$	$= 10^6$

To simplify a numerical expression containing exponents, write each factor as many times as indicated by the exponent and carry out the indicated multiplication.

$$4^3 = 4 \cdot 4 \cdot 4 = 64$$
$$2^2 \cdot 3^4 = (2 \cdot 2) \cdot (3 \cdot 3 \cdot 3 \cdot 3) = 4 \cdot 81 = 324$$

Example 1   Write $3 \cdot 3 \cdot 3 \cdot 5 \cdot 5$ in exponential notation.

Solution $3 \cdot 3 \cdot 3 \cdot 5 \cdot 5 = 3^3 \cdot 5^2$

You Try It 1   Write $2 \cdot 2 \cdot 2 \cdot 2 \cdot 3 \cdot 3 \cdot 3$ in exponential notation.

Your solution

Example 2   Write as a power of 10:
$10 \cdot 10 \cdot 10 \cdot 10$

Solution $10 \cdot 10 \cdot 10 \cdot 10 = 10^4$

You Try It 2   Write as a power of 10:
$10 \cdot 10 \cdot 10 \cdot 10 \cdot 10 \cdot 10 \cdot 10$

Your solution 10^7

Example 3   Simplify $3^2 \cdot 5^3$.

Solution $3^2 \cdot 5^3 = (3 \cdot 3) \cdot (5 \cdot 5 \cdot 5)$
$\qquad = 9 \cdot 125 = 1125$

You Try It 3   Simplify $2^3 \cdot 5^2$.

Your solution $(2 \cdot 2 \cdot 2) \cdot (5 \cdot 5) =$

Solutions on p. S3

Objective B   **To use the Order of Operations Agreement to simplify expressions** ....

More than one operation may occur in a numerical expression. The answer may be different, depending on the order in which the operations are performed. For example, consider $3 + 4 \times 5$.

Multiply first, then add.   Add first, then multiply.

$$3 + \underbrace{4 \times 5}$$ $$\underbrace{3 + 4} \times 5$$

$$\underbrace{3 + 20}$$ $$\underbrace{7 \times 5}$$

$$23$$ $$35$$

An Order of Operations Agreement is used so that only one answer is possible.

The Order of Operations Agreement

Step 1   Do all the operations inside parentheses.
Step 2   Simplify any number expressions containing exponents.
Step 3   Do multiplication and division as they occur from left to right.
Step 4   Do addition and subtraction as they occur from left to right.

➡ Simplify $3 \times (2 + 1) - 2^2 + 4 \div 2$ by using the Order of Operations Agreement.

$$3 \times \underbrace{(2 + 1)} - 2^2 + 4 \div 2$$ 1.   Perform operations in parentheses.

$$3 \times 3 - \underbrace{2^2} + 4 \div 2$$ 2.   Simplify expressions with exponents.

$$\underbrace{3 \times 3} - 4 + 4 \div 2$$ 3.   Do multiplications and divisions as they occur from left to right.

$$9 - 4 + \underbrace{4 \div 2}$$

$$\underbrace{9 - 4} + 2$$ 4.   Do additions and subtractions as they occur from left to right.

$$\underbrace{5 + 2}$$

$$7$$

One or more of the above steps may not be needed to simplify an expression. In that case, proceed to the next step in the Order of Operations Agreement.

➡ Simplify $5 + 8 \div 2$.   No parentheses or exponents. Proceed to step 3 of the Agreement.

$$5 + \underbrace{8 \div 2}$$ 3.   Do multiplication or division.

$$\underbrace{5 + 4}$$ 4.   Do addition or subtraction.

$$9$$

Example 4   Simplify: $64 \div (8 - 4)^2 \cdot 9 - 5^2$

Solution $64 \div (8 - 4)^2 \cdot 9 - 5^2$
$= 64 \div 4^2 \cdot 9 - 5^2$
$= 64 \div 16 \cdot 9 - 25$
$= 4 \cdot 9 - 25$
$= 36 - 25$
$= 11$

You Try It 4   Simplify: $5 \cdot (8 - 4)^2 \div 4 - 2$

Your solution

Solution on p. S3

1.6 Exercises

· ·

Objective A

Write the number in exponential notation.

1. $2 \cdot 2 \cdot 2 = 2^3$

2. $7 \cdot 7 \cdot 7 \cdot 7 \cdot 7 = 7^5$

3. $6 \cdot 6 \cdot 6 \cdot 7 \cdot 7 \cdot 7 \cdot 7 = 6^3 \cdot 7^4$

4. $6 \cdot 6 \cdot 9 \cdot 9 \cdot 9 \cdot 9 = 6^2 \cdot 9^4$

5. $2 \cdot 2 \cdot 2 \cdot 3 \cdot 3 \cdot 3 = 2^3 \cdot 3^3$

6. $3 \cdot 3 \cdot 10 \cdot 10 = 3^2 \cdot 10^2$

7. $5 \cdot 7 \cdot 7 \cdot 7 \cdot 7 \cdot 7 = 5 \cdot 7^5$

8. $4 \cdot 4 \cdot 4 \cdot 5 \cdot 5 \cdot 5$

9. $3 \cdot 3 \cdot 3 \cdot 6 \cdot 6 \cdot 6 \cdot 6$

10. $2 \cdot 2 \cdot 5 \cdot 5 \cdot 5 \cdot 8$

11. $3 \cdot 3 \cdot 3 \cdot 5 \cdot 9 \cdot 9 \cdot 9$

12. $2 \cdot 2 \cdot 2 \cdot 4 \cdot 7 \cdot 7 \cdot 7$

Simplify.

13. 2^3

14. $2^6 = 64$

15. $2^4 \cdot 5^2$

16. $2^6 \cdot 3^2$

17. $3^2 \cdot 10^2$

18. $2^3 \cdot 10^4 = 80,000$

19. $6^2 \cdot 3^3$

20. $4^3 \cdot 5^2 = 1,600$

21. $5 \cdot 2^3 \cdot 3 = 120$

22. $6 \cdot 3^2 \cdot 4$

23. $2^2 \cdot 3^2 \cdot 10$

24. $3^2 \cdot 5^2 \cdot 10$

25. $0^2 \cdot 4^3 = 0$

26. $6^2 \cdot 0^3 = 0$

27. $3^2 \cdot 10^4$

28. $5^3 \cdot 10^3$

29. $2^2 \cdot 3^3 \cdot 5$

30. $5^2 \cdot 7^3 \cdot 2$

31. $2 \cdot 3^4 \cdot 5^2$

32. $6 \cdot 2^6 \cdot 7^2$

33. $5^2 \cdot 3^2 \cdot 7^2$

34. $4^2 \cdot 9^2 \cdot 6^2$

35. $3^4 \cdot 2^6 \cdot 5$

36. $4^3 \cdot 6^3 \cdot 7$

37. $4^2 \cdot 3^3 \cdot 10^4$

Objective B

Simplify by using the Order of Operations Agreement.

38. $4 - 2 + 3$

39. $6 - 3 + 2$

40. $6 \div 3 + 2$

41. $8 \div 4 + 8$

42. $6 \cdot 3 + 5$

43. $5 \cdot 9 + 2$

44. $3^2 - 4$

45. $5^2 - 17$

46. $4 \cdot (5 - 3) + 2$ **47.** $3 + (4 + 2) \div 3$ **48.** $5 + (8 + 4) \div 6$ **49.** $8 - 2^2 + 4$

50. $16 \cdot (3 + 2) \div 10$ **51.** $12 \cdot (1 + 5) \div 12$ **52.** $10 - 2^3 + 4$ **53.** $5 \cdot 3^2 + 8$

54. $16 + 4 \cdot 3^2$ **55.** $12 + 4 \cdot 2^3$ **56.** $16 + (8 - 3) \cdot 2$ **57.** $7 + (9 - 5) \cdot 3$

58. $2^2 + 3 \cdot (6 - 2)^2$ **59.** $3^3 + 5 \cdot (8 - 6)^3$ **60.** $2^2 \cdot 3^2 + 2 \cdot 3$ **61.** $4 \cdot 6 + 3^2 \cdot 4^2$

62. $16 - 2 \cdot 4$ **63.** $12 + 3 \cdot 5$ **64.** $3 \cdot (6 - 2) + 4$

65. $5 \cdot (8 - 4) - 6$ **66.** $8 - (8 - 2) \div 3$ **67.** $12 - (12 - 4) \div 4$

68. $8 + 2 - 3 \cdot 2 \div 3$ **69.** $10 + 1 - 5 \cdot 2 \div 5$ **70.** $3 \cdot (4 + 2) \div 6$

71. $(7 - 3)^2 \div 2 - 4 + 8$ **72.** $20 - 4 \div 2 \cdot (3 - 1)^3$ **73.** $12 \div 3 \cdot 2^2 + (7 - 3)^2$

74. $(4 - 2) \cdot 6 \div 3 + (5 - 2)^2$ **75.** $18 - 2 \cdot 3 + (4 - 1)^3$ **76.** $100 \div (2 + 3)^2 - 8 \div 2$

APPLYING THE CONCEPTS

77. Memory in computers is measured in bytes. One kilobyte (1 K) is 2^{10} bytes. Write this number in standard form.

78. Explain the difference in the order of operations for **a.** $\frac{14 - 2}{2} \div 2 \cdot 3$ and **b.** $\frac{14 - 2}{2} \div (2 \cdot 3)$. Work the two problems. What is the difference between the larger and the smaller answer?

79. If a number is divisible by 6 and 10, is the number divisible by 30? If so, explain why. If not, give an example.

1.7 Prime Numbers and Factoring

Objective A   *To factor numbers* ....

Whole-number factors of a number divide that number evenly (there is no remainder).

1, 2, 3, and 6 are whole-number factors of 6 because they divide 6 evenly.

$$\frac{6}{1)6} \quad \frac{3}{2)6} \quad \frac{2}{3)6} \quad \frac{1}{6)6}$$

Note that both the divisor and the quotient are factors of the dividend.

To find the factors of a number, try dividing the number by 1, 2, 3, 4, 5, ... Those numbers that divide the number evenly are its factors. Continue this process until the factors start to repeat.

➡ Find all the factors of 42.

$42 \div 1 = 42$	1 and 42 are factors.
$42 \div 2 = 21$	2 and 21 are factors.
$42 \div 3 = 14$	3 and 14 are factors.
$42 \div 4$	Will not divide evenly
$42 \div 5$	Will not divide evenly
$42 \div 6 = 7$	6 and 7 are factors. ⎱ Factors are repeating; all the
$42 \div 7 = 6$	7 and 6 are factors. ⎰ factors of 42 have been found.

1, 2, 3, 6, 7, 14, 21, and 42 are factors of 42.

The following rules are helpful in finding the factors of a number.

2 is a factor of a number if the last digit of the number is 0, 2, 4, 6, or 8.

436 ends in 6; therefore, 2 is a factor of 436. ($436 \div 2 = 218$)

3 is a factor of a number if the sum of the digits of the number is divisible by 3.

The sum of the digits of 489 is $4 + 8 + 9 = 21$. 21 is divisible by 3. Therefore, 3 is a factor of 489. ($489 \div 3 = 163$)

5 is a factor of a number if the last digit of the number is 0 or 5.

520 ends in 0; therefore, 5 is a factor of 520. ($520 \div 5 = 104$)

Example 1   Find all the factors of 30.

Solution
$30 \div 1 = 30$
$30 \div 2 = 15$
$30 \div 3 = 10$
$30 \div 4$ Will not divide
               evenly
$30 \div 5 = 6$
$30 \div 6 = 5$

1 2, 3, 5, 6, 10, 15, and 30 are factors of 30.

You Try It 1   Find all the factors of 40.

Your solution
$40 \div 1 = 40$ 1,2,4,5,8,10,
$40 \div 2 = 20$ 20,40
$40 \div 3 = N_0$
$40 \div 4 = 10$
$40 \div 5 = 8$
$40 \div 6 = N_0$
$40 \div 7 = N_0$
$40 \div 8 = 5$
$40 \div 9 = N_0$
$40 \div 10 = 4$

Solution on p. S3

Objective B   *To find the prime factorization of a number* ....

A number is a **prime number** if its only whole-number factors are 1 and itself. 7 is prime because its only factors are 1 and 7. If a number is not prime, it is called a **composite number.** Because 6 has factors of 2 and 3, 6 is a composite number. The number 1 is not considered a prime number; therefore it is not included in the following list of prime numbers less than 50.

2, 3, 5, 7, 11, 13, 17, 19, 23, 29, 31, 37, 41, 43, 47

The **prime factorization** of a number is the expression of the number as a product of its prime factors. We use a "T-diagram" to find the prime factors of 60. Begin with the smallest prime number as a trial divisor, and continue with prime numbers as trial divisors until the final quotient is 1.

	60	
2	30	$60 \div 2 = 30$
2	15	$30 \div 2 = 15$
3	5	$15 \div 3 = 5$
5	1	$5 \div 5 = 1$

The prime factorization of 60 is $2 \cdot 2 \cdot 3 \cdot 5$.

Finding the prime factorization of larger numbers can be more difficult. Try each prime number as a trial divisor. Stop when the square of the trial divisor is greater than the number being factored.

➡ Find the prime factorization of 106.

	106	
2	53	
53	1	

● **53 cannot be divided evenly by 2, 3, 5, 7, or 11. Prime numbers greater than 11 need not be tested because 11^2 is greater than 53.**

The prime factorization of 106 is $2 \cdot 53$.

Example 2   Find the prime factorization of 315.

Solution

	315
3	105
3	35
5	7
7	1

$315 = 3 \cdot 3 \cdot 5 \cdot 7$

You Try It 2   Find the prime factorization of 44.

Your solution

Example 3   Find the prime factorization of 201.

Solution

	201
3	67
67	1

● **Try only 2, 3, 5, 7, and 11 because $11^2 > 67$.**

$201 = 3 \cdot 67$

You Try It 3   Find the prime factorization of 177.

Your solution

Solutions on p. S3

1.7 Exercises

Objective A

Find all the factors of the number.

1. 4

2. 6

3. 10

4. 20

5. 7

6. 12

7. 9

8. 8

9. 13

10. 17

11. 18

12. 24

13. 56

14. 36

15. 45

16. 28

17. 29

18. 33

19. 22

20. 26

21. 52

22. 49

23. 82

24. 37

25. 57

26. 69

27. 48

28. 64

29. 95

30. 46

31. 54

32. 50

33. 66

34. 77

35. 80

36. 100

37. 96

38. 85

39. 90

40. 101

Objective B

Find the prime factorization.

41. 6   **42.** 14   **43.** 17   **44.** 83

45. 24   **46.** 12   **47.** 27   **48.** 9

49. 36   **50.** 40   **51.** 19   **52.** 37

53. 90   **54.** 65   **55.** 115   **56.** 80

57. 18   **58.** 26   **59.** 28   **60.** 49

61. 31   **62.** 42   **63.** 62   **64.** 81

65. 22   **66.** 39   **67.** 101   **68.** 89

69. 66   **70.** 86   **71.** 74   **72.** 95

73. 67   **74.** 78   **75.** 55   **76.** 46

77. 120   **78.** 144   **79.** 160   **80.** 175

81. 216   **82.** 400   **83.** 625   **84.** 225

APPLYING THE CONCEPTS

85. Twin primes are two prime numbers that differ by 2. For instance, 17 and 19 are twin primes. Find three sets of twin primes, not including 17 and 19.

86. In 1742, Christian Goldbach conjectured that every even number greater than 2 could be expressed as the sum of two prime numbers. Show that this conjecture is true for 8, 24, and 72. (*Note*: Mathematicians have not yet been able to determine whether Goldbach's conjecture is true or false.)

87. Explain why 2 is the only even prime number.

88. Explain the method of finding prime numbers using the Sieve of Erathosthenes.

Focus on Problem Solving

You encounter problem-solving situations every day. Some problems are easy to solve, and you may mentally solve these problems without considering the steps you are taking in order to draw a conclusion. Others may be more challenging and require more thought and consideration.

Suppose a friend suggests that you both take a trip over spring break. You'd like to go. What questions go through your mind? You might ask yourself some of the following questions:

How much will the trip cost? What will be the cost for travel, hotel rooms, meals, etc.?

Are some costs going to be shared by both me and my friend?

Can I afford it?

How much money do I have in the bank?

How much more money than I have now do I need?

How much time is there to earn that much money?

How much can I earn in that amount of time?

How much money must I keep in the bank in order to pay the next tuition bill (or some other expense)?

These questions require different mathematical skills. Determining the cost of the trip requires **estimation**; for example, you must use your knowledge of air fares or the cost of gasoline to arrive at an estimate of these costs. If some of the costs are going to be shared, you need to **divide** those costs by 2 in order to determine your share of the expense. The question regarding how much more money you need requires **subtraction**: the amount needed minus the amount currently in the bank. To determine how much money you can earn in the given amount of time requires **multiplication**—for example, the amount you earn per week times the number of weeks to be worked. To determine if the amount you can earn in the given amount of time is sufficient, you need to use your knowledge of **order relations** to compare the amount you can earn with the amount needed.

Facing the problem-solving situation described above may not seem difficult to you. The reason may be that you have faced similar situations before and, therefore, know how to work through this one. You may feel better prepared to deal with a circumstance such as this one because you know what questions to ask. An important aspect of learning to solve problems is learning what questions to ask. As you work through application problems in this text, try to become more conscious of the mental process you are going through. You might begin the process by asking yourself the following questions whenever you are solving an application problem:

1. Have I read the problem enough times to be able to understand the situation being described?

2. Will restating the problem in different words help me to better understand the problem situation?

3. What facts are given? (You might make a list of the information contained in the problem.)

4. What information is being asked for?

5. What relationship exists between the given facts? What relationship exists between the given facts and the solution?

6. What mathematical operations are needed in order to solve the problem?

Try to focus on the problem-solving situation and not the computation or getting the answer quickly. And remember, the more problems you solve, the better able you will be to solve other problems in the future, partly because you are learning what questions to ask.

Projects and Group Activities

Order of Operations

Does your calculator use the Order of Operations Agreement? To find out, try this problem:

$$2 + 4 \cdot 7$$

If your answer is 30, then the calculator uses the Order of Operations Agreement. If your answer is 42, it does not use that agreement.

Even if your calculator does not use the Order of Operations Agreement, you can still correctly evaluate numerical expressions. The parentheses keys, ⌐(and)⌐, are used for this purpose.

Remember that $2 + 4 \cdot 7$ means $2 + (4 \cdot 7)$ because the multiplication must be completed before the addition. To evaluate this expression, enter the following:

Enter: 2 [+] [( ] 4 [×] 7 [ )] [=]
Display: 2   2   [( ] 4   4   7   28   30

When using your calculator to evaluate numerical expressions, insert parentheses around multiplications or divisions. This has the effect of forcing the calculator to do the operations in the order you want rather than in the order the calculator wants.

Evaluate.

1. $3 \cdot 8 - 5$

2. $6 + 8 \div 2$

3. $3 \cdot (8 - 2)^2$

4. $24 - (4 - 2)^2 \div 4$

5. $3 + (6 \div 2 + 4)^2 - 2$

6. $16 \div 2 + 4 \cdot (8 - 12 \div 4)^2 - 50$

7. $3 \cdot (15 - 2 \cdot 3) - 36 \div 3$

8. $4 \cdot 2^2 - (12 + 24 \div 6) + 5$

9. $16 \div 4 \cdot 3 + (3 \cdot 4 - 5) + 2$

10. $15 \cdot 3 \div 9 + (2 \cdot 6 - 3) + 4$

Patterns in Mathematics

For the circle at the left, use a straight line to connect each dot on the circle with every other dot on the circle. How many different straight lines are there?

Follow the same procedure for each of the circles shown below. How many different straight lines are there in each?

Find a pattern to describe the number of dots on a circle and the corresponding number of different lines drawn. Use the pattern to determine the number of different lines that would be drawn in a circle with 7 dots and in a circle with 8 dots.

You are arranging a tennis tournament with 9 players. Following the pattern of finding the number of lines drawn on a circle, find the number of singles matches that can be played among the 9 players if each player plays each of the other players only once.

Search the World Wide Web

Go to the Internet and find www.census.gov. Click on U.S. population count. On the next screen, click on U.S. and estimate the following to the nearest thousand. Assume that the same rate of increase in population will continue into the future.

1. What is the population of the United States today?

2. Find the increase in population for a 24-hour period.

3. Using the information found in Exercise 2, find the increase in population for a 30-day period. Find the increase in population for one year (365 days).

4. Using the information in Exercise 3, estimate the population in the United States 50 years from now.

Click back and then click on "World Population." Estimate to the nearest million.

5. How much did the population increase in the last month? In the last year?

6. Use the increase in population for the past year from Exercise 5 to estimate the population of the world 10 years from now and 50 years from now.

Chapter Summary

Key Words   The *whole numbers* are 0, 1, 2, 3, 4, 5, 6, 7, 8, 9, 10, ...

The symbol for *"is less than"* is $<$.

The symbol for *"is greater than"* is $>$.

The position of a digit in a number determines the digit's *place value.*

Giving an approximate value for an exact number is called *rounding.*

An expression of the form 4^2 is in *exponential notation,* where 4 is the *base* and 2 is the *exponent.*

A number is *prime* if its only whole-number factors are 1 and itself.

The *prime factorization* of a number is the expression of the number as a product of its prime factors.

Essential Rules

The Addition Property of Zero	Zero added to a number does not change the number. $5 + 0 = 0 + 5 = 5$
The Commutative Property of Addition	Two numbers can be added in either order. $6 + 5 = 5 + 6 = 11$
The Associative Property of Addition	Grouping numbers to be added in any order gives the same result. $(2 + 3) + 5 = 2 + (3 + 5) = 10$
The Multiplication Property of Zero	The product of a number and zero is zero. $0 \times 3 = 3 \times 0 = 0$
The Multiplication Property of One	The product of a number and one is the number. $1 \times 8 = 8 \times 1 = 8$
The Commutative Property of Multiplication	Two numbers can be multiplied in any order. $5 \times 2 = 2 \times 5 = 10$
The Associative Property of Multiplication	Grouping numbers to be multiplied in any order gives the same result. $(3 \times 2) \times 5 = 3 \times (2 \times 5) = 30$
The Properties of Zero in Division	Zero divided by any other number is zero. Division by zero is not allowed.

Order of Operations Agreement

Step 1   Perform operations inside grouping symbols.
Step 2   Simplify expressions with exponents.
Step 3   Do multiplications and divisions as they occur from left to right.
Step 4   Do additions and subtractions as they occur from left to right.

Chapter Review

1. Simplify: $3 \cdot 2^3 \cdot 5^2$

2. Write 10,327 in expanded form.

3. Find all the factors of 18.

4. Find the sum of 5894, 6301, and 298.

5. Subtract: 4926
            $- \underline{3177}$

6. Divide: $7\overline{)14{,}945}$

7. Place the correct symbol, $<$ or $>$, between the two numbers: 101   87

8. Write $5 \cdot 5 \cdot 7 \cdot 7 \cdot 7 \cdot 7 \cdot 7$ in exponential notation.

9. What is 2019 multiplied by 307?

10. What is 10,134 decreased by 4725?

11. Add: 298
           461
        $+ \underline{322}$

12. Simplify: $2^3 - 3 \cdot 2$

13. Find the prime factorization of 42.

14. Write 276,057 in words.

15. Find the quotient of 109,763 and 84.

16. Write two million eleven thousand forty-four in standard form.

17. Find all the factors of 30.

18. Simplify: $3^2 + 2^2 \cdot (5 - 3)$

19. Simplify: $8 \cdot (6 - 2)^2 \div 4$

20. Find the prime factorization of 72.

21. Write $2 \cdot 2 \cdot 2 \cdot 2 \cdot 5 \cdot 5 \cdot 5$ in exponential notation.

22. Multiply:
$$
\begin{array}{r}
843 \\
\times\ 27 \\
\hline
\end{array}
$$

23. Vincent Meyers, a sales assistant, earns $240 for working a 40-hour week. Last week Vincent worked an additional 12 hours at $12 an hour. Find Vincent's total pay for last week's work.

24. Louis Reyes, a sales executive, drove a car 351 miles on 13 gallons of gas. Find the number of miles driven per gallon of gasoline.

25. A car is purchased for $8940, with a down payment of $1500. The balance is paid in 48 equal monthly payments. Find the monthly car payment.

26. An insurance account executive received commissions of $723, $544, $812, and $488 during a 4-week period. Find the total income from commissions for the 4 weeks.

27. You had a balance of $516 in your checking account before making deposits of $88 and $213. Find the total amount deposited, and determine your new account balance.

28. You have a car payment of $123 per month. What is the total of the car payments over a 12-month period?

The graph at the right shows the amount that companies from different countries spend on business trips. Use the graph for Exercises 29 to 31.

29. How much more does a U.S. company spend for business travel and entertainment than does a company from Italy?

Big (Travel) Spenders

U.S. companies spend more than twice the amount many European companies spend for business travel and entertainment. Spending per employee:

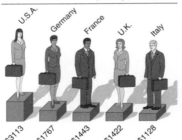

30. How much more does a U.S. company spend for business travel and entertainment than the combined spending of a company from the U.K. and a company from Italy?

31. Does a company from the United States outspend the combined spending of a company from Germany and a company from the U.K.?

Chapter Test

1. Simplify: $3^3 \cdot 4^2$

2. Write 207,068 in words.

3. Subtract:
$$\begin{array}{r} 17,495 \\ - \ 8,162 \end{array}$$

4. Find all the factors of 20.

5. Multiply:
$$\begin{array}{r} 9736 \\ \times \ 704 \end{array}$$

6. Simplify: $4^2 \cdot (4 - 2) \div 8 + 5$

7. Write 906,378 in expanded form.

8. Round 74,965 to the nearest hundred.

9. Divide: $97\overline{)108,764}$

10. Write $3 \cdot 3 \cdot 3 \cdot 7 \cdot 7$ in exponential form.

11. Find the sum of 89,756, 9094, and 37,065.

12. Find the prime factorization of 84.

13. Simplify: $16 \div 4 \cdot 2 - (7 - 5)^2$

14. Find the product of 90,763 and 8.

15. Write one million two hundred four thousand six in standard form.

16. Divide: $7\overline{)60,972}$

17. Place the correct symbol, $<$ or $>$, between the two numbers: 21   19

18. Find the quotient of 5624 and 8.

19. Add: 25,492
         +71,306

20. Find the difference between 29,736 and 9814.

The table at the right shows the estimated starting salaries of bachelor's degree candidates with engineering degrees. Use this table for Exercises 21 and 22.

21. What is the salary difference between the starting salaries of bachelor's degree candidates with degrees in mechanical engineering and industrial engineering?

Bachelor's Degree	Starting Salary
Chemical engineering	$42,758
Mechanical engineering	$39,852
Electrical engineering	$38,811
Industrial engineering	$37,732

Source: Michigan State University

22. Find the average salary of the bachelor's degree candidates with these engineering degrees. Round to the nearest whole number.

23. A farmer harvested 48,290 pounds of lemons from one grove and 23,710 pounds of lemons from another grove. The lemons were packed in boxes with 24 pounds of lemons in each box. How many boxes were needed to pack the lemons?

24. An investor receives $237 each month from a corporate bond fund. How much will the investor receive over a 12-month period?

25. A family drives 425 miles the first day, 187 miles the second day, and 243 miles the third day of their vacation. The odometer read 47,626 miles at the start of the vacation.
   a. How many miles were driven during the 3 days?
   b. What is the odometer reading at the end of the 3 days?

CHAPTER

2 | Fractions

Furniture makers can create intricate inlaid-wood patterns in, for example, dining tables or cabinets. Each piece of the pattern is precisely cut to very exact measurements. Measuring the pieces to be cut and ensuring that the pattern will fit the furniture requires that the furniture maker understand fractions and the operations on fractions.

Objectives

Section 2.1
To find the least common multiple (LCM)
To find the greatest common factor (GCF)

Section 2.2
To write a fraction that represents part of a whole
To write an improper fraction as a mixed number or a whole number, and a mixed number as an improper fraction

Section 2.3
To find equivalent fractions by raising to higher terms
To write a fraction in simplest form

Section 2.4
To add fractions with the same denominator
To add fractions with unlike denominators
To add whole numbers, mixed numbers, and fractions
To solve application problems

Section 2.5
To subtract fractions with the same denominator
To subtract fractions with unlike denominators
To subtract whole numbers, mixed numbers, and fractions
To solve application problems

Section 2.6
To multiply fractions
To multiply whole numbers, mixed numbers, and fractions
To solve application problems

Section 2.7
To divide fractions
To divide whole numbers, mixed numbers, and fractions
To solve application problems

Section 2.8
To identify the order relation between two fractions
To simplify expressions containing exponents
To use the Order of Operations Agreement to simplify expressions

Egyptian Fractions

The Rhind papyrus is one of the earliest written accounts of mathematics.[1] In the papyrus, a scribe named Ahmes gives an early account of the concept of fractions. A portion of the Rhind papyrus is rendered here with its hieroglyphic transcription.

The early Egyptians primarily used unit fractions. These are fractions in which the numerator is a 1. To write a fraction, a small oval was placed above a series of lines. The number of lines indicated the denominator. Some examples of these fractions are

$$\frac{\bigcirc}{||||} = \frac{1}{4} \qquad \bigtriangleup = \frac{1}{2}$$

In the first example, each line represents a 1. Because there are 4 lines, the fraction is $\frac{1}{4}$.

The second example is the special symbol that was used for the fraction $\frac{1}{2}$.

[1]Papyrus comes from the stem of a plant. The stem was dried and then pounded thin. The resulting material served as a primitive type of paper.

2.1 The Least Common Multiple and Greatest Common Factor

Objective A   *To find the least common multiple (LCM)* ....

The **multiples** of a number are the products of that number and the numbers 1, 2, 3, 4, 5, ...

$3 \times 1 = 3$
$3 \times 2 = 6$
$3 \times 3 = 9$
$3 \times 4 = 12$ The multiples of 3 are 3, 6, 9, 12, 15, ...
$3 \times 5 = 15$

A number that is a multiple of two or more numbers is a **common multiple** of those numbers.

The multiples of 4 are 4, 8, 12, 16, 20, 24, 28, 32, 36, ...
The multiples of 6 are 6, 12, 18, 24, 30, 36, 42, ...
Some common multiples of 4 and 6 are 12, 24, and 36.

The **least common multiple** (LCM) is the smallest common multiple of two or more numbers.

The least common multiple of 4 and 6 is 12.

Listing the multiples of each number is one way to find the LCM. Another way to find the LCM uses the prime factorization of each number.

To find the LCM of 450 and 600, find the prime factorization of each number and write the factorization of each number in a table. Circle the largest product in each column. The LCM is the product of the circled numbers.

	2	3	5
450 =	2	(3 · 3)	5 · 5
600 =	(2 · 2 · 2)	3	(5 · 5)

In the column headed by 5, the products are equal. Circle just one product.

The LCM is the product of the circled numbers.
The LCM = 2 · 2 · 2 · 3 · 3 · 5 · 5 = 1800.

Example 1   Find the LCM of 24, 36, and 50.

Solution

	2	3	5
24 =	(2 · 2 · 2)	3	
36 =	2 · 2	(3 · 3)	
50 =	2		(5 · 5)

The LCM =
2 · 2 · 2 · 3 · 3 · 5 · 5 = 1800.

You Try It 1   Find the LCM of 50, 84, and 135.

Your solution

50 = 2 · 5 · 5
84 =

Solution on p. S4

Objective B *To find the greatest common factor (GCF)* ....

Recall that a number that divides another number evenly is a factor of that number. 64 can be evenly divided by 1, 2, 4, 8, 16, 32, and 64. 1, 2, 4, 8, 16, 32, and 64 are factors of 64.

A number that is a factor of two or more numbers is a **common factor** of those numbers.

The factors of 30 are 1, 2, 3, 5, 6, 10, 15, and 30.
The factors of 105 are 1, 3, 5, 7, 15, 21, 35, and 105.
The common factors of 30 and 105 are 1, 3, 5, and 15.

The **greatest common factor** (GCF) is the largest common factor of two or more numbers.

The greatest common factor of 30 and 105 is 15.

Listing the factors of each number is one way of finding the GCF. Another way to find the GCF uses the prime factorization of each number.

To find the GCF of 126 and 180, find the prime factorization of each number and write the factorization of each number in a table. Circle the smallest product in each column that does not have a blank. The GCF is the product of the circled numbers.

	2	3	5	7
126 =	②	(3 · 3)		7
180 =	2 · 2	3 · 3	5	

In the column headed by 3, the products are equal. Circle just one product.

Columns 5 and 7 have a blank, so 5 and 7 are not common factors of 126 and 180. Do not circle any number in these columns.

The GCF is the product of the circled numbers.
The GCF = $2 \cdot 3 \cdot 3 = 18$.

Example 2 Find the GCF of 90, 168, and 420.

Solution

	2	3	5	7
90 =	②	3 · 3	5	
168 =	2 · 2 · 2	③		7
420 =	2 · 2	3	5	7

The GCF = $2 \cdot 3 = 6$.

You Try It 2 Find the GCF of 36, 60, and 72.

Your solution

Example 3 Find the GCF of 7, 12, and 20.

Solution

	2	3	5	7
7 =				7
12 =	2 · 2	3		
20 =	2 · 2		5	

Since no numbers are circled, the GCF = 1.

You Try It 3 Find the GCF of 11, 24, and 30.

Your solution

Solutions on p. S4

2.1 Exercises

· ·

Objective A

Find the LCM.

1. 5, 8 **2.** 3, 6 **3.** 3, 8 **4.** 2, 5 **5.** 5, 6

6. 5, 7 **7.** 4, 6 **8.** 6, 8 **9.** 8, 12 **10.** 12, 16

11. 5, 12 **12.** 3, 16 **13.** 8, 14 **14.** 6, 18 **15.** 3, 9

16. 4, 10 **17.** 8, 32 **18.** 7, 21 **19.** 9, 36 **20.** 14, 42

21. 44, 60 **22.** 120, 160 **23.** 102, 184 **24.** 123, 234 **25.** 4, 8, 12

26. 5, 10, 15 **27.** 3, 5, 10 **28.** 2, 5, 8 **29.** 3, 8, 12 **30.** 5, 12, 18

31. 9, 36, 64 **32.** 18, 54, 63 **33.** 16, 30, 84 **34.** 9, 12, 15

Objective B

Find the GCF.

35. 3, 5 **36.** 5, 7 **37.** 6, 9 **38.** 18, 24 **39.** 15, 25

40. 14, 49 **41.** 25, 100 **42.** 16, 80 **43.** 32, 51 **44.** 21, 44

45. 12, 80   **46.** 8, 36   **47.** 16, 140   **48.** 12, 76

49. 24, 30   **50.** 48, 144   **51.** 44, 96   **52.** 18, 32

53. 3, 5, 11   **54.** 6, 8, 10   **55.** 7, 14, 49   **56.** 6, 15, 36

57. 10, 15, 20   **58.** 12, 18, 20   **59.** 24, 40, 72   **60.** 3, 17, 51

61. 17, 31, 81   **62.** 14, 42, 84   **63.** 25, 125, 625   **64.** 12, 68, 92

65. 28, 35, 70   **66.** 1, 49, 153   **67.** 32, 56, 72   **68.** 24, 36, 48

APPLYING THE CONCEPTS

69. Define the phrase *relatively prime numbers*. List three pairs of relatively prime numbers.

70. Joe Salvo, a clerk, works three days and then has a day off. A friend works five days and then has a day off. How many days after Joe and his friend have a day off together will they have another day off together?

71. Find the LCM of the following pairs of numbers: 2 and 3, 5 and 7, and 11 and 19. Can you draw a conclusion about the LCM of two prime numbers? Suggest a way of finding the LCM of three prime numbers.

72. Find the GCF of the following pairs of numbers: 3 and 5, 7 and 11, and 29 and 43. Can you draw a conclusion about the GCF of two prime numbers? What is the GCF of three prime numbers?

73. Is the LCM of two numbers always divisible by the GCF of the two numbers? If so, explain why. If not, give an example.

74. Using the pattern for the first two triangles at the right, determine the center number of the last triangle.

75. The ancient Mayans used two calendars, a solar calendar of 365 days and a ritual calendar of 220 days. If a solar year and the ritual year begin on the same day, how many solar years and how many ritual years will pass before this situation occurs again? (*Mathematics Teacher*, November 1993, page 104)

2.2 Introduction to Fractions

Objective A *To write a fraction that represents part of a whole* ....

A **fraction** can represent the number of equal parts of a whole.

The shaded portion of the circle is represented by the fraction $\frac{4}{7}$. Four-sevenths of the circle are shaded.

Each part of a fraction has a name.

Fraction bar $\rightarrow \dfrac{4}{7}$ $\leftarrow$ Numerator
$\leftarrow$ Denominator

A **proper fraction** is a fraction less than 1. The numerator of a proper fraction is smaller than the denominator. The shaded portion of the circle can be represented by the proper fraction $\frac{3}{4}$.

A **mixed number** is a number greater than 1 with a whole-number part and a fractional part. The shaded portion of the circles can be represented by the mixed number $2\frac{1}{4}$.

An **improper fraction** is a fraction greater than or equal to 1. The numerator of an improper fraction is greater than or equal to the denominator. The shaded portion of the circles can be represented by the improper fraction $\frac{9}{4}$. The shaded portion of the square can be represented by $\frac{4}{4}$.

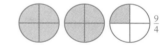

Example 1
Express the shaded portion of the circles as a mixed number.

Solution $3\frac{2}{5}$

Example 2
Express the shaded portion of the circles as an improper fraction.

Solution $\frac{17}{5}$

You Try It 1
Express the shaded portion of the circles as a mixed number.

Your solution

You Try It 2
Express the shaded portion of the circles as an improper fraction.

Your solution

Solutions on p. S4

Objective B *To write an improper fraction as a mixed number or a whole number, and a mixed number as an improper fraction* ....

Note from the diagram that the mixed number $2\frac{3}{5}$ and the improper fraction $\frac{13}{5}$ both represent the shaded portion of the circles.

$$2\frac{3}{5} = \frac{13}{5}$$

An improper fraction can be written as a mixed number.

⇒ Write $\frac{13}{5}$ as a mixed number.

Divide the numerator by the denominator.	To write the fractional part of the mixed number, write the remainder over the divisor.	Write the answer.
$$\begin{array}{r} 2 \\ 5\overline{)\ 13} \\ -10 \\ \hline 3 \end{array}$$	$$\begin{array}{r} 2\frac{3}{5} \\ 5\overline{)\ 13} \\ -10 \\ \hline 3 \end{array}$$	$$\frac{13}{5} = 2\frac{3}{5}$$

POINT OF INTEREST

Archimedes (c. 287–212 B.C.) is the person who calculated that $\pi \approx 3\frac{1}{7}$. He actually showed that $3\frac{10}{71} < \pi < 3\frac{1}{7}$. The approximation $3\frac{10}{71}$ is more accurate but more difficult to use.

To write a mixed number as an improper fraction, multiply the denominator of the fractional part by the whole-number part. The sum of this product and the numerator of the fractional part is the numerator of the improper fraction. The denominator remains the same.

⇒ Write $7\frac{3}{8}$ as an improper fraction.

$$7\frac{3}{8} = \frac{(8 \times 7) + 3}{8} = \frac{56 + 3}{8} = \frac{59}{8} \qquad 7\frac{3}{8} = \frac{59}{8}$$

Example 3 Write $\frac{21}{4}$ as a mixed number.

Solution
$$\begin{array}{r} 5 \\ 4\overline{)\ 21} \\ -20 \\ \hline 1 \end{array} \qquad \frac{21}{4} = 5\frac{1}{4}$$

You Try It 3 Write $\frac{22}{5}$ as a mixed number.

Your solution

Example 4 Write $\frac{18}{6}$ as a whole number.

Solution
$$\begin{array}{r} 3 \\ 6\overline{)\ 18} \\ -18 \\ \hline 0 \end{array} \qquad \frac{18}{6} = 3$$

Note: The remainder is zero.

You Try It 4 Write $\frac{28}{7}$ as a whole number.

Your solution

Example 5 Write $21\frac{3}{4}$ as an improper fraction.

Solution $21\frac{3}{4} = \frac{84 + 3}{4} = \frac{87}{4}$

You Try It 5 Write $14\frac{5}{8}$ as an improper fraction.

Your solution

Solutions on p. S4

2.2 Exercises

Objective A

Express the shaded portion of the circle as a fraction.

1.   **2.**   **3.**   **4.**

Express the shaded portion of the circles as a mixed number.

5. 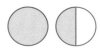   **6.**

7.    **8.**

9.    **10.**

Express the shaded portion of the circles as an improper fraction.

11. 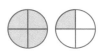   **12.**

13.    **14.**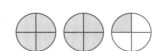

15. 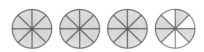   **16.**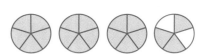

17. Shade $\frac{5}{6}$ of   **18.** Shade $\frac{3}{8}$ of

19. Shade $1\frac{2}{5}$ of 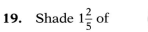   **20.** Shade $1\frac{3}{4}$ of

21. Shade $\frac{6}{5}$ of   **22.** Shade $\frac{7}{3}$ of

Objective B

Write the improper fraction as a mixed number or a whole number.

23. $\dfrac{11}{4}$ **24.** $\dfrac{16}{3}$ **25.** $\dfrac{20}{4}$ **26.** $\dfrac{18}{9}$ **27.** $\dfrac{9}{8}$ **28.** $\dfrac{13}{4}$

29. $\dfrac{23}{10}$ **30.** $\dfrac{29}{2}$ **31.** $\dfrac{48}{16}$ **32.** $\dfrac{51}{3}$ **33.** $\dfrac{8}{7}$ **34.** $\dfrac{16}{9}$

35. $\dfrac{7}{3}$ **36.** $\dfrac{9}{5}$ **37.** $\dfrac{16}{1}$ **38.** $\dfrac{23}{1}$ **39.** $\dfrac{17}{8}$ **40.** $\dfrac{31}{16}$

41. $\dfrac{12}{5}$ **42.** $\dfrac{19}{3}$ **43.** $\dfrac{9}{9}$ **44.** $\dfrac{40}{8}$ **45.** $\dfrac{72}{8}$ **46.** $\dfrac{3}{3}$

Write the mixed number as an improper fraction.

47. $2\dfrac{1}{3}$ **48.** $4\dfrac{2}{3}$ **49.** $6\dfrac{1}{2}$ **50.** $8\dfrac{2}{3}$ **51.** $6\dfrac{5}{6}$ **52.** $7\dfrac{3}{8}$

53. $9\dfrac{1}{4}$ **54.** $6\dfrac{1}{4}$ **55.** $10\dfrac{1}{2}$ **56.** $15\dfrac{1}{8}$ **57.** $8\dfrac{1}{9}$ **58.** $3\dfrac{5}{12}$

59. $5\dfrac{3}{11}$ **60.** $3\dfrac{7}{9}$ **61.** $2\dfrac{5}{8}$ **62.** $12\dfrac{2}{3}$ **63.** $1\dfrac{5}{8}$ **64.** $5\dfrac{3}{7}$

65. $11\dfrac{1}{9}$ **66.** $12\dfrac{3}{5}$ **67.** $3\dfrac{3}{8}$ **68.** $4\dfrac{5}{9}$ **69.** $6\dfrac{7}{13}$ **70.** $8\dfrac{5}{14}$

APPLYING THE CONCEPTS

71. What fraction of the states in the United States of America have names that begin with the letter M? What fraction of the states have names that begin and end with a vowel?

72. Find the business section of your local newspaper. Choose a stock and record the fluctuations in the stock price for one week. Explain the part that fractions play in reporting the price and change in price of the stock.

73. Explain in your own words the procedure for rewriting a mixed number as an improper fraction.

2.3 Writing Equivalent Fractions

Objective A   To find equivalent fractions by raising to higher terms ....

Equal fractions with different denominators are called **equivalent fractions.**

$\frac{4}{6}$ is equivalent to $\frac{2}{3}$.

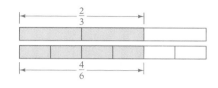

Remember that the Multiplication Property of One stated that the product of a number and one is the number. This is true for fractions as well as whole numbers. This property can be used to write equivalent fractions.

$$\frac{2}{3} \times 1 = \frac{2}{3} \times \frac{1}{1} = \frac{2 \cdot 1}{3 \cdot 1} = \frac{2}{3}$$

$$\frac{2}{3} \times 1 = \frac{2}{3} \times \boxed{\frac{2}{2}} = \frac{2 \cdot 2}{3 \cdot 2} = \frac{4}{6} \qquad \frac{4}{6} \text{ is equivalent to } \frac{2}{3}.$$

$$\frac{2}{3} \times 1 = \frac{2}{3} \times \boxed{\frac{4}{4}} = \frac{2 \cdot 4}{3 \cdot 4} = \frac{8}{12} \qquad \frac{8}{12} \text{ is equivalent to } \frac{2}{3}.$$

$\frac{2}{3}$ was rewritten as the equivalent fractions $\frac{4}{6}$ and $\frac{8}{12}$.

➡ Write a fraction that is equivalent to $\frac{5}{8}$ and has a denominator of 32.

$$8)\overline{32} \quad \overset{4}{}$$

$$\frac{5}{8} = \frac{5 \cdot 4}{8 \cdot 4} = \frac{20}{32}$$

• Divide the larger denominator by the smaller.
• Multiply the numerator and denominator of the given fraction by the quotient (4).

$\frac{20}{32}$ is equivalent to $\frac{5}{8}$.

Example 1   Write $\frac{2}{3}$ as an equivalent fraction that has a denominator of 42.

Solution $3)\overline{42} \quad \overset{14}{}$ $\frac{2}{3} = \frac{2 \cdot 14}{3 \cdot 14} = \frac{28}{42}$

$\frac{28}{42}$ is equivalent to $\frac{2}{3}$.

You Try It 1   Write $\frac{3}{5}$ as an equivalent fraction that has a denominator of 45.

Your solution

Example 2   Write 4 as a fraction that has a denominator of 12.

Solution   Write 4 as $\frac{4}{1}$.

$1)\overline{12} \quad \overset{12}{}$ $4 = \frac{4 \cdot 12}{1 \cdot 12} = \frac{48}{12}$

$\frac{48}{12}$ is equivalent to 4.

You Try It 2   Write 6 as a fraction that has a denominator of 18.

Your solution

Solutions on p. S4

***Objective B*   To write a fraction in simplest form** ....

A fraction is in **simplest form** when there are no common factors in the numerator and the denominator.

The fractions $\frac{4}{6}$ and $\frac{2}{3}$ are equivalent fractions.

$\frac{4}{6}$ has been written in simplest form as $\frac{2}{3}$.

The Multiplication Property of One can be used to write fractions in simplest form. Write the numerator and denominator of the given fraction as a product of factors. Write factors common to both the numerator and denominator as an improper fraction equivalent to 1.

$$\frac{4}{6} = \frac{2 \cdot 2}{2 \cdot 3} = \frac{2}{2} \cdot \frac{2}{3} = 1 \cdot \frac{2}{3} = \frac{2}{3}$$

The process of eliminating common factors is displayed with slashes through the common factors as shown at the right.

$$\frac{4}{6} = \frac{\overset{1}{\cancel{2}} \cdot 2}{\underset{1}{\cancel{2}} \cdot 3} = \frac{2}{3}$$

To write a fraction in simplest form, eliminate the common factors.

$$\frac{18}{30} = \frac{\overset{1}{\cancel{2}} \cdot \overset{1}{\cancel{3}} \cdot 3}{\underset{1}{\cancel{2}} \cdot \underset{1}{\cancel{3}} \cdot 5} = \frac{3}{5}$$

An improper fraction can be changed to a mixed number.

$$\frac{22}{6} = \frac{\overset{1}{\cancel{2}} \cdot 11}{\underset{1}{\cancel{2}} \cdot 3} = \frac{11}{3} = 3\frac{2}{3}$$

Example 3   Write $\frac{15}{40}$ in simplest form.

Solution $\dfrac{15}{40} = \dfrac{3 \cdot \overset{1}{\cancel{5}}}{2 \cdot 2 \cdot 2 \cdot \underset{1}{\cancel{5}}} = \dfrac{3}{8}$

You Try It 3   Write $\frac{16}{24}$ in simplest form.

Your solution

Example 4   Write $\frac{6}{42}$ in simplest form.

Solution $\dfrac{6}{42} = \dfrac{\overset{1}{\cancel{2}} \cdot \overset{1}{\cancel{3}}}{\underset{1}{\cancel{2}} \cdot \underset{1}{\cancel{3}} \cdot 7} = \dfrac{1}{7}$

You Try It 4   Write $\frac{8}{56}$ in simplest form.

Your solution

Example 5   Write $\frac{8}{9}$ in simplest form.

Solution $\dfrac{8}{9} = \dfrac{2 \cdot 2 \cdot 2}{3 \cdot 3} = \dfrac{8}{9}$

$\frac{8}{9}$ is already in simplest form because there are no common factors in the numerator and denominator.

You Try It 5   Write $\frac{15}{32}$ in simplest form.

Your solution

Example 6   Write $\frac{30}{12}$ in simplest form.

Solution $\dfrac{30}{12} = \dfrac{\overset{1}{\cancel{2}} \cdot \overset{1}{\cancel{3}} \cdot 5}{\underset{1}{\cancel{2}} \cdot 2 \cdot \underset{1}{\cancel{3}}} = \dfrac{5}{2} = 2\frac{1}{2}$

You Try It 6   Write $\frac{48}{36}$ in simplest form.

Your solution

Solutions on p. S4

2.3 Exercises

Objective A

Write an equivalent fraction with the given denominator.

1. $\dfrac{1}{2} = \dfrac{}{10}$ **2.** $\dfrac{1}{4} = \dfrac{}{16}$ **3.** $\dfrac{3}{16} = \dfrac{}{48}$ **4.** $\dfrac{5}{9} = \dfrac{}{81}$ **5.** $\dfrac{3}{8} = \dfrac{}{32}$

6. $\dfrac{7}{11} = \dfrac{}{33}$ **7.** $\dfrac{3}{17} = \dfrac{}{51}$ **8.** $\dfrac{7}{10} = \dfrac{}{90}$ **9.** $\dfrac{3}{4} = \dfrac{}{16}$ **10.** $\dfrac{5}{8} = \dfrac{}{32}$

11. $3 = \dfrac{}{9}$ **12.** $5 = \dfrac{}{25}$ **13.** $\dfrac{1}{3} = \dfrac{}{60}$ **14.** $\dfrac{1}{16} = \dfrac{}{48}$ **15.** $\dfrac{11}{15} = \dfrac{}{60}$

16. $\dfrac{3}{50} = \dfrac{}{300}$ **17.** $\dfrac{2}{3} = \dfrac{}{18}$ **18.** $\dfrac{5}{9} = \dfrac{}{36}$ **19.** $\dfrac{5}{7} = \dfrac{}{49}$ **20.** $\dfrac{7}{8} = \dfrac{}{32}$

21. $\dfrac{5}{9} = \dfrac{}{18}$ **22.** $\dfrac{11}{12} = \dfrac{}{36}$ **23.** $7 = \dfrac{}{3}$ **24.** $9 = \dfrac{}{4}$ **25.** $\dfrac{7}{9} = \dfrac{}{45}$

26. $\dfrac{5}{6} = \dfrac{}{42}$ **27.** $\dfrac{15}{16} = \dfrac{}{64}$ **28.** $\dfrac{11}{18} = \dfrac{}{54}$ **29.** $\dfrac{3}{14} = \dfrac{}{98}$ **30.** $\dfrac{5}{6} = \dfrac{}{144}$

31. $\dfrac{5}{8} = \dfrac{}{48}$ **32.** $\dfrac{7}{12} = \dfrac{}{96}$ **33.** $\dfrac{5}{14} = \dfrac{}{42}$ **34.** $\dfrac{2}{3} = \dfrac{}{42}$ **35.** $\dfrac{17}{24} = \dfrac{}{144}$

36. $\dfrac{5}{13} = \dfrac{}{169}$ **37.** $\dfrac{3}{8} = \dfrac{}{408}$ **38.** $\dfrac{9}{16} = \dfrac{}{272}$ **39.** $\dfrac{17}{40} = \dfrac{}{800}$ **40.** $\dfrac{9}{25} = \dfrac{}{1000}$

Objective B

Write the fraction in simplest form.

41. $\dfrac{4}{12}$ **42.** $\dfrac{8}{22}$ **43.** $\dfrac{22}{44}$ **44.** $\dfrac{2}{14}$ **45.** $\dfrac{2}{12}$

46. $\dfrac{50}{75}$ **47.** $\dfrac{40}{36}$ **48.** $\dfrac{12}{8}$ **49.** $\dfrac{0}{30}$ **50.** $\dfrac{10}{10}$

51. $\dfrac{9}{22}$ **52.** $\dfrac{14}{35}$ **53.** $\dfrac{75}{25}$ **54.** $\dfrac{8}{60}$ **55.** $\dfrac{16}{84}$

56. $\dfrac{20}{44}$ **57.** $\dfrac{12}{35}$ **58.** $\dfrac{8}{36}$ **59.** $\dfrac{28}{44}$ **60.** $\dfrac{12}{16}$

61. $\dfrac{16}{12}$ **62.** $\dfrac{24}{18}$ **63.** $\dfrac{24}{40}$ **64.** $\dfrac{44}{60}$ **65.** $\dfrac{8}{88}$

66. $\dfrac{9}{90}$ **67.** $\dfrac{144}{36}$ **68.** $\dfrac{140}{297}$ **69.** $\dfrac{48}{144}$ **70.** $\dfrac{32}{120}$

71. $\dfrac{60}{100}$ **72.** $\dfrac{33}{110}$ **73.** $\dfrac{36}{16}$ **74.** $\dfrac{80}{45}$ **75.** $\dfrac{32}{160}$

APPLYING THE CONCEPTS

76. Make a list of five different fractions that are equivalent to $\frac{2}{3}$.

77. Make a list of five different fractions that are equivalent to 3.

78. Show that $\frac{15}{24} = \frac{5}{8}$ by using a diagram.

79. Explain the procedure for finding equivalent fractions.

80. Explain the procedure for reducing fractions.

2.4 Addition of Fractions and Mixed Numbers

Objective A   *To add fractions with the same denominator* ....

Fractions with the same denominator are added by adding the numerators and placing the sum over the common denominator. After adding, write the sum in simplest form.

➡ Add: $\dfrac{2}{7} + \dfrac{4}{7}$

$$\begin{array}{r} \dfrac{2}{7} \\ + \dfrac{4}{7} \\ \hline \dfrac{6}{7} \end{array}$$

• Add the numerators and place the sum over the common denominator.

$$\frac{2}{7} + \frac{4}{7} = \frac{2+4}{7} = \frac{6}{7}$$

Example 1   Add: $\dfrac{5}{12} + \dfrac{11}{12}$

Solution

$$\begin{array}{r} \dfrac{5}{12} \\ + \dfrac{11}{12} \\ \hline \dfrac{16}{12} = \dfrac{4}{3} = 1\dfrac{1}{3} \end{array}$$

You Try It 1   Add: $\dfrac{3}{8} + \dfrac{7}{8}$

Your solution

Solution on p. S4

Objective B   *To add fractions with unlike denominators* ....

To add fractions with unlike denominators, first rewrite the fractions as equivalent fractions with a common denominator. The common denominator is the LCM of the denominators of the fractions.

➡ Find the total of $\dfrac{1}{2}$ and $\dfrac{1}{3}$.

The common denominator is the LCM of 2 and 3. LCM = 6. The LCM of denominators is sometimes called the **least common denominator** (LCD).

Write equivalent fractions using the LCM.

$$\begin{array}{r} \dfrac{1}{2} = \dfrac{3}{6} \\ + \dfrac{1}{3} = \dfrac{2}{6} \\ \hline \end{array}$$

Add the fractions.

$$\begin{array}{r} \dfrac{1}{2} = \dfrac{3}{6} \\ + \dfrac{1}{3} = \dfrac{2}{6} \\ \hline \dfrac{5}{6} \end{array}$$

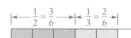

Example 2 Find $\frac{7}{12}$ more than $\frac{3}{8}$.

Solution

$$\frac{3}{8} = \frac{9}{24}$$

$$+\frac{7}{12} = \frac{14}{24}$$

$$\frac{23}{24}$$

• The LCM of 8 and 12 is 24.

You Try It 2 Find the sum of $\frac{5}{12}$ and $\frac{9}{16}$.

Your solution

Example 3 Add: $\frac{5}{8} + \frac{7}{9}$

Solution

$$\frac{5}{8} = \frac{45}{72}$$

$$+\frac{7}{9} = \frac{56}{72}$$

$$\frac{101}{72} = 1\frac{29}{72}$$

• The LCM of 8 and 9 is 72.

You Try It 3 Add: $\frac{7}{8} + \frac{11}{15}$

Your solution

Example 4 Add: $\frac{2}{3} + \frac{3}{5} + \frac{5}{6}$

Solution

$$\frac{2}{3} = \frac{20}{30}$$

$$\frac{3}{5} = \frac{18}{30}$$

$$+\frac{5}{6} = \frac{25}{30}$$

$$\frac{63}{30} = 2\frac{3}{30} = 2\frac{1}{10}$$

• The LCM of 3, 5, and 6 is 30.

You Try It 4 Add: $\frac{3}{4} + \frac{4}{5} + \frac{5}{8}$

Your solution

Solutions on pp. S4–S5

Objective C ***To add whole numbers, mixed numbers, and fractions*** ....

The sum of a whole number and a fraction is a mixed number.

➡ Add: $2 + \frac{2}{3}$

$$\boxed{2} + \frac{2}{3} = \boxed{\frac{6}{3}} + \frac{2}{3} = \frac{8}{3} = 2\frac{2}{3}$$

To add a whole number and a mixed number, write the fraction, then add the whole numbers.

➡ Add: $7\frac{2}{5} + 4$

Write the fraction.

$$7\frac{2}{5}$$
$$+4$$
$$\overline{\frac{2}{5}}$$

Add the whole numbers.

$$7\frac{2}{5}$$
$$+4$$
$$\overline{11\frac{2}{5}}$$

To add two mixed numbers, add the fractional parts and then add the whole numbers. Remember to reduce the sum to simplest form.

➡ What is $6\frac{14}{15}$ added to $5\frac{4}{9}$?

The LCM of 9 and 15 is 45.

Add the fractional parts.

$$5\frac{4}{9} = 5\frac{20}{45}$$
$$+\ 6\frac{14}{15} = 6\frac{42}{45}$$
$$\frac{62}{45}$$

Add the whole numbers.

$$5\frac{4}{9} = 5\frac{20}{45}$$
$$+\ 6\frac{14}{15} = 6\frac{42}{45}$$
$$11\frac{62}{45} = 11 + 1\frac{17}{45} = 12\frac{17}{45}$$

Example 5   Add: $5 + \frac{3}{8}$

Solution $5 + \frac{3}{8} = 5\frac{3}{8}$

You Try It 5   What is 7 added to $\frac{6}{11}$?

Your solution

Example 6   Find 17 increased by $3\frac{3}{8}$.

Solution
$$17$$
$$+\ 3\frac{3}{8}$$
$$20\frac{3}{8}$$

You Try It 6   Find the sum of 29 and $17\frac{5}{12}$.

Your solution

Example 7   Add: $5\frac{2}{3} + 11\frac{5}{6} + 12\frac{7}{9}$

Solution
$$5\frac{2}{3} = 5\frac{12}{18} \quad \bullet \text{ LCM} = 18.$$
$$11\frac{5}{6} = 11\frac{15}{18}$$
$$+\ 12\frac{7}{9} = 12\frac{14}{18}$$
$$28\frac{41}{18} = 30\frac{5}{18}$$

You Try It 7   Add: $7\frac{4}{5} + 6\frac{7}{10} + 13\frac{11}{15}$

Your solution

Example 8   Add: $11\frac{5}{8} + 7\frac{5}{9} + 8\frac{7}{15}$

Solution
$$11\frac{5}{8} = 11\frac{225}{360} \quad \bullet \text{ LCM} = 360.$$
$$7\frac{5}{9} = 7\frac{200}{360}$$
$$+\ 8\frac{7}{15} = 8\frac{168}{360}$$
$$26\frac{593}{360} = 27\frac{233}{360}$$

You Try It 8   Add: $9\frac{3}{8} + 17\frac{7}{12} + 10\frac{14}{15}$

Your solution

Solutions on p. S5

Objective D   *To solve application problems* ....

Example 9

A rain gauge collected $2\frac{1}{3}$ inches of rain in October, $5\frac{1}{2}$ inches in November, and $3\frac{3}{8}$ inches in December. Find the total rainfall for the 3 months.

Strategy

To find the total rainfall for the 3 months, add the 3 amounts of rainfall $\left(2\frac{1}{3}, 5\frac{1}{2}, \text{and } 3\frac{3}{8}\right)$.

Solution

$$2\frac{1}{3} = 2\frac{8}{24}$$
$$5\frac{1}{2} = 5\frac{12}{24}$$
$$+ 3\frac{3}{8} = 3\frac{9}{24}$$
$$10\frac{29}{24} = 11\frac{5}{24}$$

The total rainfall for the 3 months was $11\frac{5}{24}$ inches.

You Try It 9

On Monday, you spent $4\frac{1}{2}$ hours in class, $3\frac{3}{4}$ hours studying, and $1\frac{1}{3}$ hours driving. Find the number of hours spent on these three activities.

Your strategy

Your solution

Example 10

Barbara Walsh worked 4 hours, $2\frac{1}{3}$ hours, and $5\frac{2}{3}$ hours this week on a part-time job. Barbara is paid $6 an hour. How much did she earn this week?

Strategy

To find how much Barbara earned:

- Find the total number of hours worked.
- Multiply the total number of hours worked by the hourly wage ($6).

Solution

$$
\begin{array}{r}
4 \\
2\frac{1}{3} \\
+ 5\frac{2}{3} \\
\hline
11\frac{3}{3} = 12 \text{ hours worked}
\end{array}
\qquad
\begin{array}{r}
12 \\
\times \$6 \\
\hline
\$72
\end{array}
$$

Barbara earned $72 this week.

You Try It 10

Jeff Sapone, a carpenter, worked $1\frac{2}{3}$ hours of overtime on Monday, $3\frac{1}{3}$ hours of overtime on Tuesday, and 2 hours of overtime on Wednesday. At an overtime hourly rate of $24, find Jeff's overtime pay for these 3 days.

Your strategy

Your solution

Solutions on p. S5

2.4 Exercises

Objective A

Add.

1. $\dfrac{2}{7} + \dfrac{1}{7}$

2. $\dfrac{3}{11} + \dfrac{5}{11}$

3. $\dfrac{1}{2} + \dfrac{1}{2}$

4. $\dfrac{1}{3} + \dfrac{2}{3}$

5. $\dfrac{8}{11} + \dfrac{7}{11}$

6. $\dfrac{9}{13} + \dfrac{7}{13}$

7. $\dfrac{8}{5} + \dfrac{9}{5}$

8. $\dfrac{5}{3} + \dfrac{7}{3}$

9. $\dfrac{3}{5} + \dfrac{8}{5} + \dfrac{3}{5}$

10. $\dfrac{3}{8} + \dfrac{5}{8} + \dfrac{7}{8}$

11. $\dfrac{3}{4} + \dfrac{1}{4} + \dfrac{5}{4}$

12. $\dfrac{2}{7} + \dfrac{4}{7} + \dfrac{5}{7}$

13. $\dfrac{3}{8} + \dfrac{7}{8} + \dfrac{1}{8}$

14. $\dfrac{5}{12} + \dfrac{7}{12} + \dfrac{1}{12}$

15. $\dfrac{4}{15} + \dfrac{7}{15} + \dfrac{11}{15}$

16. $\dfrac{3}{4} + \dfrac{3}{4} + \dfrac{1}{4}$

17. $\dfrac{3}{16} + \dfrac{5}{16} + \dfrac{7}{16}$

18. $\dfrac{5}{18} + \dfrac{11}{18} + \dfrac{17}{18}$

19. $\dfrac{3}{11} + \dfrac{5}{11} + \dfrac{7}{11}$

20. $\dfrac{5}{7} + \dfrac{4}{7} + \dfrac{5}{7}$

21. Find the sum of $\dfrac{4}{9}$ and $\dfrac{5}{9}$.

22. Find the sum of $\dfrac{5}{12}$, $\dfrac{1}{12}$, and $\dfrac{11}{12}$.

23. Find the total of $\dfrac{5}{8}$, $\dfrac{3}{8}$, and $\dfrac{7}{8}$.

24. Find the total of $\dfrac{4}{13}$, $\dfrac{7}{13}$, and $\dfrac{11}{13}$.

Objective B

Add.

25. $\dfrac{1}{2} + \dfrac{2}{3}$

26. $\dfrac{2}{3} + \dfrac{1}{4}$

27. $\dfrac{3}{14} + \dfrac{5}{7}$

28. $\dfrac{3}{5} + \dfrac{7}{10}$

29. $\dfrac{8}{15} + \dfrac{7}{20}$

30. $\dfrac{1}{6} + \dfrac{7}{9}$

31. $\dfrac{3}{8} + \dfrac{9}{14}$

32. $\dfrac{5}{12} + \dfrac{5}{16}$

33. $\dfrac{3}{20} + \dfrac{7}{30}$

34. $\dfrac{5}{12} + \dfrac{7}{30}$

35. $\dfrac{2}{3} + \dfrac{6}{19}$

36. $\dfrac{1}{2} + \dfrac{3}{29}$

37. $\dfrac{1}{3} + \dfrac{5}{6} + \dfrac{7}{9}$ **38.** $\dfrac{2}{3} + \dfrac{5}{6} + \dfrac{7}{12}$ **39.** $\dfrac{5}{6} + \dfrac{1}{12} + \dfrac{5}{16}$ **40.** $\dfrac{2}{9} + \dfrac{7}{15} + \dfrac{4}{21}$

41. $\dfrac{2}{3} + \dfrac{1}{5} + \dfrac{7}{12}$ **42.** $\dfrac{3}{4} + \dfrac{4}{5} + \dfrac{7}{12}$ **43.** $\dfrac{1}{4} + \dfrac{4}{5} + \dfrac{5}{9}$ **44.** $\dfrac{2}{3} + \dfrac{3}{5} + \dfrac{7}{8}$

45. $\dfrac{5}{16} + \dfrac{11}{18} + \dfrac{17}{24}$ **46.** $\dfrac{3}{10} + \dfrac{14}{15} + \dfrac{9}{25}$ **47.** $\dfrac{2}{3} + \dfrac{5}{8} + \dfrac{7}{9}$ **48.** $\dfrac{1}{3} + \dfrac{2}{9} + \dfrac{7}{8}$

49. What is $\dfrac{3}{8}$ added to $\dfrac{3}{5}$?   **50.** What is $\dfrac{5}{9}$ added to $\dfrac{7}{12}$?

51. Find the sum of $\dfrac{3}{8}$, $\dfrac{5}{6}$, and $\dfrac{7}{12}$.   **52.** Find the sum of $\dfrac{11}{12}$, $\dfrac{13}{24}$, and $\dfrac{4}{15}$.

53. Find the total of $\dfrac{1}{2}$, $\dfrac{5}{8}$, and $\dfrac{7}{9}$.   **54.** Find the total of $\dfrac{5}{14}$, $\dfrac{3}{7}$, and $\dfrac{5}{21}$.

Objective C

Add.

55. $\begin{aligned} 1\tfrac{1}{2} \\ +\, 2\tfrac{1}{6} \\ \hline \end{aligned}$ **56.** $\begin{aligned} 2\tfrac{2}{5} \\ +\, 3\tfrac{3}{10} \\ \hline \end{aligned}$ **57.** $\begin{aligned} 4\tfrac{1}{2} \\ +\, 5\tfrac{7}{12} \\ \hline \end{aligned}$ **58.** $\begin{aligned} 3\tfrac{3}{8} \\ +\, 2\tfrac{5}{16} \\ \hline \end{aligned}$

59. $\begin{aligned} 4 \\ +\, 5\tfrac{2}{7} \\ \hline \end{aligned}$ **60.** $\begin{aligned} 6\tfrac{8}{9} \\ +\, 12 \\ \hline \end{aligned}$ **61.** $\begin{aligned} 3\tfrac{5}{8} \\ +\, 2\tfrac{11}{20} \\ \hline \end{aligned}$ **62.** $\begin{aligned} 4\tfrac{5}{12} \\ +\, 6\tfrac{11}{18} \\ \hline \end{aligned}$

63. $7\dfrac{5}{12} + 2\dfrac{9}{16}$ **64.** $9\dfrac{1}{2} + 3\dfrac{3}{11}$ **65.** $6\dfrac{1}{3} + 2\dfrac{3}{13}$ **66.** $8\dfrac{21}{40} + 6\dfrac{21}{32}$

67. $8\dfrac{29}{30} + 7\dfrac{11}{40}$ **68.** $17\dfrac{5}{16} + 3\dfrac{11}{24}$ **69.** $17\dfrac{3}{8} + 7\dfrac{7}{20}$ **70.** $14\dfrac{7}{12} + 29\dfrac{13}{21}$

71. $5\frac{7}{8} + 27\frac{5}{12}$

72. $7\frac{5}{6} + 3\frac{5}{9}$

73. $7\frac{5}{9} + 2\frac{7}{12}$

74. $3\frac{1}{2} + 2\frac{3}{4} + 1\frac{5}{6}$

75. $2\frac{1}{2} + 3\frac{2}{3} + 4\frac{1}{4}$

76. $3\frac{1}{3} + 7\frac{1}{5} + 2\frac{1}{7}$

77. $3\frac{1}{2} + 3\frac{1}{5} + 8\frac{1}{9}$

78. $6\frac{5}{9} + 6\frac{5}{12} + 2\frac{5}{18}$

79. $2\frac{3}{8} + 4\frac{7}{12} + 3\frac{5}{16}$

80. $2\frac{1}{8} + 4\frac{2}{9} + 5\frac{17}{18}$

81. $6\frac{5}{6} + 17\frac{2}{9} + 18\frac{5}{27}$

82. $4\frac{7}{20} + \frac{17}{80} + 25\frac{23}{60}$

83. Find the sum of $2\frac{4}{9}$ and $5\frac{7}{12}$.

84. Find $5\frac{5}{6}$ more than $3\frac{3}{8}$.

85. What is $4\frac{3}{4}$ added to $9\frac{1}{3}$?

86. What is $4\frac{8}{9}$ added to $9\frac{1}{6}$?

87. Find the total of $2\frac{2}{3}$, $4\frac{5}{8}$, and $2\frac{2}{9}$.

88. Find the total of $1\frac{5}{8}$, $3\frac{5}{6}$, and $7\frac{7}{24}$.

Objective D   *Application Problems*

89. A family with an income of $38,000 spends $\frac{1}{3}$ of its income on housing, $\frac{1}{8}$ on transportation, and $\frac{1}{4}$ on food. Find the total fractional amount of the family's income that is spent on these three items.

90. A table 30 inches high has a top that is $1\frac{1}{8}$ inches thick. Find the total thickness of the table top after a $\frac{3}{16}$-inch veneer is applied.

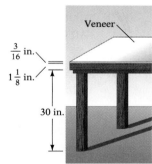

91. ◖ On June 24, 1997, the price of one share of BrMySq stock was $76\frac{3}{8}$. On June 25, 1997, the stock increased in value by $8\frac{3}{8}$ per share. Find the value of one share of the stock after this increase.

92. Find the length of the shaft.

93. Find the length of the shaft.

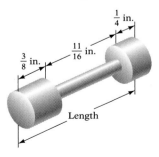

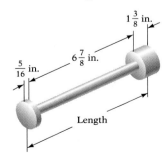

94. You are working a part-time job that pays \$7 an hour. You worked 5, $3\frac{3}{4}$, $2\frac{1}{3}$, $1\frac{1}{4}$, and $7\frac{2}{3}$ hours during the last five days.

    a. Find the total number of hours you worked during the last five days.

    b. Find your total wages for the five days.

95. Fred Thomson, a nurse, worked $2\frac{2}{3}$ hours of overtime on Monday, $1\frac{1}{4}$ hours on Wednesday, $1\frac{1}{3}$ hours on Friday, and $6\frac{3}{4}$ hours on Saturday.

    a. Find the total number of overtime hours worked during the week.

    b. At an overtime hourly wage of \$22 per hour, how much overtime pay does Fred receive?

96. Mt. Baldy had $5\frac{3}{4}$ inches of snow in December, $15\frac{1}{2}$ inches in January, and $9\frac{5}{8}$ inches in February. Find the total snowfall for the three months.

97. Alan bought 200 shares of a utility stock for $\$26\frac{3}{16}$ per share. The monthly gains for the next three months were $\$1\frac{1}{2}$, $\$\frac{5}{16}$, and $\$2\frac{5}{8}$ per share. Find the value of the stock at the end of the three months.

APPLYING THE CONCEPTS

98. What is a unit fraction? Find the sum of the three largest unit fractions. Is there a smallest unit fraction? If so, write it down. If not, explain why.

99. Use a model to illustrate and explain the addition of fractions with unlike denominators.

100. A survey was conducted to determine people's favorite color from among blue, green, red, purple, or other. The surveyor claims that $\frac{1}{3}$ of the people responded blue, $\frac{1}{6}$ responded green, $\frac{1}{8}$ responded red, $\frac{1}{12}$ responded purple, and $\frac{2}{5}$ responded some other color. Is this possible? Explain your answer.

101. The following are the average portions of each day that a person spends for each activity: sleeping, $\frac{1}{3}$; working, $\frac{1}{3}$; personal hygiene, $\frac{1}{24}$; eating, $\frac{1}{8}$; rest and relaxation, $\frac{1}{12}$. Do these five activities account for an entire day? Explain your answer.

2.5 Subtraction of Fractions and Mixed Numbers

Objective A *To subtract fractions with the same denominator* ....

Fractions with the same denominator are subtracted by subtracting the numerators and placing the difference over the common denominator. After subtracting, write the fraction in simplest form.

➡ Subtract: $\dfrac{5}{7} - \dfrac{3}{7}$

$$\begin{array}{r} \dfrac{5}{7} \\[2mm] -\dfrac{3}{7} \\ \hline \dfrac{2}{7} \end{array}$$

• Subtract the numerators and place the difference over the common denominator

$$\dfrac{5}{7} - \dfrac{3}{7} = \dfrac{5-3}{7} = \dfrac{2}{7}$$

Example 1   Find $\dfrac{17}{30}$ less $\dfrac{11}{30}$.

Solution

$$\begin{array}{r} \dfrac{17}{30} \\[2mm] -\dfrac{11}{30} \\ \hline \dfrac{6}{30} = \dfrac{1}{5} \end{array}$$

You Try It 1   Subtract: $\dfrac{16}{27} - \dfrac{7}{27}$

Your solution

Solution on p. S5

Objective B *To subtract fractions with unlike denominators* ....

To subtract fractions with unlike denominators, first rewrite the fractions as equivalent fractions with a common denominator. As with adding fractions, the common denominator is the LCM of the denominators of the fractions.

➡ Subtract: $\dfrac{5}{6} - \dfrac{1}{4}$

The common denominator is the LCM of 6 and 4. LCM = 12.

Build equivalent fractions using the LCM.

$$\dfrac{5}{6} = \dfrac{10}{12}$$
$$-\dfrac{1}{4} = \dfrac{3}{12}$$

Subtract the fractions.

$$\begin{array}{r} \dfrac{5}{6} = \dfrac{10}{12} \\[2mm] -\dfrac{1}{4} = \dfrac{3}{12} \\ \hline \dfrac{7}{12} \end{array}$$

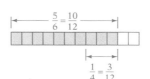

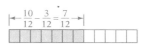

Example 2   Subtract: $\dfrac{11}{16} - \dfrac{5}{12}$

Solution

$\dfrac{11}{16} = \dfrac{33}{48}$ • LCM = 48

$-\dfrac{5}{12} = \dfrac{20}{48}$

$\phantom{-\dfrac{5}{12}} \dfrac{13}{48}$

You Try It 2   Subtract: $\dfrac{13}{18} - \dfrac{7}{24}$

Your solution

Solution on p. S5

Objective C   *To subtract whole numbers, mixed numbers, and fractions* ......

To subtract mixed numbers without borrowing, subtract the fractional parts and then subtract the whole numbers.

➡ Subtract: $5\dfrac{5}{6} - 2\dfrac{3}{4}$

Subtract the fractional parts.

$5\dfrac{5}{6} = 5\dfrac{10}{12}$ • The LCM of 6 and 4 is 12.

$-2\dfrac{3}{4} = 2\dfrac{9}{12}$

$\phantom{-2\dfrac{3}{4}} \dfrac{1}{12}$

Subtract the whole numbers.

$5\dfrac{5}{6} = 5\dfrac{10}{12}$

$-2\dfrac{3}{4} = 2\dfrac{9}{12}$

$\phantom{-2\dfrac{3}{4}} 3\dfrac{1}{12}$

Subtraction of mixed numbers sometimes involves borrowing.

➡ Subtract: $5 - 2\dfrac{5}{8}$

Borrow 1 from 5.

$5 = \overset{4}{\cancel{5}}\,1$

$-2\dfrac{5}{8} = 2\dfrac{5}{8}$

Write 1 as a fraction so that the fractions have the same denominators.

$5 = 4\dfrac{8}{8}$

$-2\dfrac{5}{8} = 2\dfrac{5}{8}$

Subtract the mixed numbers.

$5 = 4\dfrac{8}{8}$

$-2\dfrac{5}{8} = 2\dfrac{5}{8}$

$\phantom{-2\dfrac{5}{8}} 2\dfrac{3}{8}$

➡ Subtract: $7\dfrac{1}{6} - 2\dfrac{5}{8}$

Write equivalent fractions using the LCM.

$7\dfrac{1}{6} = 7\dfrac{4}{24}$

$-2\dfrac{5}{8} = 2\dfrac{15}{24}$

Borrow 1 from 7. Add the 1 to $\dfrac{4}{24}$. Write $1\dfrac{4}{24}$ as $\dfrac{28}{24}$.

$7\dfrac{1}{6} = \overset{6}{\cancel{7}}1\dfrac{4}{24} = 6\dfrac{28}{24}$

$-2\dfrac{5}{8} = 2\dfrac{15}{24} = 2\dfrac{15}{24}$

Subtract the mixed numbers.

$7\dfrac{1}{6} = 6\dfrac{28}{24}$

$-2\dfrac{5}{8} = 2\dfrac{15}{24}$

$\phantom{-2\dfrac{5}{8}} 4\dfrac{13}{24}$

Example 3   Subtract: $15\frac{7}{8} - 12\frac{2}{3}$

Solution

$$15\frac{7}{8} = 15\frac{21}{24} \qquad \text{LCM = 24.}$$
$$-\ 12\frac{2}{3} = 12\frac{16}{24}$$
$$\overline{\qquad\qquad 3\frac{5}{24}}$$

You Try It 3   Subtract: $17\frac{5}{9} - 11\frac{5}{12}$

Your solution

Example 4   Subtract: $9 - 4\frac{3}{11}$

Solution

$$9\quad\ = 8\frac{11}{11} \qquad \text{LCM = 11.}$$
$$-\ 4\frac{3}{11} = 4\frac{3}{11}$$
$$\overline{\qquad\qquad 4\frac{8}{11}}$$

You Try It 4   Subtract: $8 - 2\frac{4}{13}$

Your solution

Example 5   Find $11\frac{5}{12}$ decreased by $2\frac{11}{16}$.

Solution

$$11\frac{5}{12} = 11\frac{20}{48} = 10\frac{68}{48} \qquad \text{LCM = 48.}$$
$$-\ \ 2\frac{11}{16} = \ \ 2\frac{33}{48} = \ \ 2\frac{33}{48}$$
$$\overline{\qquad\qquad\qquad\qquad 8\frac{35}{48}}$$

You Try It 5   What is $21\frac{7}{9}$ minus $7\frac{11}{12}$?

Your solution

Solutions on p. S5

Objective D   To solve application problems ····················

Find the inside diameter of a bushing with an outside diameter of $3\frac{3}{8}$ inches and a wall thickness of $\frac{1}{4}$ inches.

$$\frac{1}{4}$$
$$+\ \frac{1}{4}$$
$$\overline{\ \ \frac{2}{4} = \frac{1}{2}}$$

• Add $\frac{1}{4}$ to $\frac{1}{4}$ to find the total thickness of the two walls.

$$3\frac{3}{8} = 3\frac{3}{8} = 2\frac{11}{8}$$
$$-\ \ \frac{1}{2} = \ \frac{4}{8} = \ \ \frac{4}{8}$$
$$\overline{\qquad\qquad\qquad 2\frac{7}{8}}$$

• Subtract the total thickness of the two walls to find the inside diameter.

The inside diameter of the bushing is $2\frac{7}{8}$ inches.

Example 6

A $2\frac{2}{3}$-inch piece is cut from a $6\frac{5}{8}$-inch board. How much of the board is left?

Strategy

To find the length remaining, subtract the length of the piece cut from the total length of the board.

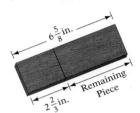

Solution
$$6\frac{5}{8} = 6\frac{15}{24} = 5\frac{39}{24}$$
$$-2\frac{2}{3} = 2\frac{16}{24} = 2\frac{16}{24}$$
$$\overline{\qquad\qquad\qquad\quad 3\frac{23}{24}}$$

$3\frac{23}{24}$ inches of the board are left.

Example 7

Two painters are staining a house. In 1 day one painter stained $\frac{1}{3}$ of the house, and the other stained $\frac{1}{4}$ of the house. How much of the job remains to be done?

Strategy

To find how much of the job remains:
- Find the total amount of the house already stained $\left(\frac{1}{3} + \frac{1}{4}\right)$.
- Subtract the amount already stained from 1, which represents the complete job.

Solution

$$\frac{1}{3} = \frac{4}{12} \qquad\qquad 1 = \frac{12}{12}$$
$$+\frac{1}{4} = \frac{3}{12} \qquad\qquad -\frac{7}{12} = \frac{7}{12}$$
$$\overline{\qquad\quad \frac{7}{12}} \qquad\qquad \overline{\qquad\quad \frac{5}{12}}$$

$\frac{5}{12}$ of the house remains to be stained.

You Try It 6

A flight from New York to Los Angeles takes $5\frac{1}{2}$ hours. After the plane has been in the air for $2\frac{3}{4}$ hours, how much flight time remains?

Your strategy

Your solution

You Try It 7

A patient is put on a diet to lose 24 pounds in 3 months. The patient lost $7\frac{1}{2}$ pounds the first month and $5\frac{3}{4}$ pounds the second month. How much weight must be lost the third month to achieve the goal?

Your strategy

Your solution

Solutions on p. S6

2.5 Exercises

Objective A

Subtract.

1. $\dfrac{9}{17}$
$-\dfrac{7}{17}$

2. $\dfrac{11}{15}$
$-\dfrac{3}{15}$

3. $\dfrac{11}{12}$
$-\dfrac{7}{12}$

4. $\dfrac{13}{15}$
$-\dfrac{4}{15}$

5. $\dfrac{9}{20}$
$-\dfrac{7}{20}$

6. $\dfrac{48}{55}$
$-\dfrac{13}{55}$

7. $\dfrac{42}{65}$
$-\dfrac{17}{65}$

8. $\dfrac{11}{24}$
$-\dfrac{5}{24}$

9. $\dfrac{23}{30}$
$-\dfrac{13}{30}$

10. $\dfrac{17}{42}$
$-\dfrac{5}{42}$

11. What is $\dfrac{5}{14}$ less than $\dfrac{13}{14}$?

12. What is $\dfrac{7}{19}$ less than $\dfrac{17}{19}$?

13. Find the difference between $\dfrac{7}{8}$ and $\dfrac{5}{8}$.

14. Find the difference between $\dfrac{7}{12}$ and $\dfrac{5}{12}$.

15. What is $\dfrac{18}{23}$ minus $\dfrac{9}{23}$?

16. What is $\dfrac{7}{9}$ minus $\dfrac{3}{9}$?

17. Find $\dfrac{17}{24}$ decreased by $\dfrac{11}{24}$.

18. Find $\dfrac{19}{30}$ decreased by $\dfrac{11}{30}$.

Objective B

Subtract.

19. $\dfrac{2}{3}$
$-\dfrac{1}{6}$

20. $\dfrac{7}{8}$
$-\dfrac{5}{16}$

21. $\dfrac{5}{8}$
$-\dfrac{2}{7}$

22. $\dfrac{5}{6}$
$-\dfrac{3}{7}$

23. $\dfrac{5}{7}$
$-\dfrac{3}{14}$

24. $\dfrac{5}{9}$
$-\dfrac{7}{15}$

25. $\dfrac{8}{15}$
$-\dfrac{7}{20}$

26. $\dfrac{7}{9}$
$-\dfrac{1}{6}$

27. $\dfrac{9}{14}$
$-\dfrac{3}{8}$

28. $\dfrac{5}{12}$
$-\dfrac{5}{16}$

Subtract.

29. $\dfrac{46}{51}$ **30.** $\dfrac{9}{16}$ **31.** $\dfrac{21}{35}$ **32.** $\dfrac{19}{40}$ **33.** $\dfrac{29}{60}$

$-\dfrac{3}{17}$ $-\dfrac{17}{32}$ $-\dfrac{5}{14}$ $-\dfrac{3}{16}$ $-\dfrac{3}{40}$

34. What is $\dfrac{3}{5}$ less than $\dfrac{11}{12}$?   **35.** What is $\dfrac{5}{9}$ less than $\dfrac{11}{15}$?

36. Find the difference between $\dfrac{11}{24}$ and $\dfrac{7}{18}$.   **37.** Find the difference between $\dfrac{9}{14}$ and $\dfrac{5}{42}$.

38. Find $\dfrac{11}{12}$ decreased by $\dfrac{11}{15}$.   **39.** Find $\dfrac{17}{20}$ decreased by $\dfrac{7}{15}$.

40. What is $\dfrac{13}{20}$ minus $\dfrac{1}{6}$?   **41.** What is $\dfrac{5}{6}$ minus $\dfrac{7}{9}$?

Objective C

Subtract.

42. $5\dfrac{7}{12}$ **43.** $16\dfrac{11}{15}$ **44.** $72\dfrac{21}{23}$ **45.** $19\dfrac{16}{17}$ **46.** $6\dfrac{1}{3}$

$-2\dfrac{5}{12}$ $-11\dfrac{8}{15}$ $-16\dfrac{17}{23}$ $-9\dfrac{7}{17}$ -2

47. $5\dfrac{7}{8}$ **48.** 10 **49.** 3 **50.** $6\dfrac{2}{5}$ **51.** $16\dfrac{3}{8}$

-1 $-6\dfrac{1}{3}$ $-2\dfrac{5}{21}$ $-4\dfrac{4}{5}$ $-10\dfrac{7}{8}$

52. $25\dfrac{4}{9}$ **53.** $8\dfrac{3}{7}$ **54.** $16\dfrac{2}{5}$ **55.** $23\dfrac{7}{8}$ **56.** 6

$-16\dfrac{7}{9}$ $-2\dfrac{6}{7}$ $-8\dfrac{4}{9}$ $-16\dfrac{2}{3}$ $-4\dfrac{3}{5}$

57. $65\dfrac{8}{35}$ **58.** $82\dfrac{4}{33}$ **59.** $101\dfrac{2}{9}$ **60.** $77\dfrac{5}{18}$ **61.** 17

$-16\dfrac{11}{14}$ $-16\dfrac{5}{22}$ -16 -61 $-7\dfrac{8}{13}$

62. What is $5\frac{3}{8}$ less than $8\frac{1}{9}$?

63. What is $7\frac{3}{5}$ less than $23\frac{3}{20}$?

64. Find the difference between $9\frac{2}{7}$ and $3\frac{1}{4}$.

65. Find the difference between $12\frac{3}{8}$ and $7\frac{5}{12}$.

66. What is $10\frac{5}{9}$ minus $5\frac{11}{15}$?

67. Find $6\frac{1}{3}$ decreased by $3\frac{3}{5}$.

Objective D   *Application Problems*

68. Find the missing dimension.

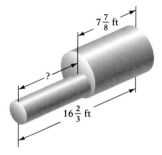

69. Find the missing dimension.

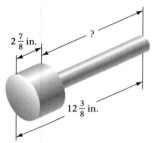

70. The horses in the Kentucky Derby run $1\frac{1}{4}$ miles. In the Belmont Stakes they run $1\frac{1}{2}$ miles, and in the Preakness Stakes they run $1\frac{3}{16}$ miles. How much farther do the horses run in the Kentucky Derby than in the Preakness Stakes? How much farther do they run in the Belmont Stakes than in the Preakness Stakes?

71. In the running high jump in the 1948 Summer Olympic Games, Alice Coachman's distance was $66\frac{1}{8}$ inches. In the same event in the 1972 Summer Olympics, Urika Meyfarth jumped $75\frac{1}{2}$ inches, and in the 1996 Olympic Games, Stefka Kostadinova jumped $80\frac{3}{4}$ inches. Find the difference between Meyfarth's distance and Coachman's distance. Find the difference between Kostadinova's distance and Meyfarth's distance.

72. Two hikers plan a 3-day $27\frac{1}{2}$-mile backpack trip carrying a total of 80 pounds. The hikers plan to travel $7\frac{3}{8}$ miles the first day and $10\frac{1}{3}$ miles the second day.
a. How many miles do the hikers plan to travel the first two days?

b. How many miles will be left to travel on the third day?

73. A 12-mile walkathon has three checkpoints. The first is $3\frac{3}{8}$ miles from the starting point. The second checkpoint is $4\frac{1}{3}$ miles from the first.

a. How many miles is it from the starting point to the second checkpoint?

b. How many miles is it from the second checkpoint to the finish line?

74. A patient with high blood pressure who weighs 225 pounds is put on a diet to lose 25 pounds in 3 months. The patient loses $8\frac{3}{4}$ pounds the first month and $11\frac{5}{8}$ pounds the second month. How much weight must be lost the third month for the goal to be achieved?

75. A wrestler is entered in the 172-pound weight class in the conference finals coming up in 3 weeks. The wrestler needs to lose $12\frac{3}{4}$ pounds. The wrestler loses $5\frac{1}{4}$ pounds the first week and $4\frac{1}{4}$ pounds the second week. **a.** Without doing the calculations, can the wrestler reach his weight class by losing less in the third week than was lost in the second week? **b.** How many pounds must be lost in the third week for the desired weight to be reached?

APPLYING THE CONCEPTS

The figure at the right shows a selected portion of the New York Stock Exchange stock prices. The figure shows the closing prices for the stocks on May 25, 1997, the 52-week high, the 52-week low, and the high and low for the day. Use this figure for Exercises 76 to 78.

	52-Week High	52-Week Low	Div	PE	Vol	High	Low	Last	Chg
Chiqt	16 1/8	11 1/8	.20	–	9308	15 3/4	15 1/4	15 3/8	– 1/4
ChokFul	6 1/4	4 1/2	–	12	1318	5 3/4	5 1/2	5 3/4	–
ChoicHt	17 5/8	12 3/4	–	–	4627	15 1/8	14 1/4	15	+ 3/4
ChrsCr	43 7/16	37 1/4	–	18	1537	42 5/8	41 5/8	41 7/8	– 1/4
Chryslr s y	36 3/8	26 1/4	1.60	6	143733	32 3/4	30 7/8	32 5/8	+1 3/8
Chubb s h	62 3/4	40 7/8	1.16	20	15371	60 3/4	58 1/8	60 1/4	+2
CIGNA	173 1/2	105 1/2	3.32	12	8857	173 3/4	170	173 3/4	+3 1/2
CinnBel	67 1/2	45 3/8	.80	20	6399	58 1/8	55 1/8	57 7/8	+2 7/8
CinMil	25 1/4	17 7/8	.36	13	3659	22 5/8	21 3/4	22 3/8	+ 1/8
CINrgy	35 3/4	29 1/8	1.80	17	12151	35 5/8	34 5/8	34 7/8	– 1/4
CIPSCO h	38 5/8	33 1/2	2.12	16	1662	34 3/4	33 1/2	34 1/4	– 1/8
CirCtyCmx	22	13 1/2	–	–	2483	15 1/4	14 5/8	15	– 1/8
CirCtyCC	40 7/8	28 5/8	.14	29	19164	40 7/8	38	39 3/8	+ 3/4
Circus	44 5/8	23 1/2	–	27	34065	28 1/2	25 7/8	26 5/8	–1 1/2
Citicorp	127 1/8	72 1/4	2.10	15	83140	120 1/4	114 5/8	117 7/8	+ 3/4
Citzcp	27 1/4	18	.20	10	145	25 1/2	24 3/4	25 1/4	– 3/8
CitzU A s	12 1/2	8 7/8	–	13	18715	9 1/2	9	9 1/4	+ 1/8
CitzU B s	12 1/2	9	–	13	20806	9 1/2	9	9 1/4	+ 1/8

76. Find the price of one share of Circus at the end of the day, May 25.

77. Find the difference between the price of one share of Citicorp stock at the 52-week low price and the price at the end of the May 25 trading day.

78. Find the 52-week difference in the price of one share of ChrsCr Stock.

79. Fill in the square to produce a true statement: $5\frac{1}{3} - \square = 2\frac{1}{2}$.

80. Fill in the square to produce a true statement: $\square - 4\frac{1}{2} = 1\frac{5}{8}$.

81. Fill in the blank squares at the right so that the sum of the numbers along any row, column, or diagonal is the same. The resulting square is called a magic square.

		$\frac{3}{4}$
	$\frac{5}{8}$	
$\frac{1}{2}$		$\frac{7}{8}$

82. If $\frac{4}{15}$ of an electrician's income is spent for housing, what fraction of the electrician's income is not spent for housing?

2.6 Multiplication of Fractions and Mixed Numbers

Objective A   *To multiply fractions* ....

The product of two fractions is the product of the numerators over the product of the denominators.

➡ Multiply: $\frac{2}{3} \times \frac{4}{5}$

$$\frac{2}{3} \times \frac{4}{5} = \frac{2 \cdot 4}{3 \cdot 5} = \frac{8}{15}$$

- Multiply the numerators.
- Multiply the denominators.

The product $\frac{2}{3} \times \frac{4}{5}$ can be read "$\frac{2}{3}$ times $\frac{4}{5}$" or "$\frac{2}{3}$ of $\frac{4}{5}$."

Reading the times sign as "of" is useful in application problems.

$\frac{4}{5}$ of the bar is shaded.

Shade $\frac{2}{3}$ of the $\frac{4}{5}$ already shaded.

$\frac{8}{15}$ of the bar is then shaded light yellow.

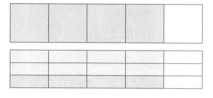

$$\frac{2}{3} \text{ of } \frac{4}{5} = \frac{2}{3} \times \frac{4}{5} = \frac{8}{15}$$

After multiplying two fractions, write the product in simplest form.

➡ Multiply: $\frac{3}{4} \times \frac{14}{15}$

$$\frac{3}{4} \times \frac{14}{15} = \frac{3 \cdot 14}{4 \cdot 15}$$

- Multiply the numerators.
- Multiply the denominators.

$$= \frac{3 \cdot 2 \cdot 7}{2 \cdot 2 \cdot 3 \cdot 5}$$

- Write the prime factorization of each number.

$$= \frac{\overset{1}{\cancel{3}} \cdot \overset{1}{\cancel{2}} \cdot 7}{\underset{1}{\cancel{2}} \cdot 2 \cdot \underset{1}{\cancel{3}} \cdot 5} = \frac{7}{10}$$

- Eliminate the common factors. Then multiply the factors of the numerator and denominator.

This example could also be worked by using the GCF.

$$\frac{3}{4} \times \frac{14}{15} = \frac{42}{60}$$

- Multiply the numerators.
- Multiply the denominators.

$$= \frac{6 \cdot 7}{6 \cdot 10}$$

- The GCF of 42 and 60 is 6. Factor 6 from 42 and 60.

$$= \frac{\overset{1}{\cancel{6}} \cdot 7}{\underset{1}{\cancel{6}} \cdot 10} = \frac{7}{10}$$

- Eliminate the GCF.

Example 1

Multiply $\frac{4}{15}$ and $\frac{5}{28}$.

Solution

$$\frac{4}{15} \times \frac{5}{28} = \frac{4 \cdot 5}{15 \cdot 28} = \frac{\overset{1}{\cancel{2}} \cdot \overset{1}{\cancel{2}} \cdot \overset{1}{\cancel{5}}}{3 \cdot \underset{1}{\cancel{5}} \cdot \underset{1}{\cancel{2}} \cdot \underset{1}{\cancel{2}} \cdot 7} = \frac{1}{21}$$

You Try It 1

Multiply $\frac{4}{21}$ and $\frac{7}{44}$.

Your solution

Example 2

Find the product of $\frac{9}{20}$ and $\frac{33}{35}$.

Solution

$$\frac{9}{20} \times \frac{33}{35} = \frac{9 \cdot 33}{20 \cdot 35} = \frac{3 \cdot 3 \cdot 3 \cdot 11}{2 \cdot 2 \cdot 5 \cdot 5 \cdot 7} = \frac{297}{700}$$

You Try It 2

Find the product of $\frac{2}{21}$ and $\frac{10}{33}$.

Your solution

Example 3

What is $\frac{14}{9}$ times $\frac{12}{7}$?

Solution

$$\frac{14}{9} \times \frac{12}{7} = \frac{14 \cdot 12}{9 \cdot 7} = \frac{2 \cdot \overset{1}{\cancel{7}} \cdot 2 \cdot 2 \cdot \overset{1}{\cancel{3}}}{3 \cdot \underset{1}{\cancel{3}} \cdot \underset{1}{\cancel{7}}} = \frac{8}{3} = 2\frac{2}{3}$$

You Try It 3

What is $\frac{16}{5}$ times $\frac{15}{24}$?

Your solution

Solutions on p. S6

Objective B   **To multiply whole numbers, mixed numbers, and fractions** ....

To multiply a whole number by a fraction or mixed number, first write the whole number as a fraction with a denominator of 1.

➡ Multiply: $4 \times \frac{3}{7}$

$$4 \times \frac{3}{7} = \frac{4}{1} \times \frac{3}{7} = \frac{4 \cdot 3}{1 \cdot 7} = \frac{2 \cdot 2 \cdot 3}{7} = \frac{12}{7} = 1\frac{5}{7}$$

• Write 4 with a denominator of 1; then multiply the fractions.

When one or more of the factors in a product is a mixed number, write the mixed number as an improper fraction before multiplying.

➡ Multiply: $2\frac{1}{3} \times \frac{3}{14}$

$$2\frac{1}{3} \times \frac{3}{14} = \frac{7}{3} \times \frac{3}{14} = \frac{7 \cdot 3}{3 \cdot 14} = \frac{\overset{1}{\cancel{7}} \cdot \overset{1}{\cancel{3}}}{\underset{1}{\cancel{3}} \cdot 2 \cdot \underset{1}{\cancel{7}}} = \frac{1}{2}$$

• Write $2\frac{1}{3}$ as an improper fraction; then multiply the fractions.

Example 4

Multiply: $4\frac{5}{6} \times \frac{12}{13}$

Solution

$4\frac{5}{6} \times \frac{12}{13} = \frac{29}{6} \times \frac{12}{13} = \frac{29 \cdot 12}{6 \cdot 13}$

$= \frac{29 \cdot \overset{1}{\cancel{2}} \cdot 2 \cdot \overset{1}{\cancel{3}}}{\underset{1}{\cancel{2}} \cdot \underset{1}{\cancel{3}} \cdot 13} = \frac{58}{13} = 4\frac{6}{13}$

You Try It 4

Multiply: $5\frac{2}{5} \times \frac{5}{9}$

Your solution

Example 5

Find $5\frac{2}{3}$ times $4\frac{1}{2}$.

Solution

$5\frac{2}{3} \times 4\frac{1}{2} = \frac{17}{3} \times \frac{9}{2} = \frac{17 \cdot 9}{3 \cdot 2}$

$= \frac{17 \cdot \overset{1}{\cancel{3}} \cdot 3}{\underset{1}{\cancel{3}} \cdot 2} = \frac{51}{2} = 25\frac{1}{2}$

You Try It 5

Multiply: $3\frac{2}{5} \times 6\frac{1}{4}$

Your solution

Example 6

Multiply: $4\frac{2}{5} \times 7$

Solution

$4\frac{2}{5} \times 7 = \frac{22}{5} \times \frac{7}{1} = \frac{22 \cdot 7}{5 \cdot 1}$

$= \frac{2 \cdot 11 \cdot 7}{5} = \frac{154}{5} = 30\frac{4}{5}$

You Try It 6

Multiply: $3\frac{2}{7} \times 6$

Your solution

Solutions on p. S6

Objective C   **To solve application problems** ⋯⋯⋯⋯⋯⋯⋯⋯⋯⋯⋯⋯⋯⋯⋯

Length (ft)	Weight (lb/ft)
$6\frac{1}{2}$	$\frac{3}{8}$
$8\frac{5}{8}$	$1\frac{1}{4}$
$10\frac{3}{4}$	$2\frac{1}{2}$
$12\frac{7}{12}$	$4\frac{1}{3}$

The table at the left lists the length of steel rods and the weight per foot. The weight per foot is measured in pounds for each foot of rod and is abbreviated as lb/ft.

➡ Find the weight of the steel bar that is $10\frac{3}{4}$ feet long.

Strategy
To find the weight of the steel bar, multiply its length by the weight per foot.

Solution $10\frac{3}{4} \times 2\frac{1}{2} = \frac{43}{4} \times \frac{5}{2} = \frac{43 \cdot 5}{4 \cdot 2} = \frac{215}{8} = 26\frac{7}{8}$

The weight of the $10\frac{3}{4}$-foot rod is $26\frac{7}{8}$ lb.

Example 7

An electrician earns $150 for each day worked. What are the electrician's earnings for working $4\frac{1}{2}$ days?

Strategy

To find the electrician's total earnings, multiply the daily earnings ($150) by the number of days worked $\left(4\frac{1}{2}\right)$.

Solution $150 \times 4\frac{1}{2} = \frac{150}{1} \times \frac{9}{2}$

$$= \frac{150 \cdot 9}{1 \cdot 2}$$

$$= 675$$

The electrician's earnings are $675.

You Try It 7

Over the last 10 years, a house increased in value by $3\frac{1}{2}$ times. The price of the house 10 years ago was $30,000. What is the value of the house today?

Your strategy

Your solution

Example 8

The value of a small office building and the land on which it is built is $90,000. The value of the land is $\frac{1}{4}$ the total value. What is the value of the building (in dollars)?

Strategy

To find the value of the building:

- Find the value of the land $\left(\frac{1}{4} \times 90,000\right)$.

- Subtract the value of the land from the total value.

Solution $\frac{1}{4} \times 90,000 = \frac{90,000}{4}$

$= 22,500$ value of
the land

$$\begin{array}{r} 90,000 \\ -\ 22,500 \\ \hline 67,500 \end{array}$$

The value of the building is $67,500.

You Try It 8

A paint company bought a drying chamber and an air compressor for spray painting. The total cost of the two items was $60,000. The drying chamber's cost was $\frac{4}{5}$ of the total cost. What was the cost of the air compressor?

Your strategy

Your solution

Solutions on p. S6

2.6 Exercises

· ·

Objective A

Multiply.

1. $\dfrac{2}{3} \times \dfrac{7}{8}$ **2.** $\dfrac{1}{2} \times \dfrac{2}{3}$ **3.** $\dfrac{5}{16} \times \dfrac{7}{15}$ **4.** $\dfrac{3}{8} \times \dfrac{6}{7}$

5. $\dfrac{1}{6} \times \dfrac{1}{8}$ **6.** $\dfrac{2}{5} \times \dfrac{5}{6}$ **7.** $\dfrac{11}{12} \times \dfrac{6}{7}$ **8.** $\dfrac{11}{12} \times \dfrac{3}{5}$

9. $\dfrac{1}{6} \times \dfrac{6}{7}$ **10.** $\dfrac{3}{5} \times \dfrac{10}{11}$ **11.** $\dfrac{1}{5} \times \dfrac{5}{8}$ **12.** $\dfrac{6}{7} \times \dfrac{14}{15}$

13. $\dfrac{8}{9} \times \dfrac{27}{4}$ **14.** $\dfrac{3}{5} \times \dfrac{3}{10}$ **15.** $\dfrac{5}{6} \times \dfrac{1}{2}$ **16.** $\dfrac{3}{8} \times \dfrac{5}{12}$

17. $\dfrac{16}{9} \times \dfrac{27}{8}$ **18.** $\dfrac{5}{8} \times \dfrac{16}{15}$ **19.** $\dfrac{3}{2} \times \dfrac{4}{9}$ **20.** $\dfrac{5}{3} \times \dfrac{3}{7}$

21. $\dfrac{7}{8} \times \dfrac{3}{14}$ **22.** $\dfrac{2}{9} \times \dfrac{1}{5}$ **23.** $\dfrac{1}{10} \times \dfrac{3}{8}$ **24.** $\dfrac{5}{12} \times \dfrac{6}{7}$

25. $\dfrac{15}{8} \times \dfrac{16}{3}$ **26.** $\dfrac{5}{6} \times \dfrac{4}{15}$ **27.** $\dfrac{1}{2} \times \dfrac{2}{15}$ **28.** $\dfrac{3}{8} \times \dfrac{5}{16}$

29. $\dfrac{5}{7} \times \dfrac{14}{15}$ **30.** $\dfrac{3}{8} \times \dfrac{15}{41}$ **31.** $\dfrac{5}{12} \times \dfrac{42}{65}$ **32.** $\dfrac{16}{33} \times \dfrac{55}{72}$

33. $\dfrac{12}{5} \times \dfrac{5}{3}$ **34.** $\dfrac{17}{9} \times \dfrac{81}{17}$ **35.** $\dfrac{16}{85} \times \dfrac{125}{84}$ **36.** $\dfrac{19}{64} \times \dfrac{48}{95}$

37. Multiply $\frac{7}{12}$ and $\frac{15}{42}$.

38. Multiply $\frac{32}{9}$ and $\frac{3}{8}$.

39. Find the product of $\frac{5}{9}$ and $\frac{3}{20}$.

40. Find the product of $\frac{7}{3}$ and $\frac{15}{14}$.

41. What is $\frac{1}{2}$ times $\frac{8}{15}$?

42. What is $\frac{3}{8}$ times $\frac{12}{17}$?

Objective B

Multiply.

43. $4 \times \frac{3}{8}$

44. $14 \times \frac{5}{7}$

45. $\frac{2}{3} \times 6$

46. $\frac{5}{12} \times 40$

47. $\frac{1}{3} \times 1\frac{1}{3}$

48. $\frac{2}{5} \times 2\frac{1}{2}$

49. $1\frac{7}{8} \times \frac{4}{15}$

50. $2\frac{1}{5} \times \frac{5}{22}$

51. $55 \times \frac{3}{10}$

52. $\frac{5}{14} \times 49$

53. $4 \times 2\frac{1}{2}$

54. $9 \times 3\frac{1}{3}$

55. $2\frac{1}{7} \times 3$

56. $5\frac{1}{4} \times 8$

57. $3\frac{2}{3} \times 5$

58. $4\frac{2}{9} \times 3$

59. $\frac{1}{2} \times 3\frac{3}{7}$

60. $\frac{3}{8} \times 4\frac{4}{5}$

61. $6\frac{1}{8} \times \frac{4}{7}$

62. $5\frac{1}{3} \times \frac{5}{16}$

63. $5\frac{1}{8} \times 5$

64. $6\frac{1}{9} \times 2$

65. $\frac{3}{8} \times 4\frac{1}{2}$

66. $\frac{5}{7} \times 2\frac{1}{3}$

67. $6 \times 2\frac{2}{3}$

68. $6\frac{1}{8} \times 0$

69. $1\frac{1}{3} \times 2\frac{1}{4}$

70. $2\frac{5}{8} \times \frac{3}{23}$

71. $2\frac{5}{8} \times 3\frac{2}{5}$

72. $5\frac{3}{16} \times 5\frac{1}{3}$

73. $3\frac{1}{7} \times 2\frac{1}{8}$

74. $16\frac{5}{8} \times 1\frac{1}{16}$

75. $2\frac{2}{5} \times 3\frac{1}{12}$ **76.** $2\frac{2}{3} \times \frac{3}{20}$ **77.** $5\frac{1}{5} \times 3\frac{1}{13}$ **78.** $3\frac{3}{4} \times 2\frac{3}{20}$

79. $10\frac{1}{4} \times 3\frac{1}{5}$ **80.** $12\frac{3}{5} \times 1\frac{3}{7}$ **81.** $5\frac{3}{7} \times 5\frac{1}{4}$ **82.** $6\frac{1}{2} \times 1\frac{3}{13}$

83. Multiply $2\frac{1}{2}$ and $3\frac{3}{5}$.

84. Multiply $4\frac{3}{8}$ and $3\frac{3}{5}$.

85. Find the product of $2\frac{1}{8}$ and $\frac{5}{17}$.

86. Find the product of $12\frac{2}{5}$ and $3\frac{7}{31}$.

87. What is $1\frac{3}{8}$ times $2\frac{1}{5}$?

88. What is $3\frac{1}{8}$ times $2\frac{4}{7}$?

Objective C   *Application Problems*

89. Salmon costs $8 per pound. Find the cost of $2\frac{3}{4}$ pounds of salmon.

90. Maria Rivera can walk $3\frac{1}{2}$ miles in 1 hour. At this rate, how far can Maria walk in $\frac{1}{3}$ hour?

91. A Honda Civic travels 38 miles on each gallon of gasoline. How many miles can the car travel on $9\frac{1}{2}$ gallons of gasoline?

92. A board costing $6 is $9\frac{1}{4}$ feet long. One-third of the board is cut off.
   a. Without doing the calculation, is the piece being cut off at least 4 feet long?
   b. What is the length of the piece cut off?

93. The F-1 engine in the first stage of the Saturn 5 rocket burns 214,000 gallons of propellant in 1 minute. The first stage burns $2\frac{1}{2}$ minutes before burnout. How much propellant is used before burnout?

94. A family budgets $\frac{2}{5}$ of its monthly income of $3200 per month for housing and utilities.
   a. What amount is budgeted for housing and utilities?
   b. What amount remains for purposes other than housing and utilities?

95. $\frac{5}{6}$ of a chemistry class of 36 has passing grades. $\frac{1}{5}$ of the students with passing grades received an A.
   a. How many students passed the chemistry course?
   b. How many students received A grades?

96. The parents of the Newton Junior High School Choir members are making robes for the choir. Each robe requires $2\frac{5}{8}$ yards of material at a cost of $8 per yard. Find the total cost of 24 choir robes.

97. A college spends $\frac{5}{8}$ of its monthly income on employee salaries. During one month the college had an income of \$712,000. How much of the monthly income remained after the employees' salaries were paid?

The table at the right shows the length of steel rods and their weight per foot. Use this table for Exercises 98 to 100.

Length (ft)	Weight (lb/ft)
$6\frac{1}{2}$	$\frac{3}{8}$
$8\frac{5}{8}$	$1\frac{1}{4}$
$10\frac{3}{4}$	$2\frac{1}{2}$
$12\frac{7}{12}$	$4\frac{1}{3}$

98. Find the weight of $6\frac{1}{2}$ feet of steel rod.

99. Find the weight of $12\frac{7}{12}$ feet of steel rod.

100. Find the total weight of $8\frac{5}{8}$ feet and $10\frac{3}{4}$ feet of steel rod.

101. A state park has reserved $\frac{4}{5}$ of its total acreage for a wildlife preserve. Three-fourths of the wildlife preserve is heavily wooded. What fraction of the state park is heavily wooded?

State Park

102. The manager of a mutual fund has one-half of the portfolio invested in bonds. Of the amount invested in bonds, $\frac{3}{8}$ is invested in corporate bonds. What fraction of the total portfolio is invested in corporate bonds?

APPLYING THE CONCEPTS

103. The product of 1 and a number is $\frac{1}{2}$. Find the number.

104. Our calendar is based on the solar year, which is $365\frac{1}{4}$ days. Use this fact to explain leap years.

105. If two positive fractions, each less than 1, are multiplied, is the product less than 1?

106. Is the product of two positive fractions always greater than either one of the two numbers? If so, explain why. If not, give an example.

107. Which of the labeled points on the number line at the right could be the graph of the product of B and C?

$$0 \quad A \quad B \quad C \quad 1 \quad D \quad \quad 2 \quad E \quad 3$$

108. Fill in the circles on the square at the right with the fractions $\frac{1}{6}$, $\frac{5}{18}$, $\frac{4}{9}$, $\frac{5}{9}$, $\frac{2}{3}$, $\frac{3}{4}$, $1\frac{1}{9}$, $1\frac{1}{2}$, and $2\frac{1}{4}$ so that the product of any row is equal to $\frac{5}{18}$. (*Note:* There is more than one answer.)

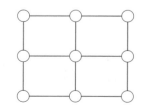

2.7 Division of Fractions and Mixed Numbers

Objective A   To divide fractions ....

The **reciprocal** of a fraction is the fraction with the numerator and denominator interchanged. For instance, the reciprocal of $\frac{2}{3}$ is $\frac{3}{2}$. The process of interchanging the numerator and denominator of a fraction is called **inverting** the fraction.

To find the reciprocal of a whole number, first write the whole number as a fraction with a denominator of 1; then find the reciprocal of that fraction.

The reciprocal of 5 is $\frac{1}{5}$. $\left(\text{Think } 5 = \frac{5}{1}\right)$

Reciprocals are used to rewrite division problems as related multiplication problems. Look at the following two problems:

$$8 \div 2 = 4 \qquad\qquad 8 \times \frac{1}{2} = 4$$

8 divided by 2 is **4**.   8 times the reciprocal of 2 is **4**.

"Divided by" means the same as "times the reciprocal of." Thus "÷ 2" can be replaced with "$\times \frac{1}{2}$," and the answer will be the same. Fractions are divided by making this replacement.

➡ Divide: $\frac{2}{3} \div \frac{3}{4}$ $\frac{2}{3} \div \frac{3}{4} = \frac{2}{3} \times \frac{4}{3} = \frac{2 \cdot 4}{3 \cdot 3} = \frac{2 \cdot 2 \cdot 2}{3 \cdot 3} = \frac{8}{9}$

Example 1   Divide: $\frac{5}{8} \div \frac{4}{9}$

Solution $\frac{5}{8} \div \frac{4}{9} = \frac{5}{8} \times \frac{9}{4} = \frac{5 \cdot 9}{8 \cdot 4}$

$= \frac{5 \cdot 3 \cdot 3}{2 \cdot 2 \cdot 2 \cdot 2 \cdot 2} = \frac{45}{32} = 1\frac{13}{32}$

You Try It 1   Divide: $\frac{3}{7} \div \frac{2}{3}$

Your solution

Example 2   Find the quotient of $\frac{3}{5}$ and $\frac{12}{25}$.

Solution $\frac{3}{5} \div \frac{12}{25} = \frac{3}{5} \times \frac{25}{12} = \frac{3 \cdot 25}{5 \cdot 12}$

$= \frac{\overset{1}{\cancel{3}} \cdot \overset{1}{\cancel{5}} \cdot 5}{\underset{1}{\cancel{5}} \cdot 2 \cdot 2 \cdot \underset{1}{\cancel{3}}} = \frac{5}{4} = 1\frac{1}{4}$

You Try It 2   Divide: $\frac{3}{4} \div \frac{9}{10}$

Your solution

Solutions on p. S7

Objective B   To divide whole numbers, mixed numbers, and fractions ....

To divide a fraction and a whole number, first write the whole number as a fraction with a denominator of 1.

➡ Divide: $\frac{3}{7} \div 5$

$\frac{3}{7} \div \boxed{5} = \frac{3}{7} \div \boxed{\frac{5}{1}} = \frac{3}{7} \times \frac{1}{5} = \frac{3 \cdot 1}{7 \cdot 5} = \frac{3}{35}$ • Write 5 with a denominator of 1; then divide the fractions.

When one of the numbers in a quotient is a mixed number, write the mixed number as an improper fraction before dividing.

➡ Divide: $4\frac{2}{3} \div \frac{8}{15}$

Write $4\frac{2}{3}$ as an improper fraction; then divide the fractions.

$$4\frac{2}{3} \div \frac{8}{15} = \frac{14}{3} \div \frac{\mathbf{8}}{\mathbf{15}} = \frac{14}{3} \times \frac{\mathbf{15}}{\mathbf{8}} = \frac{14 \cdot 15}{3 \cdot 8} = \frac{\overset{1}{\cancel{2}} \cdot 7 \cdot \overset{1}{\cancel{3}} \cdot 5}{\underset{1}{\cancel{3}} \cdot 2 \cdot \underset{1}{\cancel{2}} \cdot 2} = \frac{35}{4} = 8\frac{3}{4}$$

➡ Divide: $1\frac{13}{15} \div 4\frac{4}{5}$

Write the mixed numbers as improper fractions. Then divide the fractions.

$$1\frac{13}{15} \div 4\frac{4}{5} = \frac{28}{15} \div \frac{24}{5} = \frac{28}{15} \times \frac{5}{24} = \frac{28 \cdot 5}{15 \cdot 24} = \frac{\overset{1}{\cancel{2}} \cdot \overset{1}{\cancel{2}} \cdot 7 \cdot \overset{1}{\cancel{5}}}{3 \cdot \underset{1}{\cancel{5}} \cdot \underset{1}{\cancel{2}} \cdot \underset{1}{\cancel{2}} \cdot 2 \cdot 3} = \frac{7}{18}$$

Example 3 Divide $\frac{4}{9}$ by 5.

Solution

$$\frac{4}{9} \div 5 = \frac{4}{9} \div \frac{5}{1} = \frac{4}{9} \times \frac{1}{5}$$

$$= \frac{4 \cdot 1}{9 \cdot 5} = \frac{2 \cdot 2}{3 \cdot 3 \cdot 5} = \frac{4}{45}$$

You Try It 3 Divide $\frac{5}{7}$ by 6.

Your solution

Example 4

Find the quotient of $\frac{3}{8}$ and $2\frac{1}{10}$.

Solution

$$\frac{3}{8} \div 2\frac{1}{10} = \frac{3}{8} \div \frac{21}{10} = \frac{3}{8} \times \frac{10}{21}$$

$$= \frac{3 \cdot 10}{8 \cdot 21} = \frac{\overset{1}{\cancel{3}} \cdot \overset{1}{\cancel{2}} \cdot 5}{\underset{1}{\cancel{2}} \cdot 2 \cdot 2 \cdot \underset{1}{\cancel{3}} \cdot 7} = \frac{5}{28}$$

You Try It 4

Find the quotient of $12\frac{3}{5}$ and 7.

Your solution

Example 5 Divide: $2\frac{3}{4} \div 1\frac{5}{7}$

Solution

$$2\frac{3}{4} \div 1\frac{5}{7} = \frac{11}{4} \div \frac{12}{7} = \frac{11}{4} \times \frac{7}{12} = \frac{11 \cdot 7}{4 \cdot 12}$$

$$= \frac{11 \cdot 7}{2 \cdot 2 \cdot 2 \cdot 2 \cdot 3} = \frac{77}{48} = 1\frac{29}{48}$$

You Try It 5 Divide: $3\frac{2}{3} \div 2\frac{2}{5}$

Your solution

Solutions on p. S7

Example 6   Divide: $1\frac{13}{15} \div 4\frac{1}{5}$

Solution

$$1\frac{13}{15} \div 4\frac{1}{5} = \frac{28}{15} \div \frac{21}{5} = \frac{28}{15} \times \frac{5}{21} = \frac{28 \cdot 5}{15 \cdot 21}$$

$$= \frac{2 \cdot 2 \cdot \overset{1}{\cancel{7}} \cdot \overset{1}{\cancel{5}}}{3 \cdot \underset{1}{\cancel{5}} \cdot 3 \cdot \underset{1}{\cancel{7}}} = \frac{4}{9}$$

You Try It 6   Divide: $2\frac{5}{6} \div 8\frac{1}{2}$

Your solution

Example 7   Divide: $4\frac{3}{8} \div 7$

Solution

$$4\frac{3}{8} \div 7 = \frac{35}{8} \div \frac{7}{1} = \frac{35}{8} \times \frac{1}{7}$$

$$= \frac{35 \cdot 1}{8 \cdot 7} = \frac{5 \cdot \overset{1}{\cancel{7}}}{2 \cdot 2 \cdot 2 \cdot \underset{1}{\cancel{7}}} = \frac{5}{8}$$

You Try It 7   Divide: $6\frac{2}{5} \div 4$

Your solution

Solutions on p. S7

Objective C   *To solve application problems* ....

Example 8

A car used $15\frac{1}{2}$ gallons of gasoline on a 310-mile trip. How many miles can this car travel on 1 gallon of gasoline?

Strategy
To find the number of miles, divide the number of miles traveled by the number of gallons of gasoline used.

Solution

$$310 \div 15\frac{1}{2} = \frac{310}{1} \div \frac{31}{2}$$

$$= \frac{310}{1} \times \frac{2}{31}$$

$$= \frac{310 \cdot 2}{1 \cdot 31}$$

$$= 20$$

The car travels 20 miles on 1 gallon of gasoline.

You Try It 8

Leon Dern purchased a $\frac{1}{2}$-ounce gold coin for \$195. What would the price of a 1-ounce coin be?

Your strategy

Your solution

Solution on p. S7

Example 9

A 12-foot board is cut into pieces $2\frac{1}{4}$ feet long for use as bookshelves. What is the length of the remaining piece after as many shelves as possible are cut?

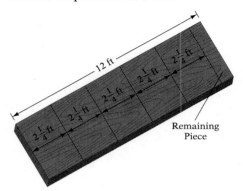

Strategy

To find the length of the remaining piece:

• Divide the total length by the length of each shelf $\left(2\frac{1}{4}\right)$. This will give you the number of shelves cut, with a certain fraction of a shelf left over.

• Multiply the fraction left over by the length of one shelf to determine the length of the remaining piece.

Solution

$$12 \div 2\frac{1}{4} = \frac{12}{1} \div \frac{9}{4} = \frac{12}{1} \times \frac{4}{9}$$

$$= \frac{12 \cdot 4}{1 \cdot 9} = \frac{16}{3} = 5\frac{1}{3}$$

5 pieces $2\frac{1}{4}$ feet long

1 piece $\frac{1}{3}$ of $2\frac{1}{4}$ feet long

$$\frac{1}{3} \times 2\frac{1}{4} = \frac{1}{3} \times \frac{9}{4} = \frac{1 \cdot 9}{3 \cdot 4} = \frac{3}{4}$$

The length of the piece remaining is $\frac{3}{4}$ foot.

You Try It 9

A 16-foot board is cut into pieces $3\frac{1}{3}$ feet long for shelves for a bookcase. What is the length of the remaining piece after as many shelves as possible are cut?

Your strategy

Your solution

Solution on p. S7

2.7 Exercises

Objective A

Divide.

1. $\dfrac{1}{3} \div \dfrac{2}{5}$

2. $\dfrac{3}{7} \div \dfrac{3}{2}$

3. $\dfrac{3}{7} \div \dfrac{3}{7}$

4. $0 \div \dfrac{1}{2}$

5. $0 \div \dfrac{3}{4}$

6. $\dfrac{16}{33} \div \dfrac{4}{11}$

7. $\dfrac{5}{24} \div \dfrac{15}{36}$

8. $\dfrac{11}{15} \div \dfrac{1}{12}$

9. $\dfrac{15}{16} \div \dfrac{16}{39}$

10. $\dfrac{2}{15} \div \dfrac{3}{5}$

11. $\dfrac{8}{9} \div \dfrac{4}{5}$

12. $\dfrac{11}{15} \div \dfrac{5}{22}$

13. $\dfrac{1}{9} \div \dfrac{2}{3}$

14. $\dfrac{10}{21} \div \dfrac{5}{7}$

15. $\dfrac{2}{5} \div \dfrac{4}{7}$

16. $\dfrac{3}{8} \div \dfrac{5}{12}$

17. $\dfrac{1}{2} \div \dfrac{1}{4}$

18. $\dfrac{1}{3} \div \dfrac{1}{9}$

19. $\dfrac{1}{5} \div \dfrac{1}{10}$

20. $\dfrac{4}{15} \div \dfrac{2}{5}$

21. $\dfrac{7}{15} \div \dfrac{14}{5}$

22. $\dfrac{5}{8} \div \dfrac{15}{2}$

23. $\dfrac{14}{3} \div \dfrac{7}{9}$

24. $\dfrac{7}{4} \div \dfrac{9}{2}$

25. $\dfrac{5}{9} \div \dfrac{25}{3}$

26. $\dfrac{5}{16} \div \dfrac{3}{8}$

27. $\dfrac{2}{3} \div \dfrac{1}{3}$

28. $\dfrac{4}{9} \div \dfrac{1}{9}$

29. $\dfrac{5}{7} \div \dfrac{2}{7}$

30. $\dfrac{5}{6} \div \dfrac{1}{9}$

31. $\dfrac{2}{3} \div \dfrac{2}{9}$

32. $\dfrac{5}{12} \div \dfrac{5}{6}$

33. $4 \div \dfrac{2}{3}$

34. $\dfrac{2}{3} \div 4$

35. $\dfrac{3}{2} \div 3$

36. $3 \div \dfrac{3}{2}$

37. Divide $\frac{7}{8}$ by $\frac{3}{4}$.

38. Divide $\frac{7}{12}$ by $\frac{3}{4}$.

39. Find the quotient of $\frac{5}{7}$ and $\frac{3}{14}$.

40. Find the quotient of $\frac{6}{11}$ and $\frac{9}{32}$.

Objective B

Divide.

41. $\frac{5}{6} \div 25$

42. $22 \div \frac{3}{11}$

43. $6 \div 3\frac{1}{3}$

44. $5\frac{1}{2} \div 11$

45. $6\frac{1}{2} \div \frac{1}{2}$

46. $\frac{3}{8} \div 2\frac{1}{4}$

47. $\frac{5}{12} \div 4\frac{4}{5}$

48. $1\frac{1}{2} \div 1\frac{3}{8}$

49. $8\frac{1}{4} \div 2\frac{3}{4}$

50. $3\frac{5}{9} \div 32$

51. $4\frac{1}{5} \div 21$

52. $6\frac{8}{9} \div \frac{31}{36}$

53. $\frac{11}{12} \div 2\frac{1}{3}$

54. $\frac{7}{8} \div 3\frac{1}{4}$

55. $\frac{5}{16} \div 5\frac{3}{8}$

56. $\frac{9}{14} \div 3\frac{1}{7}$

57. $35 \div \frac{7}{24}$

58. $\frac{3}{8} \div 2\frac{3}{4}$

59. $\frac{11}{18} \div 2\frac{2}{9}$

60. $\frac{21}{40} \div 3\frac{3}{10}$

61. $2\frac{1}{16} \div 2\frac{1}{2}$

62. $7\frac{3}{5} \div 1\frac{7}{12}$

63. $1\frac{2}{3} \div \frac{3}{8}$

64. $16 \div \frac{2}{3}$

65. $1\frac{5}{8} \div 4$

66. $13\frac{3}{8} \div \frac{1}{4}$

67. $16 \div 1\frac{1}{2}$

68. $9 \div \frac{7}{8}$

69. $16\frac{5}{8} \div 1\frac{2}{3}$

70. $24\frac{4}{5} \div 2\frac{3}{5}$

71. $1\frac{1}{3} \div 5\frac{8}{9}$

72. $13\frac{2}{3} \div 0$

2.8 Order, Exponents, and the Order of Operations Agreement

Objective A *To identify the order relation between two fractions* ....

Recall that whole numbers can be graphed as points on the number line. Fractions can also be graphed as points on the number line.

The graph of $\dfrac{3}{4}$ on the number line

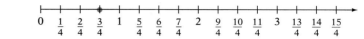

The number line can be used to determine the order relation between two fractions. A fraction that appears to the left of a given fraction is less than the given fraction. A fraction that appears to the right of a given fraction is greater than the given fraction.

$$\dfrac{1}{8} < \dfrac{3}{8} \qquad \dfrac{6}{8} > \dfrac{3}{8}$$

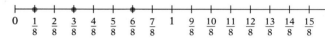

To find the order relation between two fractions with the same denominator, compare numerators. The fraction that has the smaller numerator is the smaller fraction. When the denominators are different, begin by writing equivalent fractions with a common denominator; then compare numerators.

➡ Find the order relation between $\dfrac{11}{18}$ and $\dfrac{5}{8}$.

The LCM of 18 and 8 is 72.

$\dfrac{11}{18} = \dfrac{44}{72}$ ← Smaller numerator

$\dfrac{5}{8} = \dfrac{45}{72}$ ← Larger numerator

$\dfrac{11}{18} < \dfrac{5}{8}$ or $\dfrac{5}{8} > \dfrac{11}{18}$

Example 1 Place the correct symbol, < or >, between the two numbers.

$\dfrac{5}{12} \qquad \dfrac{7}{18}$

Solution $\dfrac{5}{12} = \dfrac{15}{36} \qquad \dfrac{7}{18} = \dfrac{14}{36}$

$\dfrac{5}{12} > \dfrac{7}{18}$

You Try It 1 Place the correct symbol, < or >, between the two numbers.

$\dfrac{9}{14} \qquad \dfrac{13}{21}$

Your solution

Solution on p. S7

Objective B *To simplify expressions containing exponents* ....

Repeated multiplication of the same fraction can be written in two ways:

$$\dfrac{1}{2} \cdot \dfrac{1}{2} \cdot \dfrac{1}{2} \cdot \dfrac{1}{2} \qquad \text{or} \qquad \left(\dfrac{1}{2}\right)^4 \leftarrow \text{Exponent}$$

The exponent indicates how many times the fraction occurs as a factor in the multiplication. The expression $\left(\dfrac{1}{2}\right)^4$ is in exponential notation.

Example 2   Simplify: $\left(\dfrac{5}{6}\right)^3 \cdot \left(\dfrac{3}{5}\right)^2$

Solution $\left(\dfrac{5}{6}\right)^3 \cdot \left(\dfrac{3}{5}\right)^2 = \left(\dfrac{5}{6} \cdot \dfrac{5}{6} \cdot \dfrac{5}{6}\right) \cdot \left(\dfrac{3}{5} \cdot \dfrac{3}{5}\right)$

$= \dfrac{\overset{1}{\cancel{5}} \cdot \overset{1}{\cancel{5}} \cdot 5 \cdot \overset{1}{\cancel{3}} \cdot \overset{1}{\cancel{3}}}{2 \cdot \underset{1}{\cancel{3}} \cdot 2 \cdot \underset{1}{\cancel{3}} \cdot 2 \cdot 3 \cdot \underset{1}{\cancel{5}} \cdot \underset{1}{\cancel{5}}} = \dfrac{5}{24}$

You Try It 2   Simplify: $\left(\dfrac{7}{11}\right)^2 \cdot \left(\dfrac{2}{7}\right)$

Your solution

Solution on p. S7

Objective C   To use the Order of Operations Agreement to simplify expressions ....

The Order of Operations Agreement is used for fractions as well as whole numbers.

Step 1	Do all operations inside parentheses.
Step 2	Simplify any number expressions containing exponents.
Step 3	Do multiplications and divisions as they occur from left to right.
Step 4	Do additions and subtractions as they occur from left to right.

➡ Simplify $\dfrac{14}{15} - \left(\dfrac{1}{2}\right)^2 \times \left(\dfrac{2}{3} + \dfrac{4}{5}\right)$ by using the Order of Operations Agreement.

$\dfrac{14}{15} - \left(\dfrac{1}{2}\right)^2 \times \left(\dfrac{2}{3} + \dfrac{4}{5}\right)$ **1.** Perform operations in parentheses.

$\dfrac{14}{15} - \left(\dfrac{1}{2}\right)^2 \times \dfrac{22}{15}$ **2.** Simplify expressions with exponents.

$\dfrac{14}{15} - \dfrac{1}{4} \times \dfrac{22}{15}$ **3.** Do multiplications and divisions as they occur from left to right.

$\dfrac{14}{15} - \dfrac{11}{30}$ **4.** Do additions and subtractions as they occur from left to right.

$\dfrac{17}{30}$

One or more of the above steps may not be needed to simplify an expression. In that case, proceed to the next step in the Order of Operations Agreement.

Example 3   Simplify: $\left(\dfrac{3}{4}\right)^2 \div \left(\dfrac{3}{8} - \dfrac{1}{12}\right)$

Solution $\left(\dfrac{3}{4}\right)^2 \div \left(\dfrac{3}{8} - \dfrac{1}{12}\right)$

$= \left(\dfrac{3}{4}\right)^2 \div \left(\dfrac{7}{24}\right) = \dfrac{9}{16} \div \dfrac{7}{24}$

$= \dfrac{9}{16} \cdot \dfrac{24}{7} = \dfrac{27}{14} = 1\dfrac{13}{14}$

You Try It 3   Simplify:
$\left(\dfrac{1}{13}\right)^2 \cdot \left(\dfrac{1}{4} + \dfrac{1}{6}\right) \div \dfrac{5}{13}$

Your solution

Solution on p. S8

2.8 Exercises

Objective A

Place the correct symbol, $<$ or $>$, between the two numbers.

1. $\dfrac{11}{40}$ $\dfrac{19}{40}$

2. $\dfrac{92}{103}$ $\dfrac{19}{103}$

3. $\dfrac{2}{3}$ $\dfrac{5}{7}$

4. $\dfrac{2}{5}$ $\dfrac{3}{8}$

5. $\dfrac{5}{8}$ $\dfrac{7}{12}$

6. $\dfrac{11}{16}$ $\dfrac{17}{24}$

7. $\dfrac{7}{9}$ $\dfrac{11}{12}$

8. $\dfrac{5}{12}$ $\dfrac{7}{15}$

9. $\dfrac{13}{14}$ $\dfrac{19}{21}$

10. $\dfrac{13}{18}$ $\dfrac{7}{12}$

11. $\dfrac{7}{24}$ $\dfrac{11}{30}$

12. $\dfrac{13}{36}$ $\dfrac{19}{48}$

Objective B

Simplify.

13. $\left(\dfrac{3}{8}\right)^2$

14. $\left(\dfrac{5}{12}\right)^2$

15. $\left(\dfrac{2}{9}\right)^3$

16. $\left(\dfrac{1}{2}\right)\cdot\left(\dfrac{2}{3}\right)^2$

17. $\left(\dfrac{2}{3}\right)\cdot\left(\dfrac{1}{2}\right)^4$

18. $\left(\dfrac{1}{3}\right)^2\cdot\left(\dfrac{3}{5}\right)^3$

19. $\left(\dfrac{2}{5}\right)^3\cdot\left(\dfrac{5}{7}\right)^2$

20. $\left(\dfrac{5}{9}\right)^3\cdot\left(\dfrac{18}{25}\right)^2$

21. $\left(\dfrac{1}{3}\right)^4\cdot\left(\dfrac{9}{11}\right)^2$

22. $\left(\dfrac{1}{2}\right)^6\cdot\left(\dfrac{32}{35}\right)^2$

23. $\left(\dfrac{2}{3}\right)^4\cdot\left(\dfrac{81}{100}\right)^2$

24. $\left(\dfrac{1}{6}\right)\cdot\left(\dfrac{6}{7}\right)^2\cdot\left(\dfrac{2}{3}\right)$

25. $\left(\dfrac{2}{7}\right)\cdot\left(\dfrac{7}{8}\right)^2\cdot\left(\dfrac{8}{9}\right)$

26. $3\cdot\left(\dfrac{3}{5}\right)^3\cdot\left(\dfrac{1}{3}\right)^2$

27. $4\cdot\left(\dfrac{3}{4}\right)^3\cdot\left(\dfrac{4}{7}\right)^2$

28. $11\cdot\left(\dfrac{3}{8}\right)^3\cdot\left(\dfrac{8}{11}\right)^2$

29. $5\cdot\left(\dfrac{3}{5}\right)^3\cdot\left(\dfrac{2}{3}\right)^4$

30. $\left(\dfrac{2}{7}\right)^2\cdot\left(\dfrac{7}{9}\right)^2\cdot\left(\dfrac{9}{11}\right)^2$

Objective C

Simplify using the Order of Operations Agreement.

31. $\dfrac{1}{2} - \dfrac{1}{3} + \dfrac{2}{3}$

32. $\dfrac{2}{5} + \dfrac{3}{10} - \dfrac{2}{3}$

33. $\dfrac{1}{3} \div \dfrac{1}{2} + \dfrac{3}{4}$

34. $\dfrac{3}{5} \div \dfrac{6}{7} + \dfrac{4}{5}$

35. $\dfrac{4}{5} + \dfrac{3}{7} \cdot \dfrac{14}{15}$

36. $\dfrac{2}{3} + \dfrac{5}{8} \cdot \dfrac{16}{35}$

37. $\left(\dfrac{3}{4}\right)^2 - \dfrac{5}{12}$

38. $\left(\dfrac{3}{5}\right)^3 - \dfrac{3}{25}$

39. $\dfrac{5}{6} \cdot \left(\dfrac{2}{3} - \dfrac{1}{6}\right) + \dfrac{7}{18}$

40. $\dfrac{3}{4} \cdot \left(\dfrac{11}{12} - \dfrac{7}{8}\right) + \dfrac{5}{16}$

41. $\dfrac{7}{12} - \left(\dfrac{2}{3}\right)^2 + \dfrac{5}{8}$

42. $\dfrac{11}{16} - \left(\dfrac{3}{4}\right)^2 + \dfrac{7}{12}$

43. $\dfrac{3}{4} \cdot \left(\dfrac{4}{9}\right)^2 + \dfrac{1}{2}$

44. $\dfrac{9}{10} \cdot \left(\dfrac{2}{3}\right)^3 + \dfrac{2}{3}$

45. $\left(\dfrac{1}{2} + \dfrac{3}{4}\right) \div \dfrac{5}{8}$

46. $\left(\dfrac{2}{3} + \dfrac{5}{6}\right) \div \dfrac{5}{9}$

47. $\dfrac{3}{8} \div \left(\dfrac{5}{12} + \dfrac{3}{8}\right)$

48. $\dfrac{7}{12} \div \left(\dfrac{2}{3} + \dfrac{5}{9}\right)$

49. $\left(\dfrac{3}{8}\right)^2 \div \left(\dfrac{3}{7} + \dfrac{3}{14}\right)$

50. $\left(\dfrac{5}{6}\right)^2 \div \left(\dfrac{5}{12} + \dfrac{2}{3}\right)$

51. $\dfrac{2}{5} \div \dfrac{3}{8} \cdot \dfrac{4}{5}$

APPLYING THE CONCEPTS

52. $\dfrac{2}{3} < \dfrac{3}{4}$. Is $\dfrac{2+3}{3+4}$ less than $\dfrac{2}{3}$, greater than $\dfrac{2}{3}$, or between $\dfrac{2}{3}$ and $\dfrac{3}{4}$?

53. A farmer died and left 17 horses to be divided among 3 children. The first child was to receive $\dfrac{1}{2}$ of the horses, the second child $\dfrac{1}{3}$ of the horses, and the third child $\dfrac{1}{9}$ of the horses. The executor for the family's estate realized that 17 horses could not be divided by halves, thirds, or ninths and so added a neighbor's horse to the farmer's. With 18 horses, the executor gave 9 horses to the first child, 6 horses to the second child, and 2 horses to the third child. This accounted for the 17 horses, so the executor returned the borrowed horse to the neighbor. Explain why this worked.

Focus on Problem Solving

An application problem may not provide all the information that is needed to solve the problem. Sometimes, however, the necessary information is common knowledge.

You are traveling by bus from Boston to New York. The trip is 4 hours long. If the bus leaves Boston at 10 A.M., what time should you arrive in New York?

What other information do you need to solve this problem?

You need to know that, using a 12-hour clock, the hours run

10 A.M.
11 A.M.
12 P.M.
1 P.M.
2 P.M.

Four hours after 10 A.M. is 2 P.M.

You should arrive in New York at 2 P.M.

You purchase a 32¢ stamp at the post office and hand the clerk a one-dollar bill. How much change do you receive?

What information do you need to solve this problem?

You need to know that there are 100¢ in one dollar.

Your change is 100¢ − 32¢.

$$100 - 32 = 68$$

You receive 68¢ in change.

What information do you need to know to solve each of the following problems?

1. You sell a dozen tickets to a fundraiser. Each ticket costs $10. How much money do you collect?

2. The weekly lab period for your science course is one hour and twenty minutes long. Find the length of the science lab period in minutes.

3. An employee's monthly salary is $1750. Find the employee's annual salary.

4. A survey revealed that eighth graders spend an average of 3 hours each day watching television. Find the total time an eighth grader spends watching TV each week.

5. You want to buy a carpet for a room that is 15 feet wide and 18 feet long. Find the amount of carpet that you need.

Projects and Group Activities

Music   In musical notation, notes are printed on a staff, which is a set of five horizontal lines and the spaces between them. The notes of a musical composition are grouped into measures, or bars. Vertical lines separate measures on a staff. The shape of a note indicates how long it should be held. The whole note has the longest time value of any note. Each time value is divided by 2 in order to find the next smallest note value.

The time signature is a fraction that appears at the beginning of a piece of music. The numerator of the fraction indicates the number of beats in a measure. The denominator indicates what kind of note receives one beat. For example, music written in $\frac{2}{4}$ time has 2 beats to a measure, and a quarter note receives one beat. One measure in $\frac{2}{4}$ time may have 1 half note, 2 quarter notes, 4 eighth notes, or any other combination of notes totaling 2 beats. Other common time signatures include $\frac{4}{4}$, $\frac{3}{4}$, and $\frac{6}{8}$.

1. Explain the meaning of the 6 and the 8 in the time signature $\frac{6}{8}$. Give some possible combinations of notes in one measure of a piece written in $\frac{4}{4}$ time.

2. What does a dot at the right of a note indicate? What is the effect of a dot at the right of a half note? A quarter note? An eighth note?

3. Symbols called rests are used to indicate periods of silence in a piece of music. What symbols are used to indicate the different time values of rests?

4. Find some examples of musical compositions written in different time signatures. Use a few measures from each to show that the sum of the time values of the notes and rests in each measure equals the numerator of the time signature.

Construction

Suppose you are involved in building your own home. Design a stairway from the first floor of the house to the second floor. Some of the questions you will need to answer follow.

What is the distance from the floor of the first story to the floor of the second story?

Typically, what is the number of steps in a stairway?

What is a reasonable length for the run of each step?

What width wood is being used to build the staircase?

In designing the stairway, remember that each riser should be the same height and each run should be the same length. And the width of the wood used for the steps will have to be incorporated in the calculation.

Search the World Wide Web

There are many addresses on the Web where you can find investment information. Some of these addresses are:

http://www.streeteye.com/
http://www.brill.com/fundlink
http://www.thegroup.net:80/invest
http://stocks.com/

Below is a printout for MCN Energy Group Inc. for June 9, 1997, from http://www.streeteye.com/.

Last Sale	29 1/4	52 Week High	32 5/8
Tick	Down	52 Week Low	22 3/4
Net Change	+ 1/8	Volatility	20.06
Percent Change	+ 0.43	Ex-Dividend Date	08/06/97
Exchange	New York	Dividend Amount	0.243
Time of Last Sale	14:03	Dividend Frequency	Quarterly
Size of Last Sale	300	Earnings per Share	1.71
Open	29 1/8	P/E Ratio	17.10
High	29 3/8	Yield	3.32
Low	29	Shares Outstanding	67197
Volume	28200		
Previous Close	29 1/8		

Search these Web sites until you become familiar with the information that can be found on each site. Assume that you have $10,000 to invest. Make up some rules, such as that you cannot own more than 5 stocks at one time and you cannot make more than two trades a week. Assume that there are no commission fees. Keep a weekly record of your stocks and see if you can beat the professional investors.

Chapter Summary

Key Words

The *least common multiple* (LCM) is the smallest common multiple of two or more numbers.

The *greatest common factor* (GCF) is the largest common factor of two or more numbers.

A *fraction* can represent the number of equal parts of a whole.

A *proper fraction* is a fraction less than 1.

A *mixed number* is a number greater than 1 with a whole-number part and a fractional part.

An *improper fraction* is a fraction greater than or equal to 1.

Equal fractions with different denominators are called *equivalent fractions.*

A fraction is in *simplest form* when there are no common factors in the numerator and the denominator.

The *reciprocal* of a fraction is the fraction with the numerator and denominator interchanged.

Inverting is the process of finding the reciprocal of a fraction.

Essential Rules

Addition of Fractions with Like Denominators
To add fractions with like denominators, add the numerators and place the sum over the common denominator.

Addition of Fractions with Unlike Denominators
To add fractions with unlike denominators, first rewrite the fractions as equivalent fractions with the same denominator. Then add the numerators and place the sum over the common denominator.

Subtraction of Fractions with Like Denominators
To subtract fractions with like denominators, subtract the numerators and place the difference over the common denominator.

Subtraction of Fractions with Unlike Denominators
To subtract fractions with unlike denominators, rewrite the fractions as equivalent fractions with the same denominator. Then subtract the numerators and place the difference over the common denominator.

Multiplication of Fractions
To multiply two fractions, multiply the numerators and place the product over the product of the denominators.

Division of Fractions
To divide two fractions, multiply by the reciprocal of the divisor.

Chapter Review

1. Write $\frac{30}{45}$ in simplest form.

2. Simplify: $\left(\frac{3}{4}\right)^3 \cdot \frac{20}{27}$

3. Express the shaded portion of the circles as an improper fraction.

4. Find the total of $\frac{2}{3}$, $\frac{5}{6}$, and $\frac{2}{9}$.

5. Place the correct symbol, $<$ or $>$, between the two numbers.

    $\frac{11}{18} \quad \frac{17}{24}$

6. Subtract: $18\frac{1}{6}$
    $-\ 3\frac{5}{7}$
    $\overline{}$

7. Simplify: $\frac{2}{7}\left(\frac{5}{8} - \frac{1}{3}\right) \div \frac{3}{5}$

8. Multiply: $2\frac{1}{3} \times 3\frac{7}{8}$

9. Divide: $1\frac{1}{3} \div \frac{2}{3}$

10. Find $\frac{17}{24}$ decreased by $\frac{3}{16}$.

11. Divide: $8\frac{2}{3} \div 2\frac{3}{5}$

12. Find the GCF of 20 and 48.

13. Write an equivalent fraction with the given denominator.

    $\frac{2}{3} = \frac{}{36}$

14. What is $\frac{15}{28}$ divided by $\frac{5}{7}$?

15. Write an equivalent fraction with the given denominator.

    $\frac{8}{11} = \frac{}{44}$

16. Multiply: $2\frac{1}{4} \times 7\frac{1}{3}$

17. Find the LCM of 18 and 12.

18. Write $\frac{16}{44}$ in simplest form.

19. Add: $\frac{3}{8} + \frac{5}{8} + \frac{1}{8}$

20. Subtract: $\begin{array}{r} 16 \\ -\ 5\frac{7}{8} \\ \hline \end{array}$

21. Add: $4\frac{4}{9} + 2\frac{1}{6} + 11\frac{17}{27}$

22. Find the GCF of 15 and 25.

23. Write $\frac{17}{5}$ as a mixed number.

24. Simplify: $\left(\frac{4}{5} - \frac{2}{3}\right)^2 \div \frac{4}{15}$

25. Add: $\frac{3}{8} + 1\frac{2}{3} + 3\frac{5}{6}$

26. Find the LCM of 18 and 27.

27. Subtract: $\frac{11}{18} - \frac{5}{18}$

28. Write $2\frac{5}{7}$ as an improper fraction.

29. Divide: $\frac{5}{6} \div \frac{5}{12}$

30. Multiply: $\frac{5}{12} \times \frac{4}{25}$

31. What is $\frac{11}{50}$ multiplied by $\frac{25}{44}$?

32. Express the shaded portion of the circles as a mixed number.

33. During three months of the rainy season, $5\frac{7}{8}$, $6\frac{2}{3}$, and $8\frac{3}{4}$ inches of rain fell. Find the total rainfall for the three months.

34. A home building contractor bought $4\frac{2}{3}$ acres for \$168,000. What was the cost of each acre?

35. A 15-mile race has three checkpoints. The first checkpoint is $4\frac{1}{2}$ miles from the starting point. The second checkpoint is $5\frac{3}{4}$ miles from the first checkpoint. How many miles is the second checkpoint from the finish line?

36. A compact car gets 36 miles on each gallon of gasoline. How many miles can the car travel on $6\frac{3}{4}$ gallons of gasoline?

Chapter Test

1. Multiply: $\dfrac{9}{11} \times \dfrac{44}{81}$

2. Find the GCF of 24 and 80.

3. Divide: $\dfrac{5}{9} \div \dfrac{7}{18}$

4. Simplify: $\left(\dfrac{3}{4}\right)^2 \div \left(\dfrac{2}{3} + \dfrac{5}{6}\right) - \dfrac{1}{12}$

5. Write $9\dfrac{4}{5}$ as an improper fraction.

6. What is $5\dfrac{2}{3}$ multiplied by $1\dfrac{7}{17}$?

7. Write $\dfrac{40}{64}$ in simplest form.

8. Place the correct symbol, $<$ or $>$, between the two numbers.
   $\dfrac{3}{8}$ $\dfrac{5}{12}$

9. Simplify: $\left(\dfrac{1}{4}\right)^3 \div \left(\dfrac{1}{8}\right)^2 - \dfrac{1}{6}$

10. Find the LCM of 24 and 40.

11. Subtract: $\dfrac{17}{24} - \dfrac{11}{24}$

12. Write $\dfrac{18}{5}$ as a mixed number.

13. Find the quotient of $6\dfrac{2}{3}$ and $3\dfrac{1}{6}$.

14. Write an equivalent fraction with the given denominator.
    $\dfrac{5}{8} = \dfrac{}{72}$

15. Add: $\dfrac{5}{6}$

$\dfrac{7}{9}$

$+\ \dfrac{1}{15}$

_____

16. Subtract: $23\dfrac{1}{8}$

$-\ \ 9\dfrac{9}{44}$

_____

17. What is $\dfrac{9}{16}$ minus $\dfrac{5}{12}$?

18. Simplify: $\left(\dfrac{2}{3}\right)^4 \cdot \dfrac{27}{32}$

19. Add: $\dfrac{7}{12} + \dfrac{11}{12} + \dfrac{5}{12}$

20. What is $12\dfrac{5}{12}$ more than $9\dfrac{17}{20}$?

21. Express the shaded portion of the circles as an improper fraction.

22. An electrician earns $120 for each day worked. What is the total of the electrician's earnings for working $3\dfrac{1}{2}$ days?

23. Grant Miura bought $7\dfrac{1}{4}$ acres of land for a housing project. One and three-fourths acres were set aside for a park, and the remaining land was developed into $\dfrac{1}{2}$-acre lots. How many lots were available for sale?

24. Chris Aguilar bought 100 shares of a utility stock at $\$24\dfrac{1}{2}$ per share. The stock gained $\$5\dfrac{5}{8}$ during the first month of ownership and lost $\$2\dfrac{1}{4}$ during the second month. Find the value of 1 share of the utility stock at the end of the second month.

25. The rainfall for a 3-month period was $11\dfrac{1}{2}$ inches, $7\dfrac{5}{8}$ inches, and $2\dfrac{1}{3}$ inches. Find the total rainfall for the 3 months.

Cumulative Review

1. Round 290,496 to the nearest thousand.

2. Subtract: 390,047
                 − 98,769

3. Find the product of 926 and 79.

4. Divide: $57\overline{)30{,}792}$

5. Simplify: $4 \cdot (6 - 3) \div 6 - 1$

6. Find the prime factorization of 44.

7. Find the LCM of 30 and 42.

8. Find the GCF of 60 and 80.

9. Write $7\frac{2}{3}$ as an improper fraction.

10. Write $\frac{25}{4}$ as a mixed number.

11. Write an equivalent fraction with the given denominator.

$$\frac{5}{16} = \frac{}{48}$$

12. Write $\frac{24}{60}$ in simplest form.

13. What is $\frac{9}{16}$ more than $\frac{7}{12}$?

14. Add: $3\frac{7}{8}$

$7\frac{5}{12}$

$+ \ 2\frac{15}{16}$

15. Find $\frac{3}{8}$ less than $\frac{11}{12}$.

16. Subtract: $5\frac{1}{6}$

$- \ 3\frac{7}{18}$

17. Multiply: $\dfrac{3}{8} \times \dfrac{14}{15}$

18. Multiply: $3\dfrac{1}{8} \times 2\dfrac{2}{5}$

19. Divide: $\dfrac{7}{16} \div \dfrac{5}{12}$

20. Find the quotient of $6\dfrac{1}{8}$ and $2\dfrac{1}{3}$.

21. Simplify: $\left(\dfrac{1}{2}\right)^3 \cdot \dfrac{8}{9}$

22. Simplify: $\left(\dfrac{1}{2} + \dfrac{1}{3}\right) \div \left(\dfrac{2}{5}\right)^2$

23. Molly O'Brien had $1359 in a checking account. During the week, Molly wrote checks of $128, $54, and $315. Find the amount in the checking account at the end of the week.

24. The tickets for a movie were $5 for an adult and $2 for a student. Find the total income from the sale of 87 adult tickets and 135 student tickets.

25. Find the total weight of three packages that weigh $1\dfrac{1}{2}$ pounds, $7\dfrac{7}{8}$ pounds, and $2\dfrac{2}{3}$ pounds.

26. A board $2\dfrac{5}{8}$ feet long is cut from a board $7\dfrac{1}{3}$ feet long. What is the length of the remaining piece?

27. A car travels 27 miles on each gallon of gasoline. How many miles can the car travel on $8\dfrac{1}{3}$ gallons of gasoline?

28. Jimmy Santos purchased $10\dfrac{1}{3}$ acres of land to build a housing development. Jimmy donated 2 acres for a park. How many $\dfrac{1}{3}$-acre parcels can be sold from the remaining land?

3

Decimals

Bookkeepers record the transactions of a business. This requires excellent skills in adding, subtracting, multiplying, and dividing decimal numbers, which are the numbers we use to represent amounts of money.

Objectives

Section 3.1
To write decimals in standard form and in words
To round a decimal to a given place value

Section 3.2
To add decimals
To solve application problems

Section 3.3
To subtract decimals
To solve application problems

Section 3.4
To multiply decimals
To solve application problems

Section 3.5
To divide decimals
To solve application problems

Section 3.6
To convert fractions to decimals
To convert decimals to fractions
To identify the order relation between two decimals or between a decimal and a fraction

Decimal Fractions

How would you like to add $\frac{37{,}544}{23{,}465} + \frac{5184}{3456}$? These two fractions are very cumbersome, and it would take even a mathematician some time to get the answer.

Well, around 1550, help with such problems arrived with the publication of a book called *La Disme* (*The Tenth*), which urged the use of decimal fractions. A decimal fraction is one in which the denominator is 10, 100, 1000, 10,000, and so on.

This book suggested that all whole numbers were "units" and when written would end with the symbol ⓪. For example, the number 294⓪ would be the number two hundred ninety-four. This is very much like the way numbers are currently written (except for the ⓪).

For a fraction between 0 and 1, a unit was broken down into parts called "primes." The fraction three-tenths would be written

$$\frac{3}{10} = 3①$$ The ① was used to mean the end of the primes, or what are now called tenths.

Each prime was further broken down into "seconds," and each second was broken down into "thirds," and so on. The primes ended with ①, the seconds ended with ②, and the thirds ended with ③.

Examples of these numbers in our modern fraction notation and the old notation are shown below.

$$\frac{37}{100} = 3①7②\qquad\qquad \frac{257}{1000} = 2①5②7③$$

After completing this chapter, you might come back to the problem in the first line. Use decimals instead of fractions to find the answer. The answer is 3.1.

3.1 Introduction to Decimals

Objective A   **To write decimals in standard form and in words** ....

The smallest human bone is found in the middle ear and measures 0.135 inch in length. The number 0.135 is in **decimal notation.**

Note the relationship between fractions and numbers written in decimal notation.

Three-tenths   Three-hundredths   Three-thousandths

$$\frac{3}{10} = 0.\underline{3}$$ $$\frac{3}{100} = 0.0\underline{3}$$ $$\frac{3}{1000} = 0.00\underline{3}$$

1 zero 1 decimal place   2 zeros 2 decimal places   3 zeros 3 decimal places

A number written in decimal nota-   351   .   7089
tion has three parts.   **Whole-number   Decimal   Decimal**
                                               part   point   part

A number written in decimal notation is often called simply a **decimal.** The position of a digit in a decimal determines the digit's place value.

In the decimal 351.7089, the position of the digit 9 determines that its place value is ten-thousandths.

When writing a decimal in words, write the decimal part as if it were a whole number; then name the place value of the last digit.

0.6481   Six thousand four hundred eighty-one ten-thousandths
549.238   Five hundred forty-nine and two hundred thirty-eight thousandths
            (The decimal point is read as "and.")

To write a decimal in standard form, zeros may have to be inserted after the decimal point so that the last digit is in the given place-value position.

Five and thirty-eight <u>hundredths</u>

    8 is in the hundredths' place.   5.3<u>8</u>

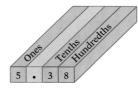

Nineteen and four <u>thousandths</u>

    4 is in the thousandths' place.   19.00<u>4</u>
    Insert two zeros so that 4 is in
    the thousandths' place.

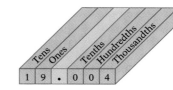

Seventy-one <u>ten-thousandths</u>

    1 is in the ten-thousandths'   0.007<u>1</u>
    place. Insert two zeros so that
    1 is in the ten-thousandths' place.

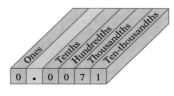

Example 1	Write 307.4027 in words.	**You Try It 1**	Write 209.05838 in words.
Solution	Three hundred seven and four thousand twenty-seven ten-thousandths	**Your solution**	
Example 2	Write six hundred seven and seven hundred eight hundred-thousandths in standard form.	**You Try It 2**	Write forty-two thousand and two hundred seven millionths in standard form.
Solution	607.00708	**Your solution**	

Solutions on p. S8

Objective B   To round a decimal to a given place value ······················· 5   CT

Rounding decimals is similar to rounding whole numbers except that the digits to the right of the given place value are dropped instead of being replaced by zeros.

If the digit to the right of the given place value is less than 5, drop that digit and all digits to the right. If the digit to the right of the given place value is greater than or equal to 5, increase the number in the given place value by 1 and drop all digits to its right.

➡ Round 26.3799 to the nearest hundredth.

Given place value

26.3799

9 > 5   Increase 7 by 1 and drop all digits to the right of 7.

26.3799 rounded to the nearest hundredth is 26.38.

Example 3	Round 0.39275 to the nearest ten-thousandth.	**You Try It 3**	Round 4.349254 to the nearest hundredth.
Solution	Given place value 0.39275 5 = 5 0.3928	**Your solution**	
Example 4	Round 42.0237412 to the nearest hundred-thousandth.	**You Try It 4**	Round 3.290532 to the nearest hundred-thousandth.
Solution	Given place value 42.0237412 1 < 5 42.02374	**Your solution**	

Solutions on p. S8

3.1 Exercises

Objective A

Write each decimal in words.

1. 0.27

2. 0.92

3. 1.005

4. 3.067

5. 36.4

6. 59.7

7. 0.00035

8. 0.00092

9. 10.007

10. 20.009

11. 52.00095

12. 64.00037

13. 0.0293

14. 0.0717

15. 6.324

16. 8.916

17. 276.3297

18. 418.3115

19. 216.0729

20. 976.0317

21. 4625.0379

22. 2986.0925

23. 1.00001

24. 3.00003

Write each decimal in standard form.

25. Seven hundred sixty-two thousandths

26. Two hundred ninety-five thousandths

27. Sixty-two millionths

28. Forty-one millionths

29. Eight and three hundred four ten-thousandths

30. Four and nine hundred seven ten-thousandths

31. Three hundred four and seven hundredths

32. Eight hundred ninety-six and four hundred seven thousandths

Write each decimal in standard form.

33. Three hundred sixty-two and forty-eight thousandths

34. Seven hundred eighty-four and eighty-four thousandths

35. Three thousand forty-eight and two thousand two ten-thousandths

36. Seven thousand sixty-one and nine thousand one ten-thousandths

---

Objective B

Round each decimal to the given place value.

37. 7.359 Tenths

38. 6.405 Tenths

39. 23.009 Tenths

40. 89.19204 Tenths

41. 22.68259 Hundredths

42. 16.30963 Hundredths

43. 7.072854 Thousandths

44. 1946.3745 Thousandths

45. 62.009435 Thousandths

46. 0.029876 Ten-thousandths

47. 0.012346 Ten-thousandths

48. 1.702596 Nearest whole number

49. 2.079239 Hundred-thousandths

50. 0.0102903 Millionths

51. 0.1009754 Millionths

APPLYING THE CONCEPTS

52. To what decimal place value are timed events in the Olympics recorded? Provide some specific examples of events and the winning times in each.

53. Provide an example of a situation in which a decimal is always rounded up, even if the digit to the right is less than 5. Provide an example of a situation in which a decimal is always rounded down, even if the digit to the right is 5 or greater than 5. (*Hint*: Think about situations in which money changes hands.)

54. Indicate which zeros of the number, if any, need not be entered on a calculator.
   a. 23.500   **b.** 0.000235   **c.** 300.0005   **d.** 0.004050

55. **a.** A decimal number was rounded to 6. Between what two numbers, to the nearest tenth, was the number?
   b. A decimal number was rounded to 10.2. Between what two numbers, to the nearest hundredth, was the number?

3.2   **Addition of Decimals**

Objective A   *To add decimals* ··

To add decimals, write the numbers so that the decimal points are on a vertical line. Add as for whole numbers, and write the decimal point in the sum directly below the decimal points in the addends.

➡ Add: 0.237 + 4.9 + 27.32

Note that by placing the decimal points on a vertical line, we make sure that digits of the same place value are added.

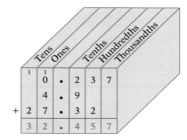

	Example 1	Find the sum of 42.3, 162.903, and 65.0729.	**You Try It 1**	Add: 4.62 + 27.9 + 0.62054

Solution

$$\begin{array}{r}\scriptstyle 1\ 1\ 1\\ 42.3\\ 162.903\\ +\ \ 65.0729\\ \hline 270.2759\end{array}$$

Your solution

Example 2   Add: 0.83 + 7.942 + 15

You Try It 2   Add: 6.05 + 12 + 0.374

Solution

$$\begin{array}{r}\scriptstyle 1\ 1\\ 0.83\\ 7.942\\ +\ 15.\\ \hline 23.772\end{array}$$

Your solution

Solutions on p. S8

ESTIMATION

Estimating the Sum of Two or More Decimals

Estimate and then use your calculator to find 23.037 + 16.7892.

To estimate the sum of two or more numbers, round each number to the same place value. In this case, we will round to the nearest whole number. Then add. The estimated answer is 40.

$$\begin{array}{r}23.037\ \approx\ \ \ 23\\ +\ 16.7892\ \approx\ +\ 17\\ \hline 40\end{array}$$

Now use your calculator to find the exact result. The exact answer is 39.8262.

23.037 ☐+☐ 16.7892 ☐=☐ 39.8262

Objective B **To solve application problems** ·································   CT

The 1997 catalog from the West Shore Acres Display Garden has the accompanying list of prices for different numbers and kinds of tulip bulbs. Use this price list for Example 3 and You Try It 3.

		10	20	30
Single Early		$5.80	$11.10	$16.40

One of the first to bloom. 14"-18"; Late March-Early April
CAT.#

050	**APRICOT BEAUTY** Soft salmon rose. Long lasting. 18".
051	**BESTSELLER** Rich golden orange. 14".
052	**CHRISTMAS DREAM** Warm rosy pink. 14".
053	**CHRISTMAS MARVEL** Glowing fuchsia pink, Eleanor's favorite. 14".
054	**GENERAL DEWET** Orange, hint of yellow. Fragrant. 14".
055	**MERRY CHRISTMAS** Cheery "Holiday" red. 14".
056	**PRINCESS IRENE** Salmon orange, purple flame. 14".

Source: 1997 catalog, West Shores Acres Display Garden Catalog, 956 Downey Road, Mount Vernon, WA 98273

Example 3
Find the cost of the following order:

10 Apricot Beauty
30 General Dewet
10 Princess Irene

Strategy
To find the total cost of the order, add the cost of each kind and number of tulips ($5.80, $16.40, $5.80).

Solution
$5.80 + $16.40 + $5.80 = $28.00

The cost of the tulips was $28.00.

Example 4
Dan Burhoe earned a salary of $138.50 for working 5 days this week as a food server. He also received $22.92, $15.80, $19.65, $39.20, and $27.70 in tips during the 5 days. Find his total income for the week.

Strategy
To find the total income, add the tips ($22.92, $15.80, $19.65, $39.20, and $27.70) to the salary ($138.50).

Solution
```
   $138.50
     22.92
     15.80
     19.65
     39.20
  +  27.70
   $263.77
```

Dan's total income for the week was $263.77.

You Try It 3
Find the cost of the following order:

30 Christmas Dream
10 Christmas Marvel
30 Merry Christmas

Your strategy

Your solution

You Try It 4
Anita Khavari, an insurance executive, earns a salary of $425 every four weeks. During the past 4-week period, she received commissions of $485.60, $599.46, $326.75, and $725.42. Find her total income for the past 4-week period.

Your strategy

Your solution

Solutions on p. S8

3.2 Exercises

· ·

Objective A

Add.

1. 16.008 + 2.0385 + 132.06

2. 17.32 + 1.0579 + 16.5

3. 1.792 + 67 + 27.0526

4. 8.772 + 1.09 + 26.5027

5. 3.02 + 62.7 + 3.924

6. 9.06 + 4.976 + 59.6

7. 82.006 + 9.95 + 0.927

8. 0.826 + 8.76 + 79.005

9. 4.307 + 99.82 + 9.078

10.   0.3
      + 0.07

11.   0.29
      + 0.4

12.   1.007
      + 2.1

13.   7.3
      + 9.005

14.   4.9257
       27.05
     + 9.0063

15.   8.72
       99.073
     + 2.9736

16.   62.4
       9.827
     + 692.44

17.   8
       89.43
     + 7.0659

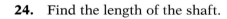 Estimate by rounding to the nearest whole number. Then use your calculator to add.

18.   342.42
       89.625
     + 176.2

19.   219.9
       0.872
     + 13.42

20.   823.9
       82.65
     + 46.923

21.   678.92
       97.6
     + 5.423

Objective B   *Application Problems*

22. A family has a mortgage of $814.72, a Visa bill of $216.40, and an electric bill of $87.32. Estimate by rounding the numbers to the nearest hundred dollars, and then calculate the exact amount of the three payments.

23. Find the length of the shaft.

24. Find the length of the shaft.

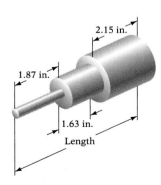

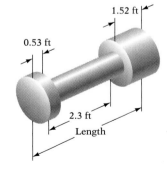

25. Commuting, Mae Chan used 12.4 gallons of gas the first week, 9.8 gallons the second week, 15.2 gallons the third week, and 10.4 gallons the fourth week. Find the total amount of gas she used during the 4 weeks.

26. The odometer on a family's car reads 24,835.9 miles. The car was driven 8.2 miles on Friday, 82.6 miles on Saturday, and 133.6 miles on Sunday.
a. How many miles was the car driven during the three days?
b. Find the odometer reading at the end of the three days.

27. You have $2143.57 in your checking account. You make deposits of $210.98, $45.32, $1236.34, and $27.99. Find the amount in your checking account after you have made the deposits if no money has been withdrawn.

The figure at the right shows the number of viewers who watch television each night of the week. Use this figure for Exercises 28 to 30.

How Many People Tune In Prime-Time TV	
Mon	91.9
Tues	89.8
Wed	90.6
Thu	93.9
Fri	78.0
Sat	77.1
Sun	87.7
Viewers (Millions) 0 50 100	

28. Find the total number of television viewers for Friday, Saturday, and Sunday nights.

29. Find the total number of television viewers for Monday, Tuesday, Wednesday, and Thursday nights.

30. Find the total number of television viewers for the week.

APPLYING THE CONCEPTS

The table at the right gives the prices for selected products in a grocery store. Use this table for Exercises 31 and 32.

Product	Cost
Raisin bran	3.45
Butter	2.69
Bread	1.23
Popcorn	.89
Potatoes	1.09
Cola (6-pack)	.98
Mayonnaise	2.25
Lunch meat	3.31
Milk	2.18
Toothpaste	2.45

31. Does a customer with $10 have enough money to purchase raisin bran, bread, milk, lunch meat, and butter?

32. Name three items that would cost more than $6 but less than $7. (There is more than one answer.)

33. Can a piece of rope 4 feet long be wrapped around the box shown at the right?

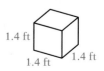

1.4 ft
1.4 ft
1.4 ft

3.3 Subtraction of Decimals

Objective A   To subtract decimals ....

To subtract decimals, write the numbers so that the decimal points are on a vertical line. Subtract as for whole numbers, and write the decimal point in the difference directly below the decimal point in the subtrahend.

➡ Subtract 21.532 − 9.875 and check.

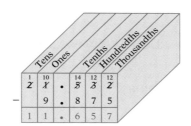

Placing the decimal points on a vertical line ensures that digits of the same place value are subtracted.

Check:

		$\overset{1\ 1\ \ 11}{9.875}$
	Subtrahend	
	+ Difference	+ 11.657
	= Minuend	21.532

➡ Subtract 4.3 − 1.7942 and check.

$$\overset{3\quad 12\ 9\ 9\ 10}{\cancel{4}.\cancel{3}\,\cancel{0}\,\cancel{0}\,\cancel{0}}$$
$$-\ 1.7942$$
$$\underline{\quad 2.5058}$$

If necessary, insert zeros in the minuend before subtracting.

Check:

$$\overset{1\ \ 1\ 1\ 1}{1.7942}$$
$$+\ 2.5058$$
$$\underline{\quad 4.3000}$$

Example 1   Subtract 39.047 − 7.96 and check.

Solution
$$\overset{8\quad 9\ 14}{3\cancel{9}.\cancel{0}\,\cancel{4}\,7}$$
$$-\ \ 7.96$$
$$\underline{\ \ 31.087}$$

Check:
$$\overset{1\ 1}{7.96}$$
$$+\ 31.087$$
$$\underline{\ \ 39.047}$$

You Try It 1   Subtract 72.039 − 8.47 and check.

Your solution

Example 2   Find 9.23 less than 29 and check.

Solution
$$\overset{1\ 18\quad 9\ 10}{\cancel{2}\,\cancel{9}.\cancel{0}\,\cancel{0}}$$
$$-\ \ 9.23$$
$$\underline{\ \ 19.77}$$

Check:
$$\overset{1\ 1\ 1}{9.23}$$
$$+\ 19.77$$
$$\underline{\ \ 29.00}$$

You Try It 2   Subtract 35 − 9.67 and check.

Your solution

Example 3   Subtract 1.2 − 0.8235 and check.

Solution
$$\overset{0\quad 11\ 9\ 9\ 10}{\cancel{1}.\cancel{2}\,\cancel{0}\,\cancel{0}\,\cancel{0}}$$
$$-\ 0.8235$$
$$\underline{\ \ 0.3765}$$

Check:
$$\overset{1\ 1\,1\,1}{0.8235}$$
$$+\ 0.3765$$
$$\underline{\ \ 1.2000}$$

You Try It 3   Subtract 3.7 − 1.9715 and check.

Your solution

Solutions on p. S8

ESTIMATION

Estimating the Difference Between Two Decimals

Estimate and then use your calculator to find 820.2306 − 475.74815.

To estimate the difference between two numbers, round each number to the same place value. In this case we will round to the nearest ten. Then subtract. The estimated answer is 340.

$$
\begin{array}{r}
820.2306 \approx 820 \\
- \; 475.74815 \approx - \; 480 \\
\hline
340
\end{array}
$$

Now use your calculator to find the exact result. The exact answer is 344.48245.

820.2306 ⊟ 475.74815 ⊟ 344.48245

Objective B   *To solve application problems* ....

Example 4
You bought a book for $15.87. How much change did you receive from a $20.00 bill?

Strategy
To find the amount of change, subtract the cost of the book ($15.87) from $20.00.

Solution
$$
\begin{array}{r}
\$20.00 \\
- \; 15.87 \\
\hline
\$ \; 4.13
\end{array}
$$

You received $4.13 in change.

You Try It 4
Your breakfast cost $3.85. How much change did you receive from a $5.00 bill?

Your strategy

Your solution

Example 5
You had a balance of $62.41 in your checking account. You then bought a cassette for $8.95, film for $3.17, and a skateboard for $39.77. After paying for these items with a check, how much do you have left in your checking account?

Strategy
To find the new balance:
• Find the total cost of the three items ($8.95 + $3.17 + $39.77).
• Subtract the total cost from the old balance ($62.41).

Solution
$$
\begin{array}{r}
\$ \; 8.95 \\
3.17 \\
+ \; 39.77 \\
\hline
\$51.89 \text{ total cost}
\end{array}
\qquad
\begin{array}{r}
\$62.41 \\
- \; 51.89 \\
\hline
\$10.52
\end{array}
$$

The new balance is $10.52.

You Try It 5
You had a balance of $2472.69 in your checking account. You then wrote checks for $1025.60, $79.85, and $162.47. Find the new balance in your checking account.

Your strategy

Your solution

Solutions on pp. S8–S9

3.3 Exercises

. . . .

Objective A

Subtract and check.

1. 24.037 − 18.41   **2.** 26.029 − 19.31   **3.** 123.07 − 9.4273   **4.** 214 − 7.143

5. 16.5 − 9.7902   **6.** 13.2 − 8.6205   **7.** 235.79 − 20.093   **8.** 463.27 − 40.095

9. 63.005 − 9.1274   **10.** 23.004 − 7.2175   **11.** 92 − 19.2909   **12.** 41.2405 − 25.2709

13. 7.01 − 2.325   **14.** 8.07 − 5.392   **15.** 19.0035 − 8.967

16.   0.32   **17.**   0.78   **18.**   3.005   **19.**   6.007
   − 0.0058   − 0.0073   − 1.982   − 2.734

20.   352.16   **21.**   872   **22.**   724.32   **23.**   625.46
   − 90.994   − 80.753   − 69   − 77.509

24.   362.394   **25.**   421.385   **26.**   19   **27.**   23.4
   − 19.4672   − 17.5293   − 10.372   − 0.921

Estimate by rounding to the nearest ten. Then use your calculator to subtract.

28.   620.59   **29.**   835.07   **30.**   67.3   **31.**   84.1
   − 132.79   − 244.82   − 19.793   − 48.906

Estimate by rounding to the nearest whole number (nearest one). Then use your calculator to subtract.

32.   93.079256   **33.**   3.7529   **34.**   76.53902   **35.**   9.07325
   − 66.09249   − 1.00784   − 45.73005   − 1.924

Objective B *Application Problems*

36. The manager of the Edgewater Cafe takes a reading of the cash register tape each hour. At 1:00 P.M. the tape read $967.54; at 2:00 P.M. the tape read $1437.15. Find the amount of sales between 1:00 P.M. and 2:00 P.M.

37. Find the missing dimension.

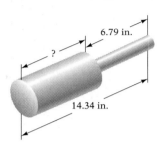

38. Find the missing dimension.

39. You had a balance of $1029.74 in your checking account. You then wrote checks for $67.92, $43.10, and $496.34.
  a. Find the total amount of the checks written.
  b. Find the new balance in your checking account.

40. The price of gasoline is $1.22 per gallon after the price rose $.07 one month and $.12 the next month. Find the price of gasoline before these increases in price.

41. Rainfall for the last 3 months of the year was 1.42 inches, 5.39 inches, and 3.55 inches. The normal rainfall for the last 3 months of the year is 11.22 inches. How many inches below normal was the rainfall?

42. Grace Herrera owned 357.448 shares of a mutual fund on January 1, 1998. On December 31, 1998, she had 439.917 shares. What was the increase in the number of shares for the year?

43. In 1990, Arie Luyendyk set the Indianapolis 500 average speed record of 185.981 mph. Luyendyk won the race again in 1997 with an average speed of 145.857 mph. Find the difference in average speed for those two races.

The table at the right shows the amount that an investor would pay to buy stocks from various companies. Use the table for Exercises 44 and 45.

44. How much would be saved from buying 200 shares at $25 dollars a share from Quick & Reilly instead of from Schwab?

45. How much would be saved from buying 500 shares at $18 per share from Waterhouse Securities instead of from Merrill Lynch?

Retail Broker	200 sh. @ $25	300 sh. @ $20	500 sh. @ $18
Merrill Lynch	$129.50	$164.85	$225.23
Schwab	$89.00	$95.60	$106.60
Fidelity	$88.50	$95.10	$106.10
Quick & Reilly	$60.50	$65.00	$81.50
Waterhouse Securities	$35.00	$40.82	$57.62

Source: Kiplinger Magazine, March 1997, page 100

APPLYING THE CONCEPTS

46. Find the largest amount by which the estimate of the sum of two decimals with tenths, hundredths, and thousandths places could differ from the exact sum.

3.4 Multiplication of Decimals

Objective A *To multiply decimals* .... ⟨ 5 ⟩ CT

Decimals are multiplied as if they were whole numbers; then the decimal point is placed in the product. Writing the decimals as fractions shows where to write the decimal point in the product.

$$0.\underline{3} \times 5 = \frac{3}{10} \times \frac{5}{1} = \frac{15}{10} = 1.\underline{5}$$

1 decimal place   1 decimal place

$$0.\underline{3} \times 0.\underline{5} = \frac{3}{10} \times \frac{5}{10} = \frac{15}{100} = 0.\underline{15}$$

1 decimal place   1 decimal place   2 decimal places

$$0.\underline{3} \times 0.\underline{05} = \frac{3}{10} \times \frac{5}{100} = \frac{15}{1000} = 0.\underline{015}$$

1 decimal place   2 decimal places   3 decimal places

To multiply decimals, multiply the numbers as in whole numbers. Write the decimal point in the product so that the number of decimal places in the product is the sum of the decimal places in the factors.

➡ Multiply: 21.4×0.36

$$
\begin{array}{r}
21.4 \\
\times\, 0.36 \\
\hline
1284 \\
642 \\
\hline
7.704
\end{array}
$$

21.4 — 1 decimal place
× 0.36 — 2 decimal places
7.704 — 3 decimal places

➡ Multiply: 0.037×0.08

$$
\begin{array}{r}
0.037 \\
\times\, 0.08 \\
\hline
0.00296
\end{array}
$$

0.037 — 3 decimal places
× 0.08 — 2 decimal places
0.00296 — 5 decimal places

• **Two zeros must be inserted between the 2 and the decimal point so that there are 5 decimal places in the product.**

To multiply a decimal by a power of 10 (10, 100, 1000, . . .), move the decimal point to the right the same number of places as there are zeros in the power of 10.

$3.8925 \times 1\underline{0}$ $= 38.925$

1 zero   1 decimal place

$3.8925 \times 1\underline{00}$ $= 389.25$

2 zeros   2 decimal places

$3.8925 \times 1\underline{000}$ $= 3892.5$

3 zeros   3 decimal places

$3.8925 \times 1\underline{0,000}$ $= 38,925.$

4 zeros   4 decimal places

$3.8925 \times 1\underline{00,000} = 389,250.$

5 zeros   5 decimal places

Note that a zero must be inserted before the decimal point.

Note that if the power of 10 is written in exponential notation, the exponent indicates how many places to move the decimal point.

$$3.8925 \times 10^1 = 38.925$$
1 decimal place

$$3.8925 \times 10^2 = 389.25$$
2 decimal places

$$3.8925 \times 10^3 = 3892.5$$
3 decimal places

$$3.8925 \times 10^4 = 38{,}925.$$
4 decimal places

$$3.8925 \times 10^5 = 389{,}250.$$
5 decimal places

Example 1   Multiply: 920×3.7

Solution
$$
\begin{array}{r}
920 \\
\times\ \ 3.7 \\
\hline
644\ 0 \\
2760\ \ \\
\hline
3404.0
\end{array}
$$

You Try It 1   Multiply: 870×4.6

Your solution

Example 2   Find 0.00079 multiplied by 0.025.

Solution
$$
\begin{array}{r}
0.00079 \\
\times\ \ \ \ 0.025 \\
\hline
395 \\
158\ \ \\
\hline
0.00001975
\end{array}
$$

You Try It 2   Find 0.000086 multiplied by 0.057.

Your solution

Example 3   Find the product of 3.69 and 2.07.

Solution
$$
\begin{array}{r}
3.69 \\
\times\ 2.07 \\
\hline
2583 \\
7380\ \ \\
\hline
7.6383
\end{array}
$$

You Try It 3   Find the product of 4.68 and 6.03.

Your solution

Example 4   Multiply: $42.07 \times 10{,}000$

Solution $42.07 \times 10{,}000 = 420{,}700$

You Try It 4   Multiply: 6.9×1000

Your solution

Example 5   Find 3.01 times 10^3.

Solution $3.01 \times 10^3 = 3010$

You Try It 5   Find 4.0273 times 10^2.

Your solution

Solutions on p. S9

ESTIMATION

Estimating the Product of Two Decimals

Estimate and then use your calculator to find 28.259×0.029.

To estimate a product, round each
number so that there is one nonzero
digit. Then multiply.

$$
\begin{array}{r}
28.259 \approx \quad 30 \\
\times \ 0.029 \approx \times 0.03 \\
\hline
0.90
\end{array}
$$

The estimated answer is 0.90.

Now use your calculator to find the
exact answer.

28.259 ⨯ 0.029 ═ 0.819511

The exact answer is 0.819511.

Objective B   *To solve application problems* ....

The tables below list water rates and meter fees for a city. These tables are used
for Example 6 and You Try It 6.

Water Charges		Meter Charges	
Commercial	$1.39/1000 gal	*Meter*	*Meter Fee*
Comm Restaurant	$1.39/1000 gal	5/8" & 3/4"	$13.50
Industrial	$1.39/1000 gal	1"	$21.80
Institutional	$1.39/1000 gal	1-1/2"	$42.50
Res—No Sewer		2"	$67.20
Residential—SF		3"	$133.70
>0 <200 gal. per day	$1.15/1000 gal	4"	$208.20
>200 <1500 gal. per day	$1.39/1000 gal	6"	$415.10
>1500 gal. per day	$1.54/1000 gal	8"	$663.70

Example 6

Find the total bill for an industrial water
user with a 6-inch meter that uses 152,000
gallons of water for July and August.

Strategy

To find the total cost of water:

• Find the cost of water by multiplying the
cost per 1000 gallons ($1.39) by the
number of 1000-gallon units used.
• Add the cost of the water to the meter fee
($415.10).

Solution

Cost of water $= \dfrac{152{,}000}{1000} \cdot 1.39 = 211.28$

Total cost $= 211.28 + 415.10 = 626.38$

The total cost is $626.38.

You Try It 6

Find the total bill for a commercial user
that used 5000 gallons of water per day for
July and August. The user has a 3-inch
meter.

Your strategy

Your solution

Solution on p. S9

Example 7

It costs $.036 an hour to operate an electric motor. How much does it cost to operate the motor for 120 hours?

Strategy

To find the cost of running the motor for 120 hours, multiply the hourly cost ($.036) by the number of hours the motor is run (120).

Solution

$$
\begin{array}{r}
\$.036 \\
\times \quad 120 \\
\hline
720 \\
36 \quad \\
\hline
\$4.320
\end{array}
$$

The cost of running the motor for 120 hours is $4.32.

You Try It 7

The cost of electricity to run a freezer for 1 hour is $.035. This month the freezer has run for 210 hours. Find the total cost of running the freezer this month.

Your strategy

Your solution

Example 8

Jason Ng earns a salary of $280 for a 40-hour work week. This week he worked 12 hours of overtime at a rate of $10.50 for each hour of overtime worked. Find his total income for the week.

Strategy

To find Jason's total income for the week:

- Find the overtime pay by multiplying the hourly overtime rate ($10.50) by the number of hours of overtime worked (12).
- Add the overtime pay to the weekly salary ($280).

Solution

$$
\begin{array}{r}
\$10.50 \\
\times \quad 12 \\
\hline
21\ 00 \\
105\ 0 \quad \\
\hline
\$126.00 \text{ overtime pay}
\end{array}
\qquad
\begin{array}{r}
\$280.00 \\
+\ 126.00 \\
\hline
\$406.00
\end{array}
$$

Jason's total income for this week is $406.00.

You Try It 8

You make a down payment of $175 on a stereo and agree to make payments of $37.18 a month for the next 18 months to repay the remaining balance. Find the total cost of the stereo.

Your strategy

Your solution

Solutions on p. S9

3.4 Exercises

· ·

Objective A

Multiply.

1.	0.9 × 0.4	2.	0.7 × 0.9	3.	0.5 × 0.6	4.	0.3 × 0.7	5.	0.5 × 0.5

6.	0.7 × 0.7	7.	0.9 × 0.5	8.	0.2 × 0.6	9.	7.7 × 0.9	10.	3.4 × 0.4

11.	9.2 × 0.2	12.	2.6 × 0.7	13.	7.2 × 0.6	14.	6.8 × 0.4	15.	7.4 × 0.1

16.	3.8 × 0.1	17.	7.9 × 5	18.	9.3 × 7	19.	0.68 × 4	20.	0.83 × 9

21.	0.67 × 0.9	22.	0.84 × 0.3	23.	0.16 × 0.6	24.	0.47 × 0.8	25.	2.5 × 5.4

26.	3.9 × 1.9	27.	8.4 × 9.5	28.	7.6 × 5.8	29.	0.83 × 5.2	30.	0.24 × 2.7

31.	0.46 × 3.9	32.	0.78 × 6.8	33.	0.2 × 0.3	34.	0.3 × 0.3	35.	0.24 × 0.3

36.	0.17 × 0.5	37.	1.47 × .09	38.	6.37 × 0.05	39.	8.92 × 0.004	40.	6.75 × 0.007

Multiply.

41. 0.49
$\times$ 0.16

42. 0.38
$\times$ 0.21

43. 7.6
$\times$ 0.01

44. 5.1
$\times$ 0.01

45. 8.62
$\times$ 4

46. 5.83
$\times$ 7

47. 64.5
$\times$ 9

48. 37.8
$\times$ 8

49. 2.19
$\times$ 9.2

50. 1.25
$\times$ 5.6

51. 1.85
$\times$ 0.023

52. 37.8
$\times$ 0.052

53. 0.478
$\times$ 0.37

54. 0.526
$\times$ 0.22

55. 48.3
$\times$ 0.0041

56. 67.2
$\times$ 0.0086

57. 2.437
$\times$ 6.1

58. 4.237
$\times$ 0.54

59. 0.413
$\times$ 0.0016

60. 0.517
$\times$ 0.0029

61. 94.73
$\times$ 0.57

62. 89.23
$\times$ 0.62

63. 8.005
$\times$ 0.067

64. 9.032
$\times$ 0.019

65. 4.29×0.1

66. 6.78×0.1

67. 5.29×0.4

68. 6.78×0.5

69. 0.68×0.7

70. 0.56×0.9

71. 1.4×0.73

72. 6.3×0.37

73. 5.2×7.3

74. 7.4×2.9

75. 3.8×0.61

76. 7.2×0.72

77. 0.32×10

78. 6.93×10

79. 0.065×100

80. 0.039×100

81. 6.2856×1000

Multiply.

82. 3.2954 × 1000   **83.** 3.2 × 1000   **84.** 0.006 × 10,000   **85.** 3.57 × 10,000

86. 8.52×10^1 **87.** 0.63×10^1 **88.** 82.9×10^2

89. 0.039×10^2 **90.** 6.8×10^3 **91.** 4.9×10^4

92. 6.83×10^4 **93.** 0.067×10^2 **94.** 0.052×10^2

95. Find the product of 0.0035 and 3.45.   **96.** Find the product of 237 and 0.34.

97. Multiply 3.005 by 0.00392.   **98.** Multiply 20.34 by 1.008.

99. Multiply 1.348 by 0.23.   **100.** Multiply 0.000358 by 3.56.

101. Find the product of 23.67 and 0.0035.   **102.** Find the product of 0.00346 and 23.1.

103. Find the product of 5, 0.45, and 2.3.   **104.** Find the product of 0.03, 23, and 9.45.

Estimate and then use your calculator to multiply.

105. 28.5 × 3.2   **106.** 86.3 × 4.4   **107.** 2.38 × 0.44   **108.** 9.82 × 0.77

109. 0.866 × 4.5   **110.** 0.239 × 8.2   **111.** 4.34 × 2.59   **112.** 6.87 × 9.98

113. 8.434 × 0.044   **114.** 7.037 × 0.094   **115.** 28.44 × 1.12   **116.** 86.57 × 7.33

117. 49.6854 × 39.0672   **118.** 2.00547 × 9.672   **119.** 0.00456 × 0.009542   **120.** 7.00637 × 0.0128

Objective B *Application Problems*

121. It costs $8 a day and $.28 per mile to rent a car. Find the cost to rent a car for five days if the car is driven 530 miles.

122. An electric motor costing $315.45 has an operating cost of $.027 for 1 hour of operation. Find the cost to run the motor for 56 hours. Round to the nearest cent.

123. Four hundred empty soft drink cans weigh 18.75 pounds. A recycling center pays $.75 per pound for the cans. Find the amount received for the 400 cans. Round to the nearest cent.

124. A recycling center pays $.045 per pound for newspapers. Estimate the payment for recycling 520 pounds of newspapers. Find the actual amount received from recycling the newspapers.

125. A broker's fee for buying stock is 0.045 times the price of the stock. An investor bought 100 shares of stock at $38.50 per share. Find the broker's fee.

126. A broker's fee for buying a stock is 0.028 times the price of the stock. An investor bought 100 shares of stock at $54.25 per share. Estimate the broker's fee. Calculate the actual broker's fee.

127. You bought a car for $2000 down and made payments of $127.50 each month for 36 months.
a. Find the amount of the payments over the 36 months.
b. Find the total cost of the car.

128. As a nurse, Rob Martinez earns a salary of $344 for a 40-hour work week. This week he worked 15 hours of overtime at a rate of $12.90 for each hour of overtime worked.
a. Find the amount of overtime pay.
b. Find Rob's total income for the week.

129. Bay Area Rental Cars charges $12 a day and $.12 per mile for renting a car. You rented a car for 3 days and drove 235 miles. Find the total cost of renting the car.

130. A taxi costs $1.50 and $.20 for each $\frac{1}{8}$ mile driven. Find the cost of hiring a taxi to get from the airport to the hotel—a distance of 5.5 miles.

The table at the right lists three pieces of steel required for a repair project. Use this table for Exercises 131 and 132.

Grade of Steel	Weight (Pounds per Foot)	Required Number of Feet	Cost per Pound
1	2.2	8	$1.20
2	3.4	6.5	1.35
3	6.75	15.4	1.94

131. Find the total cost of each of the grades of steel.

132. Find the total cost of the three pieces of steel.

A confectioner ships holiday packs of candy and nuts anywhere in the United States. At the right is a price list for nuts and candy, and below is a table of shipping charges to zones in the United States. For any fraction of a pound, use the next higher weight. Use these tables for Exercise 133. (16 oz = 1 lb)

Code	Description	Price
112	Almonds 16 oz	4.75
116	Cashews 8 oz	2.90
117	Cashews 16 oz	5.50
130	Macadamias 7 oz	5.25
131	Macadamias 16 oz	9.95
149	Pecan halves 8 oz	6.25
155	Mixed nuts 10 oz	4.80
160	Cashew brittle 8 oz	1.95
182	Pecan roll 8 oz	3.70
199	Chocolate peanuts 8 oz	1.90

Pounds	Zone 1	Zone 2	Zone 3	Zone 4
1–3	6.55	6.85	7.25	7.75
4–6	7.10	7.40	7.80	8.30
7–9	7.50	7.80	8.20	8.70
10–12	7.90	8.20	8.60	9.10

133. Find the cost of sending the following orders to the given mail zone.

a. Code	Quantity	b. Code	Quantity	c. Code	Quantity
116	2	112	1	117	3
130	1	117	4	131	1
149	3	131	2	155	2
182	4	160	3	160	4
Mail to zone 4.		182	5	182	1
		Mail to zone 3.		199	3
				Mail to zone 2.	

APPLYING THE CONCEPTS

134. An emissions test for cars requires that of the total engine exhaust, less than 1 part per thousand $\left(\frac{1}{1000} = 0.001 \right)$ be hydrocarbon emissions. Using this figure, determine which of the cars in the table below would fail the emissions test.

Car	Total Engine Exhaust	Hydrocarbon Emission
1	367,921	36
2	401,346	42
3	298,773	21
4	330,045	32
5	432,989	45

Chris works at B & W Garage as an auto mechanic and has just completed an engine overhaul for a customer. To determine the cost of the repair job, Chris keeps a list of times worked and parts used. A parts list and a list of the times worked are shown below. Use these tables for Exercises 135 to 138.

| Parts Used | | Time Spent | | Price List | | |
Item	Quantity	Day	Hours	Item Number	Description	Unit Price
Gasket set	1	Monday	7.0	27345	Valve spring	$1.85
Ring set	1	Tuesday	7.5	41257	Main bearing	3.40
Valves	8	Wednesday	6.5	54678	Valve	4.79
Wrist pins	8	Thursday	8.5	29753	Ring set	33.98
Valve springs	16	Friday	9.0	45837	Gasket set	48.99
Rod bearings	8			23751	Timing chain	42.95
Main bearings	5			23765	Fuel pump	77.59
Valve seals	16			28632	Wrist pin	2.71
Timing chain	1			34922	Rod bearing	2.67
				2871	Valve seal	0.42

135. Organize a table of data showing the parts used, the unit price for each, and the price of the quantity used. *Hint:* Use the following headings for the table.

Quantity   *Item Number*   *Description*   *Unit Price*   *Total*

136. Add up the numbers in the "Total" column to find the total cost of the parts.

137. If the charge for labor is $26.75 per hour, compute the cost of labor.

138. What is the total cost for parts and labor?

139. Explain how the decimal point is placed when a number is multiplied by 10, 100, 1000, 10,000, etc.

140. Explain how the decimal point is placed in the product of two decimals.

141. Show how the decimal is placed in the product of 1.3×2.31 by first writing each number as a fraction and then multiplying. Now change back to decimal notation.

3.5 Division of Decimals

Objective A   *To divide decimals*

To divide decimals, move the decimal point in the divisor to the right to make the divisor a whole number. Move the decimal point in the dividend the same number of places to the right. Place the decimal point in the quotient directly over the decimal point in the dividend, and then divide as in whole numbers.

➡ Divide: $3.25\overline{)15.275}$

$$3.25.\overline{)15.27.5}$$

Move the decimal point 2 places to the right in the divisor and then in the dividend. Place the decimal point in the quotient.

$$
\begin{array}{r}
4.7 \\
325.\overline{)\ 1527.5} \\
-1300 \\
\hline
227\ 5 \\
-227\ 5 \\
\hline
0
\end{array}
$$

Moving the decimal point the same number of decimal places in the divisor and dividend does not change the value of the quotient, because this process is the same as multiplying the numerator and denominator of a fraction by the same number. In the example above,

$$3.25\overline{)15.275} = \frac{15.275}{3.25} = \frac{15.275 \times 100}{3.25 \times 100} = \frac{1527.5}{325} = 325\overline{)1527.5}$$

When dividing decimals, we usually round the quotient off to a specified place value, rather than writing the quotient with a remainder.

➡ Divide: $0.3\overline{)0.56}$
Round to the nearest hundredth.

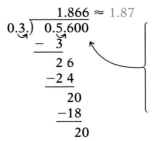

The division must be carried to the thousandths' place to round the quotient to the nearest hundredth. Therefore, zeros must be inserted in the dividend so that the quotient has a digit in the thousandths' place.

➡ Divide: 57.93 ÷ 3.24
Round to the nearest thousandth.

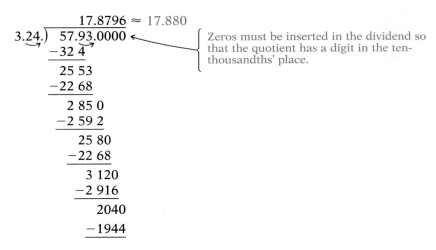

$$17.8796 \approx 17.880$$

3.24.) 57.93.0000
  −32 4
   25 53
  −22 68
    2 85 0
   −2 59 2
     25 80
    −22 68
      3 120
     −2 916
       2040
      −1944

Zeros must be inserted in the dividend so that the quotient has a digit in the ten-thousandths' place.

To divide a decimal by a power of 10 (10, 100, 1000, . . .), move the decimal point to the left the same number of places as there are zeros in the power of 10.

$34.65 \div 1\underline{0}$ $= 3.465$

1 zero   1 decimal place

$34.65 \div 1\underline{00}$ $= 0.3465$

2 zeros   2 decimal places

$34.65 \div 1\underline{000}$ $= 0.03465$ Note that a zero must be inserted between the 3 and the decimal point.

3 zeros   3 decimal places

$34.65 \div 1\underline{0,000}$ $= 0.003465$ Note that two zeros must be inserted between the 3 and the decimal point.

4 zeros   4 decimal places

If the power of 10 is written in exponential notation, the exponent indicates how many places to move the decimal point.

$34.65 \div 10^1 = 3.465$ 1 decimal place

$34.65 \div 10^2 = 0.3465$ 2 decimal places

$34.65 \div 10^3 = 0.03465$ 3 decimal places

$34.65 \div 10^4 = 0.003465$ 4 decimal places

Example 1   Divide: $0.1344 \div 0.032$

Solution

$$4.2$$

0.032.)0.134.4
   −128
     6 4
    −6 4
       0

You Try It 1   Divide: $0.1404 \div 0.052$

Your solution

Solution on p. S9

Example 2   Divide: 58.092 ÷ 82
Round to the nearest
thousandth.

Solution

$$
\begin{array}{r}
0.7084 \approx 0.708 \\
82\overline{)\ 58.0920} \\
-57\ 4 \\
\hline
69 \\
-\ \ 0 \\
\hline
692 \\
-656 \\
\hline
360 \\
-328 \\
\hline
\end{array}
$$

You Try It 2   Divide: 37.042 ÷ 76
Round to the nearest
thousandth.

Your solution

Example 3   Divide: 420.9 ÷ 7.06
Round to the nearest tenth.

Solution

$$
\begin{array}{r}
59.61 \approx 59.6 \\
7.06.\overline{)\ 420.90.00} \\
-353\ 0 \\
\hline
67\ 90 \\
-63\ 54 \\
\hline
4\ 36\ 0 \\
-4\ 23\ 6 \\
\hline
12\ 40 \\
-\ 7\ 06 \\
\hline
\end{array}
$$

You Try It 3   Divide: 370.2 ÷ 5.09
Round to the nearest tenth.

Your solution

Example 4   Divide: 402.75 ÷ 1000

Solution $402.75 \div 1000 = 0.40275$

You Try It 4   Divide: 309.21 ÷ 10,000

Your solution

Example 5   What is 0.625 divided by 10^2?

Solution $0.625 \div 10^2 = 0.00625$

You Try It 5   What is 42.93 divided by 10^4?

Your solution

Solutions on pp. S9–S10

ESTIMATION

Estimating the Quotient of Two Decimals

Estimate and then use your calculator to find 282.18 ÷ 0.48.

To estimate a quotient, round each number so
that there is one nonzero digit. Then divide.

$$282.18 \div 0.48 \approx$$
$$300 \div 0.5 = 600$$

The estimated answer is 600.

Now use your calculator to find the
exact answer.

$$282.18\ \boxed{\div}\ 0.48\ \boxed{=}\ 587.875$$

The exact answer is 587.875.

Objective B   To solve application problems

The table at the right shows the total amount of gasoline tax paid per gallon of gas in selected states in 1997. Use this table for Example 6 and You Try It 6.

Gasoline Taxes in Dollars per Gallon

State	Taxes
Arkansas	$.185
Georgia	$.875
Nebraska	$.264
Wisconsin	$.237

Example 6

Eddie Lopez's car gets 28 miles per gallon of gas. In 1997, he drove 12,110 miles in Arkansas. How much did he pay, to the nearest dollar, in gasoline taxes?

Strategy

To find the amount he paid in gasoline taxes:

- Find the total number of gallons of gas he used by dividing his total miles traveled (12,110) by the number of miles traveled per gallon (28).
- Multiply the state tax ($.185) by the total number of gallons of gasoline used.

Solution

$12,110 \div 28 = 432.5$

$0.185 \times 432.5 = 80.0125$

Eddie paid approximately $80 for gasoline taxes.

You Try It 6

In 1997, Susan Beckman drove her car 9675 miles in Georgia. If her car gets 22.5 miles per gallon, how much did she pay in gasoline taxes?

Your strategy

Your solution

Example 7

In 1996, AFLAG had earnings before income taxes of $650,001,000. AFLAG paid total income taxes of $255,638,000. Find the net earnings per share. AFLAG had issued 144,512,000 shares of stock. Round to the nearest cent.

Strategy

To find the net earnings per share:

- Find the net income by subtracting $255,638,000 from $650,001,000.
- Divide the difference by the number of shares (144,512,000).

Solution

$650,001,000 - 255,638,000 = 394,363,000$

$394,363,000 \div 144,512,000 \approx 2.728929$

Net earnings per share are approximately $2.73.

You Try It 7

A Nielsen survey of the number of people (in millions) who watch television during the week is given in the table below.

Mon.	Tues.	Wed.	Thu.	Fri.	Sat.	Sun.
91.9	89.8	90.6	93.9	78.0	77.1	87.7

Find the average number of people watching television per day.

Your strategy

Your solution

Solutions on p. S10

3.5 Exercises

Objective A

Divide.

1. 3)2.46

2. 7)3.71

3. 0.8)3.84

4. 0.9)6.93

5. 0.7)62.3

6. 0.4)52.8

7. 0.4)24

8. 0.5)65

9. 0.7)59.01

10. 0.9)8.721

11. 0.5)16.15

12. 0.8)77.6

13. 0.7)3.542

14. 0.6)2.436

15. 6.3)8.19

16. 3.2)7.04

17. 3.6)0.396

18. 2.7)0.648

19. 6.9)26.22

20. 1.7)84.66

Divide. Round to the nearest tenth.

21. 55.62 ÷ 8.8

22. 25.43 ÷ 5.4

23. 5.427 ÷ 9.5

24. 1.837 ÷ 1.4

25. 18.4 ÷ 7.3

26. 52.9 ÷ 8.1

27. 0.183 ÷ 0.17

28. 0.381 ÷ 0.47

29. 6.924 ÷ 0.053

Divide. Round to the nearest hundredth.

30. 4.817 ÷ 16

31. 6.467 ÷ 8

32. 0.0418 ÷ 0.53

33. 19.08 ÷ 0.45

34. 21.792 ÷ 0.96

35. 38.665 ÷ 0.95

36. 13.97 ÷ 25.4

37. 27.738 ÷ 60.3

38. 3.171 ÷ 45.3

Divide. Round to the nearest thousandth.

39. $1.028 \div 54$

40. $6.729 \div 27$

41. $0.0437 \div 0.5$

42. $75.469 \div 77.8$

43. $34.31 \div 95.3$

44. $0.2695 \div 2.67$

45. $0.4871 \div 4.72$

46. $0.1142 \div 17.2$

47. $0.2307 \div 26.7$

Divide. Round to the nearest whole number.

48. $16.5 \div 4$

49. $89.76 \div 90$

50. $1.94 \div 0.3$

51. $1.0478 \div 0.413$

52. $2.148 \div 0.519$

53. $0.79 \div 0.778$

54. $3.092 \div 0.075$

55. $392 \div 6.9$

56. $8.729 \div 0.075$

Divide.

57. $4.07 \div 10$

58. $0.039 \div 10$

59. $42.67 \div 10$

60. $389.7 \div 100$

61. $1.037 \div 100$

62. $237.835 \div 100$

63. $8.295 \div 1000$

64. $82,547 \div 1000$

65. $825.37 \div 1000$

66. $8.35 \div 10$

67. $0.32 \div 10$

68. $87.65 \div 10$

69. $23.627 \div 10^2$

70. $2.954 \div 10^2$

71. $0.0053 \div 10^2$

72. $289.32 \div 10^3$

73. $1.8932 \div 10^3$

74. $0.139 \div 10^3$

75. Divide 44.208 by 2.4.

76. Divide 0.04664 by 0.44.

77. Find the quotient of 723.15 and 45.

78. Find the quotient of 3.3463 and 3.07.

79. Divide 13.5 by 10^3.

80. Divide 0.045 by 10^5.

81. Find the quotient of 23.678 and 1000.

82. Find the quotient of 7.005 and 10,000.

83. What is 0.0056 divided by 0.05?

84. What is 123.8 divided by 0.02?

Estimate and then use your calculator to divide. Round your calculated answer to the nearest ten-thousandth.

85. 42.42 ÷ 3.8

86. 69.8 ÷ 7.2

87. 389 ÷ 0.44

88. 642 ÷ 0.83

89. 6.394 ÷ 3.5

90. 8.429 ÷ 4.2

91. 1.235 ÷ 0.021

92. 7.456 ÷ 0.072

93. 95.443 ÷ 1.32

94. 423.0925 ÷ 4.0927

95. 1.000523 ÷ 429.07

96. 0.03629 ÷ 0.00054

Objective B   *Application Problems*

97. Ramon, a high school football player, gained 162 yards on 26 carries in a high school football game. Find the average number of yards gained per carry. Round to the nearest hundredth.

98. Ross Lapointe earns $39,440.64 for 12 months' work as a park ranger. How much does he earn in 1 month?

99. A car with an odometer reading of 17,814.2 is filled with 9.4 gallons of gas. At an odometer reading of 18,130.4, the tank is empty and the car is filled with 12.4 gallons of gas. How many miles does the car travel on 1 gallon of gasoline?

100. A case of diet cola costs $6.79. If there are 24 cans in a case, find the cost per can. Round to the nearest cent.

101. Anne is building bookcases that are 3.4 feet long. How many complete shelves can be cut from a 12-foot board?

102. Earl is 52 years old and is buying $70,000 of life insurance for an annual premium of $703.80. If he pays each annual premium in 12 equal installments, how much is each monthly payment?

103. An oil company has issued 3,541,221,500 shares of stock. The company paid $6,090,990,120 in dividends. Find the dividend for each share of stock. Round to the nearest cent.

104. The total budget for the United States in 1996 was $1.63 trillion. If each person in the United States were to pay the same amount of taxes, how much would each person pay to raise that amount of money? Assume that there are 267 million people in the United States. Round to the nearest dollar.

105. You buy a home entertainment center for $1242.58. The down payment is $400, and the balance is to be paid in 18 equal monthly payments.
a. Find the amount to be paid in monthly payments.
b. Find the amount of each monthly payment.

APPLYING THE CONCEPTS

106. Explain how the decimal point is moved when dividing a number by 10, 100, 1,000, 10,000, etc.

107. A ball point pen priced at 50¢ was not selling. When the price was reduced to a different whole number of cents, the entire stock sold for $31.95. How many cents were charged per pen when the price was reduced? (*Hint*: There is more than one possible answer.)

108. Explain how baseball batting averages are determined. Then find Tony Gwynn's batting average with 175 hits out of 489 at bats. Round to the nearest thousandth.

109. Explain how the decimal point is placed in the quotient when dividing by a decimal.

For each of the problems below, insert a +, −, ×, or ÷ into the square so that the statement is true.

110. 3.45 ☐ 0.5 = 6.9   **111.** 3.46 ☐ 0.24 = 0.8304   **112.** 6.009 ☐ 4.68 = 1.329

113. 0.064 ☐ 1.6 = 0.1024   **114.** 9.876 ☐ 23.12 = 32.996   **115.** 3.0381 ☐ 1.23 = 2.47

Fill in the square to make a true statement.

116. 6.47 − ☐ = 1.253   **117.** 6.47 + ☐ = 9   **118.** 0.009 ÷ ☐ = 0.36

3.6 Comparing and Converting Fractions and Decimals

Objective A   *To convert fractions to decimals* ⋯⋯⋯⋯⋯⋯⋯⋯⋯⋯⋯⋯

Every fraction can be written as a decimal. To write a fraction as a decimal, divide the numerator of the fraction by the denominator. The quotient can be rounded to the desired place value.

➡ Convert $\frac{3}{7}$ to a decimal.

$$\begin{array}{r} 0.42857 \\ 7\overline{)3.00000} \end{array}$$ $\frac{3}{7}$ rounded to the nearest hundredth is 0.43.

$\frac{3}{7}$ rounded to the nearest thousandth is 0.429.

$\frac{3}{7}$ rounded to the nearest ten-thousandth is 0.4286.

➡ Convert $3\frac{2}{9}$ to a decimal. Round to the nearest thousandth.

$$3\frac{2}{9} = \frac{29}{9} \qquad \begin{array}{r} 3.2222 \\ 9\overline{)29.0000} \end{array} \qquad 3\frac{2}{9} \text{ rounded to the nearest thousandth is } 3.222.$$

Example 1   Convert $\frac{3}{8}$ to a decimal. Round to the nearest hundredth.

Solution $\begin{array}{r} 0.375 \\ 8\overline{)3.000} \end{array} \approx 0.38$

You Try It 1   Convert $\frac{9}{16}$ to a decimal. Round to the nearest tenth.

Your solution

Example 2   Convert $2\frac{3}{4}$ to a decimal. Round to the nearest tenth.

Solution $2\frac{3}{4} = \frac{11}{4} \qquad \begin{array}{r} 2.75 \\ 4\overline{)11.00} \end{array} \approx 2.8$

You Try It 2   Convert $4\frac{1}{6}$ to a decimal. Round to the nearest hundredth.

Your solution

Solutions on p. S10

Objective B   *To convert decimals to fractions* ⋯⋯⋯⋯⋯⋯⋯⋯⋯⋯⋯⋯

To convert a decimal to a fraction, remove the decimal point and place the decimal part over a denominator equal to the place value of the last digit in the decimal.

$$0.47 \overset{\text{hundredths}}{=} \frac{47}{100} \qquad\qquad 7.45 \overset{\text{hundredths}}{=} 7\frac{45}{100} = 7\frac{9}{20}$$

$$0.275 \overset{\text{thousandths}}{=} \frac{275}{1000} = \frac{11}{40} \qquad 0.16\frac{2}{3} \overset{\text{hundredths}}{=} \frac{16\frac{2}{3}}{100} = 16\frac{2}{3} \div 100 = \frac{50}{3} \times \frac{1}{100} = \frac{1}{6}$$

Example 3   Convert 0.82 and 4.75 to fractions.

Solution $0.82 = \dfrac{82}{100} = \dfrac{41}{50}$

$4.75 = 4\dfrac{75}{100} = 4\dfrac{3}{4}$

You Try It 3   Convert 0.56 and 5.35 to fractions.

Your solution

Example 4   Convert $0.15\dfrac{2}{3}$ to a fraction.

Solution $0.15\dfrac{2}{3} = \dfrac{15\dfrac{2}{3}}{100} = 15\dfrac{2}{3} \div 100$

$= \dfrac{47}{3} \times \dfrac{1}{100} = \dfrac{47}{300}$

You Try It 4   Convert $0.12\dfrac{7}{8}$ to a fraction.

Your solution

Solutions on p. S10

Objective C   ***To identify the order relation between two decimals or between a decimal and a fraction*** ....

Decimals, like whole numbers and fractions, can be graphed as points on the number line. The number line can be used to show the order of decimals. A decimal that appears to the right of a given number is greater than the given number. A decimal that appears to the left of a given number is less than the given number.

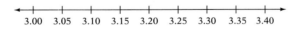

3.00   3.05   3.10   3.15   3.20   3.25   3.30   3.35   3.40

Note that 3, 3.0, and 3.00 represent the same number.

➡ Find the order relation between $\dfrac{3}{8}$ and 0.38.

$\dfrac{3}{8} = 0.375$ $0.38 = 0.380$

$0.375 < 0.380$

$\dfrac{3}{8} < 0.38$

Example 5   Place the correct symbol, < or >, between the numbers.

$\dfrac{5}{16}$ 0.32

Solution $\dfrac{5}{16} \approx 0.313$

$0.313 < 0.32$

$\dfrac{5}{16} < 0.32$

You Try It 5   Place the correct symbol, < or >, between the numbers.

0.63 $\dfrac{5}{8}$

Your solution

Solution on p. S10

3.6 Exercises

· ·

Objective A

Convert the fraction to a decimal. Round to the nearest thousandth.

1. $\dfrac{5}{8}$ 2. $\dfrac{7}{12}$ 3. $\dfrac{2}{3}$ 4. $\dfrac{5}{6}$ 5. $\dfrac{1}{6}$ 6. $\dfrac{7}{8}$

7. $\dfrac{5}{12}$ 8. $\dfrac{9}{16}$ 9. $\dfrac{7}{4}$ 10. $\dfrac{5}{3}$ 11. $1\dfrac{1}{2}$ 12. $2\dfrac{1}{3}$

13. $\dfrac{16}{4}$ 14. $\dfrac{36}{9}$ 15. $\dfrac{3}{1000}$ 16. $\dfrac{5}{10}$ 17. $7\dfrac{2}{25}$ 18. $16\dfrac{7}{9}$

19. $37\dfrac{1}{2}$ 20. $87\dfrac{1}{2}$ 21. $\dfrac{3}{8}$ 22. $\dfrac{11}{16}$ 23. $\dfrac{5}{24}$ 24. $\dfrac{4}{25}$

25. $3\dfrac{1}{3}$ 26. $8\dfrac{2}{5}$ 27. $5\dfrac{4}{9}$ 28. $3\dfrac{1}{12}$ 29. $\dfrac{5}{16}$ 30. $\dfrac{11}{12}$

Objective B

Convert the decimal to a fraction.

31. 0.8   32. 0.4   33. 0.32   34. 0.48   35. 0.125

36. 0.485   37. 1.25   38. 3.75   39. 16.9   40. 17.5

41. 8.4   42. 10.7   43. 8.437   44. 9.279   45. 2.25

46. 7.75   47. $0.15\dfrac{1}{3}$ 48. $0.17\dfrac{2}{3}$ 49. $0.87\dfrac{7}{8}$ 50. $0.12\dfrac{5}{9}$

Convert the decimal to a fraction.

51. 1.68   **52.** 7.38   **53.** 0.045   **54.** 0.085   **55.** 16.72

56. 82.32   **57.** 0.33   **58.** 0.57   **59.** $0.33\frac{1}{3}$ **60.** $0.66\frac{2}{3}$

Objective C

Place the correct symbol, < or >, between the numbers.

61. 0.15   0.5   **62.** 0.6   0.45   **63.** 6.65   6.56   **64.** 3.89   3.98

65. 2.504   2.054   **66.** 0.025   0.105   **67.** $\frac{3}{8}$ 0.365   **68.** $\frac{4}{5}$ 0.802

69. $\frac{2}{3}$ 0.65   **70.** 0.85 $\frac{7}{8}$ **71.** $\frac{5}{9}$ 0.55   **72.** $\frac{7}{12}$ 0.58

73. 0.62 $\frac{7}{15}$ **74.** $\frac{11}{12}$ 0.92   **75.** 0.161 $\frac{1}{7}$ **76.** 0.623   0.6023

77. 0.86   0.855   **78.** 0.87   0.087   **79.** 1.005   0.5   **80.** 0.033   0.3

APPLYING THE CONCEPTS

81. Which of the following is true?
     a. $\frac{137}{300} = 0.456666667$ **b.** $\frac{137}{300} < 0.45666666$ **c.** $\frac{137}{300} > 0.45666666$

82. Is $\frac{7}{13}$ in decimal form a repeating decimal? Why or why not?

83. If a number is rounded to the nearest thousandth, is it always greater than if it was rounded to the nearest hundredth? Give examples to support your answer.

84. Convert $\frac{1}{9}, \frac{2}{9}, \frac{3}{9}$, and $\frac{4}{9}$ to decimals. Describe the pattern. Use the pattern to convert $\frac{5}{9}, \frac{7}{9}$, and $\frac{8}{9}$ to decimals.

85. Explain the difference between terminating, repeating, and nonrepeating decimals. Give an example of each kind of decimal.

Focus on Problem Solving

Problems in mathematics or real life involve a question or a need and information or circumstances about that need. Solving problems in the sciences usually involves a question, observation, and measurements of some kind.

One of the challenges of problem solving in the sciences is to separate the relevant information about a problem from other information. Following is an example from the physical sciences in which some relevant information was omitted.

Hooke's Law states that the distance that a weight will stretch a spring is directly proportional to the weight on the spring. That is, $d = kF$, where d is the distance the spring is stretched and F is the force. In an experiment to verify this law, some physics students were continually getting inconsistent results. Finally, the instructor discovered that the heat produced when the lights were turned on was affecting the experiment. In this case, relevant information was omitted, namely that the temperature of the spring can affect the distance it will stretch.

A lawyer drove 8 miles to the train station. After a 35-minute ride of 18 miles, the lawyer walked 10 minutes to the office. Find the total time it took the lawyer to get to work.

From this situation, answer the following before reading on.

a. What is asked for?

b. Is there enough information to answer the question?

c. Is information given that is not necessary?

Here are the answers.

a. We want the total time for the lawyer to get to work.

b. No. We do not know the time it takes the lawyer to get to the train station.

c. Yes. The distance to the train station and the distance of the train ride are not necessary to answer the question.

In the following problems,

a. What is asked for?

b. Is there enough information to answer the question?

c. Is some information not needed?

1. A customer bought 6 boxes of strawberries and paid with a $20 bill. What was the change?

2. A board is cut into two pieces. One piece is 3 feet longer than the other piece. What is the length of the original board?

3. A family rented a car for their vacation and drove 680 miles. The cost of the rental car was $21 per day with 150 free miles per day and $.15 for each mile driven above the number of free miles allowed. How many miles did the family drive per day?

4. An investor bought 8 acres of land for $80,000. One and one-half acres were set aside for a park, and the remainder was developed into one-half-acre lots. How many lots were available for sale?

5. You wrote checks of $43.67, $122.88, and $432.22 after making a deposit of $768.55. How much do you have left in your checking account?

Projects and Group Activities

Topographical Maps Carpenters use fractions to measure the wood that is used to frame a house. For instance, a door entry may measure $42\frac{1}{2}$ inches. However, the grading contractor (the person who levels the lot on which the house is built) measures the height of the lot by using decimals. For instance, a certain place may be at a height of 554.2 feet. (The height is measured above sea level; so 554.2 means 554.2 feet above sea level.)

A surveyor provides the grading contractor with a grading plan that shows the elevation of each point of the lot. These plans are drawn so that the house can be sited in such a way that water will drain away from it. The diagram below is a **topographical map** for a lot. Along each closed curve, called a **contour curve,** the lot is at the same height above sea level. For instance, every point on the curve in red is 556.5 feet above sea level.

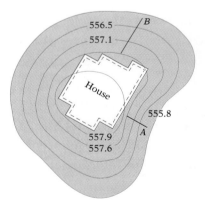

Using the map above, answer the following questions.

1. What is the elevation of the highest point on the lot?

2. What is the elevation of the lowest point on the lot?

3. What is the difference in elevation between the highest and lowest points on the lot?

4. Assuming that this map is drawn to scale, describe the significance of how steep the slope is along line *A* compared to that along line *B*.

5. Acquire a forest service contour map of a portion of a national forest and plan a cross-country hike, choosing a trail that avoids the steepest ascent.

Search the World Wide Web

Go to the Internet and find http://globe3.gsfc.nasa.gov/cgi-bin/show.cgi/&page= help-contour-productn.ht. This explains the basics of contour maps, and shows how to produce a simple contour map from scattered data points as shown below. This contour map is different from the one above in that in this exercise, the contour line connects data values of equal temperature. Complete the contour map that is shown on the Web site and discuss the meaning of (a) contour line, (b) data point, and (c) interpolation.

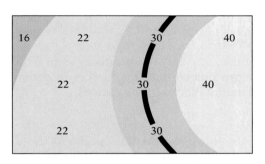

The data points that show 30 degrees are connected by a dotted line in the diagram at the left. To the right of the dotted line the temperature is warmer, and to the left of the dotted line the temperature is colder.

The Internet also provides a great deal of information on booking airline flights, car rentals, ocean cruises, and hotel reservations. Information can also be found on renting recreational vehicles, campground costs, and Amtrak vacation packages.

Go to the Internet and find http://www.cruiseweb.com/. Here you can find information on many cruises to all parts of the world. Plan a cruise of your choice and calculate the cost, including the airfare. Airfare cost can be found at http://www.airfare.com.

Go to the Internet and find http://www.amtrak.com. This address provides the history of Amtrak and vacation packages. Plan a vacation, include the route, with stopovers at two different hotels. Hotel information can be found at http://www.webscope.com/travel/chains.html.

If you like camping, the cost of renting recreational vehicles can be found at http://www.rvlink.com. If seeing Europe by rail is interesting, you can check out the cost on the Internet. The Internet address is http://www.starnetinc.com/eurorail/railindx.htm.

Chapter Summary

Key Words

A number written in *decimal notation* has three parts: a whole-number part, a decimal point, and a decimal part.

The position of a digit in a *decimal* determines the digit's *place value*.

Essential Rules

To Write a Decimal in Words

To write a decimal in words, write the decimal part as if it were a whole number. Then name the place value of the last digit.

To Add Decimals

To add decimals, write the numbers so that the decimal points are on a vertical line. Add as in whole numbers, and place the decimal point in the sum directly below the decimal point in the addends.

To Subtract Decimals

To subtract decimals, place the numbers so that the decimal points are on a vertical line. Subtract as for whole numbers, and write the decimal point in the difference directly below the decimal point in the subtrahend.

To Multiply Decimals

To multiply decimals, multiply the numbers as in whole numbers. Place the decimal point in the product so that the number of decimal places in the product is the sum of the decimal places in the factors.

To Divide Decimals

To divide decimals, move the decimal point in the divisor to make it a whole number. Move the decimal point in the dividend the same number of places to the right. Place the decimal point in the quotient directly over the decimal point in the dividend. Then divide as in whole numbers.

To Write a Fraction as a Decimal

To write a fraction as a decimal, divide the numerator of the fraction by the denominator. Round the quotient to the desired number of places.

To Convert a Decimal to a Fraction

To convert a decimal to a fraction, remove the decimal point and place the decimal part over a denominator equal to the place value of the last digit in the decimal.

Chapter Review

1. Find the quotient of 3.6515 and 0.067.

2. Find the sum of 369.41, 88.3, 9.774, and 366.474.

3.. Place the correct symbol, $<$ or $>$, between the two numbers.
0.055   0.1

4. Write 22.0092 in words.

5. Round 0.05678235 to the nearest hundred-thousandth.

6. Convert $2\frac{1}{3}$ to a decimal. Round to the nearest hundredth.

7. Convert 0.375 to a fraction.

8. Add: 3.42 + 0.794 + 32.5

9. Write thirty-four and twenty-five thousandths in standard form.

10. Place the correct symbol, $<$ or $>$, between the two numbers.
$\frac{5}{8}$ 0.62

11. Convert $\frac{7}{9}$ to a decimal. Round to the nearest thousandth.

12. Convert $0.66\frac{2}{3}$ to a fraction.

13. Subtract: 27.31 − 4.4465

14. Round 7.93704 to the nearest hundredth.

15. Find the product of 3.08 and 2.9.

16. Write 342.37 in words.

17. Write three and six thousand seven hundred fifty-three hundred-thousandths in standard form.

18. Multiply: 34.79
                   × 0.74

19. Divide: $0.053\overline{)0.349482}$

20. What is 7.796 decreased by 2.9175?

21. You had a balance of $895.68 in your checking account. You then wrote checks of $145.72 and $88.45. Find the new balance in your checking account.

22. The state income tax on the business you own is $560 plus 0.08 times your profit. You made a profit of $63,000 last year. Find the amount of income tax you paid last year.

23. A car costing $5944.20 is bought with a down payment of $1500 and 36 equal monthly payments. Find the amount of each monthly payment.

24. You have $237.44 in your checking account. You make deposits of $56.88, $127.40, and $56.30. Find the amount in your checking account after you make the deposits.

Chapter Test

1. Place the correct symbol, < or >, between the two numbers.
   0.66   0.666

2. Subtract: 13.027
            − 8.94

3. Write 45.0302 in words.

4. Convert $\frac{9}{13}$ to a decimal. Round to the nearest thousandth.

5. Convert 0.825 to a fraction.

6. Round 0.07395 to the nearest ten-thousandth.

7. Find 0.0569 divided by 0.037. Round to the nearest thousandth.

8. Find 9.23674 less than 37.003.

9. Round 7.0954625 to the nearest thousandth.

10. Divide: 0.006)‾1.392

11. Add: 270.93
            97.
            1.976
         + 88.675

12. Find the missing dimension.

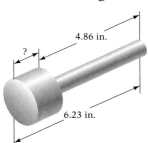

4.86 in.

?

6.23 in.

13. Multiply: 1.37
               $\times$ 0.004

14. What is the total of 62.3, 4.007, and 189.65?

15. Write two hundred nine and seven thousand eighty-six hundred-thousandths in standard form.

16. A car was bought for $6392.60, with a down payment of $1250. The balance was paid in 36 monthly payments. Find the amount of each monthly payment.

17. You received a salary of $363.75, a commission of $954.82, and a bonus of $225. Find your total income.

18. A long-distance telephone call costs $.85 for the first 3 minutes and $.42 for each additional minute. Find the cost of a 12-minute long-distance telephone call.

The table at the right shows the postal rates for selected weights for single-piece and presorted mail. Use this table for Exercises 19 and 20.

Weight (not over)	Single piece	Presorted
1 ounce	$.32	$.295
2 ounces	.55	.525
3 ounces	.78	.709
4 ounces	1.01	.939
5 ounces	1.24	1.69

19. Find the total cost of mailing the following:
3 single pieces weighing 2.3 ounces each
2 single pieces weighing 3.4 ounces each
50 presorted pieces weighing 0.3 ounce each

20. Find the total cost of mailing the following:
7 single pieces weighing 0.4 ounce each
30 presorted pieces weighing 0.7 ounce each
75 presorted pieces weighing 1.4 ounces each
Round to the nearest cent.

Cumulative Review

1. Divide: $89\overline{)20{,}932}$

2. Simplify: $2^3 \cdot 4^2$

3. Simplify: $2^2 - (7 - 3) \div 2 + 1$

4. Find the LCM of 9, 12, and 24.

5. Write $\frac{22}{5}$ as a mixed number.

6. Write $4\frac{5}{8}$ as an improper fraction.

7. Build an equivalent fraction with the given denominator.

$$\frac{5}{12} = \frac{}{60}$$

8. Add: $\frac{3}{8} + \frac{5}{12} + \frac{9}{16}$

9. What is $5\frac{7}{12}$ increased by $3\frac{7}{18}$?

10. Subtract: $9\frac{5}{9} - 3\frac{11}{12}$

11. Multiply: $\frac{9}{16} \times \frac{4}{27}$

12. Find the product of $2\frac{1}{8}$ and $4\frac{5}{17}$.

13. Divide: $\frac{11}{12} \div \frac{3}{4}$

14. What is $2\frac{3}{8}$ divided by $2\frac{1}{2}$?

15. Simplify: $\left(\frac{2}{3}\right)^2 \cdot \left(\frac{3}{4}\right)^3$

16. Simplify: $\left(\frac{2}{3}\right)^2 - \left(\frac{2}{3} - \frac{1}{2}\right) + 2$

17. Write 65.0309 in words.

18. Add: 379.006
            27.523
            9.8707
        + 88.2994

19. What is 29.005 decreased by 7.9286?

20. Multiply: $\begin{array}{r} 9.074 \\ \times\ \ 6.09 \\ \hline \end{array}$

21. Divide: $8.09\overline{)17.42963}$. Round to the nearest thousandth.

22. Convert $\frac{11}{15}$ to a decimal. Round to the nearest thousandth.

23. Convert $0.16\frac{2}{3}$ to a fraction.

24. Place the correct symbol, $<$ or $>$, between the two numbers.
$\frac{8}{9}$ 0.98

25. An airplane had 204 passengers aboard. During a stop, 97 passengers got off the plane and 127 passengers got on the plane. How many passengers were on the continuing flight?

26. An investor purchased stock at $\$32\frac{1}{8}$ per share. During the first 2 months of ownership, the stock lost $\$\frac{3}{4}$ and then gained $\$1\frac{1}{2}$. Find the value of each share of stock at the end of the 2 months.

27. You have a checking account balance of $814.35. You then write checks for $42.98, $16.43, and $137.56. Find your checking account balance after you write the checks.

28. A machine lathe takes 0.017 inch from a brass bushing that is 1.412 inches thick. Find the resulting thickness of the bushing.

29. The state income tax on your business is $820 plus 0.08 times your profit. You made a profit of $64,860 last year. Find the amount of income tax you paid last year.

30. You bought a camera costing $210.96. The down payment is $20, and the balance is to be paid in 8 equal monthly payments. Find the monthly payment.

4

Ratio and Proportion

Objectives

Section 4.1
To write the ratio of two quantities in simplest form
To solve application problems

Section 4.2
To write rates
To write unit rates
To solve application problems

Section 4.3
To determine whether a proportion is true
To solve proportions
To solve application problems

Special-effects artists use scale models to create dinosaurs, exploding spaceships, and aliens that inhabit the spaceships. A scale model of a dinosaur is produced by using ratios and proportions to determine each dimension of the scale model.

Musical Scales

When a metal wire is stretched tight and then plucked, a sound is heard. Guitars, banjos, and violins are examples of instruments that use this principle to produce music. A piano is another example of this principle, but the sound is produced by the string's being struck by a small, hammerlike object.

After the string is plucked or struck, the string begins to vibrate. The number of times the string vibrates in 1 second is called the frequency of the vibration, or the pitch. Normally humans can hear vibrations as low as 16 cps (cycles per second) and as high as 20,000 cps. The longer the string, the lower the pitch of the sound; the shorter the string, the higher the pitch of the sound. In fact, a string half as long as another string vibrates twice as fast.

Vibrating String

Most music that is heard today is based on what is called the chromatic or twelve-tone scale. For this scale, a vibration of 261 cps is called middle C. A string half as long as the string for middle C vibrates twice as fast and produces a musical note one octave higher.

To produce the notes between the two C's, strings of different lengths that produce the desired pitches are chosen. Recall that as the string gets shorter, the pitch increases. A top view of a grand piano illustrates how the strings vary in length.

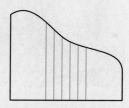

A well-tempered chromatic scale is one in which the string lengths are chosen so that the ratios of the frequencies of adjacent notes are the same.

$$\frac{C}{C\#} = \frac{C\#}{D} = \frac{D}{D\#} = \frac{D\#}{E} = \frac{E}{F} = \frac{F}{F\#} = \frac{F\#}{G} = \frac{G}{G\#} = \frac{G\#}{A} =$$

$$\frac{A}{A\#} = \frac{A\#}{B} = \frac{B}{C}$$

The common ratio for the chromatic scale is approximately $\frac{1}{1.0595}$.

4.1 Ratio

Objective A   **To write the ratio of two quantities in simplest form** ....

Quantities such as 3 feet, 12 cents, and 9 cars are number quantities written with **units**.

3 feet
12 cents   These are only some examples of units. Shirts, dollars, trees,
9 cars   miles, and gallons are further examples.
↑
units

A **ratio** is a comparison of two quantities that have the *same* units. This comparison can be written three different ways:

1. As a fraction

2. As two numbers separated by a colon (:)

3. As two numbers separated by the word *to*

The ratio of the lengths of two boards, one 8 feet long and the other 10 feet long, can be written as

1. $\dfrac{8 \text{ feet}}{10 \text{ feet}} = \dfrac{8}{10} = \dfrac{4}{5}$

2. 8 feet : 10 feet $= 8:10 = 4:5$

3. 8 feet to 10 feet $=$ 8 to 10 $=$ 4 to 5

A ratio is in **simplest form** when the two numbers do not have a common factor. Note that in a ratio, the units are not written.

This ratio means that the smaller board is $\dfrac{4}{5}$ the length of the longer board.

Example 1
Write the comparison $6 to $8 as a ratio in simplest form using a fraction, a colon, and the word *to*.

Solution $\dfrac{\$6}{\$8} = \dfrac{6}{8} = \dfrac{3}{4}$
$6 : $8 $= 6:8 = 3:4$
$6 to $8 $=$ 6 to 8 $=$ 3 to 4

You Try It 1
Write the comparison 20 pounds to 24 pounds as a ratio in simplest form using a fraction, a colon, and the word *to*.

Your solution

Example 2
Write the comparison 18 quarts to 6 quarts as a ratio in simplest form using a fraction, a colon, and the word *to*.

Solution $\dfrac{18 \text{ quarts}}{6 \text{ quarts}} = \dfrac{18}{6} = \dfrac{3}{1}$
18 quarts : 6 quarts $=$
  $18:6 = 3:1$
18 quarts to 6 quarts $=$
  18 to 6 $=$ 3 to 1

You Try It 2
Write the comparison 64 miles to 8 miles as a ratio in simplest form using a fraction, a colon, and the word *to*.

Your solution

Solutions on p. S10

Objective B   To solve application problems .... (7) [CT]

Use the table below for Example 3 and You Try It 3.

Board Feet of Wood at a Lumber Store			
Pine	*Ash*	*Oak*	*Cedar*
20,000	18,000	10,000	12,000

Example 3
Find, as a fraction in simplest form, the ratio of the number of board feet of pine to the number of board feet of oak.

Strategy
To find the ratio, write the ratio of board feet of pine (20,000) to board feet of oak (10,000) in simplest form.

Solution
$$\frac{20{,}000}{10{,}000} = \frac{2}{1}$$

The ratio is $\frac{2}{1}$.

You Try It 3
Find, as a fraction in simplest form, the ratio of the number of board feet of cedar to board feet of ash.

Your strategy

Your solution

Example 4
The cost of building a patio cover was $250 for labor and $350 for materials. What, as a fraction in simplest form, is the ratio of the cost of materials to the total cost for labor and materials?

Strategy
To find the ratio, write the ratio of the cost of materials ($350) to the total cost ($250 + $350) in simplest form.

Solution
$$\frac{\$350}{\$250 + \$350} = \frac{350}{600} = \frac{7}{12}$$

The ratio is $\frac{7}{12}$.

You Try It 4
A company spends $20,000 a month for television advertising and $15,000 a month for radio advertising. What, as a fraction in simplest form, is the ratio of the cost of radio advertising to the total cost of radio and television advertising?

Your strategy

Your solution

Solutions on p. S10

4.1 Exercises

· ·

Objective A

Write the comparison as a ratio in simplest form using a fraction, a colon (:), and the word *to*.

1. 3 pints to 15 pints

2. 6 pounds to 8 pounds

3. $40 to $20

4. 10 feet to 2 feet

5. 3 miles to 8 miles

6. 2 hours to 3 hours

7. 37 hours to 24 hours

8. 29 inches to 12 inches

9. 6 minutes to 6 minutes

10. 8 days to 12 days

11. 35 cents to 50 cents

12. 28 inches to 36 inches

13. 30 minutes to 60 minutes

14. 25 cents to 100 cents

15. 32 ounces to 16 ounces

16. 12 quarts to 4 quarts

17. 3 cups to 4 cups

18. 6 years to 7 years

19. $5 to $3

20. 30 yards to 12 yards

21. 12 quarts to 18 quarts

22. $20 to $28

23. 14 days to 7 days

24. 9 feet to 3 feet

Objective B   *Application Problems*

Write ratios in simplest form using a fraction.

			Family Budget			
Housing	Food	Transportation	Taxes	Utilities	Miscellaneous	Total
$800	$400	$300	$350	$150	$400	$2400

25. Use the table to find the ratio of housing cost to total expenses.

26. Use the table to find the ratio of food cost to total expenses.

27. Use the table to find the ratio of utilities cost to food cost.

28. Use the table to find the ratio of transportation cost to housing cost.

29. National Collegiate Athletic Association (NCAA) statistics show that for every 154,000 high school seniors playing basketball, only 4000 will play college basketball as first-year students. Write the ratio of the number of first-year students playing college basketball to the number of high school seniors playing basketball.

30. NCAA statistics show that for every 2800 college seniors playing college basketball, only 50 will play as rookies in the National Basketball Association. Write the ratio of the number of National Basketball Association rookies to the number of college seniors playing basketball.

31. A transformer has 40 turns in the primary coil and 480 turns in the secondary coil. State the ratio of the number of turns in the primary coil to the number of turns in the secondary coil.

32. Rita Sterling bought a computer system for $2400. Five years later she sold the computer for $900. Find the ratio of the amount she received for the computer to the cost of the computer.

33. A house with an original value of $90,000 increased in value to $110,000 in 5 years.
     a. Find the increase in the value of the house.
     b. What is the ratio of the increase in value to the original value of the house?

34. A decorator bought a box of ceramic floor tile for $21 and a box of wood tile for $33.
     a. What was the total cost of the box of ceramic tile and the box of wood tile?
     b. What is the ratio of the cost of the box of wood tile to the total cost?

35. The price of gasoline jumped from $.96 to $1.26 in 1 year. What is the ratio of the increase in price to the original price?

APPLYING THE CONCEPTS

A bank uses the ratio of a borrower's total monthly debts to the borrower's total monthly income to determine the maximum monthly payment for a potential homeowner. This ratio is called the debt–income ratio. Use the homeowner's income–debt table at the right for Exercises 36 and 37.

Income	Debts
$3500	$900
250	170
140	160
	95

36. Compute the debt–income ratio for the potential homeowner.

37. If Central Trust Bank will make a loan to a customer whose debt–income ratio is less than $\frac{1}{3}$, will the potential homeowner qualify? Explain your answer.

38. To make a home loan, First National Bank requires a debt–income ratio that is less than $\frac{2}{5}$. Would the homeowner whose income–debt table is given at the right qualify for a loan using these standards?

Income		Debts	
Salary	3400	Mortgage	1800
Interest	83	Property tax	104
Rent	650	Insurance	35
Dividends	34	Liabilities	120
		Credit card	234
		Car loan	197

39. Is the value of a ratio always less than 1? Explain.

Rates

Objective A   To write rates ....

A **rate** is a comparison of two quantities that have *different* units. A rate is written as a fraction.

A distance runner ran 26 miles in 4 hours. The distance-to-time rate is written

$$\frac{26 \text{ miles}}{4 \text{ hours}} = \frac{13 \text{ miles}}{2 \text{ hours}}$$ A rate is in **simplest form** when the numbers that form the rate have no common factors. Note that the units are written as part of the rate.

Example 1   Write "6 roof supports for every 9 feet" as a rate in simplest form.

Solution $\dfrac{6 \text{ supports}}{9 \text{ feet}} = \dfrac{2 \text{ supports}}{3 \text{ feet}}$

You Try It 1   Write "15 pounds of fertilizer for 12 trees" as a rate in simplest form.

Your solution

Solution on p. S11

Objective B   To write unit rates ....

A **unit rate** is a rate in which the number in the denominator is 1.

$$\frac{\$3.25}{1 \text{ pound}}$$ or \$3.25/pound is read "\$3.25 per pound."

To find unit rates, divide the number in the numerator of the rate by the number in the denominator of the rate.

A car traveled 344 miles on 16 gallons of gasoline. To find the miles per gallon (unit rate), divide the numerator of the rate by the denominator of the rate.

$$\frac{344 \text{ miles}}{16 \text{ gallons}}$$ is the rate.

$$16\overline{)344.0}$$ 21.5 miles/gallon is the unit rate.

Example 2   Write "300 feet in 8 seconds" as a unit rate.

Solution $\dfrac{300 \text{ feet}}{8 \text{ seconds}}$ $8\overline{)300.0}$

37.5 feet/second

You Try It 2   Write "260 miles in 8 hours" as a unit rate.

Your solution

Solution on p. S11

Objective C **To solve application problems** .... (7) [CT]

The table at the right shows typical air fare costs for long routes.

Long Routes	Miles	Fare
New York–Los Angeles	2475	$683
San Francisco–Dallas	1464	$536
Denver–Pittsburgh	1302	$525
Minneapolis–Hartford	1050	$483

⇒ Find the cost per mile for the four routes. Which route is the most expensive, and which is the least expensive, for each mile flown?

Strategy
To find the cost per mile, divide the fare for each route by the miles flown. Compare the costs per mile to determine the most expensive and least expensive routes per mile.

Solution New York–Los Angeles $\dfrac{683}{2475} \approx 0.28$

San Francisco–Dallas $\dfrac{536}{1464} \approx 0.37$

Denver–Pittsburgh $\dfrac{525}{1302} \approx 0.40$

Minneapolis–Hartford $\dfrac{483}{1050} = 0.46$

The Minneapolis–Hartford route is the most expensive per mile, and the New York–Los Angeles route is the least expensive per mile.

Example 3
As an investor, Jung Ho purchased 100 shares of stock for $1500. One year later, Jung sold the 100 shares for $1800. What was his profit per share?

Strategy
To find Jung's profit per share:

- Find the total profit by subtracting the original cost ($1500) from the selling price ($1800).
- Find the profit per share (unit rate) by dividing the total profit by the number of shares of stock (100).

Solution

$$
\begin{array}{r}
\$1800 \\
-\ 1500 \\
\hline
\$300 \quad \text{total profit}
\end{array}
\qquad
\begin{array}{r}
\$3 \\
100\overline{)\$300}
\end{array}
$$

Jung Ho's profit per share was $3.

You Try It 3
Erik Peltier, a jeweler, purchased 5 ounces of gold for $1625. Later, he sold the 5 ounces for $1720. What was Erik's profit per ounce?

Your strategy

Your solution

Solution on p. S11

4.2 Exercises

- -

Objective A

Write as a rate in simplest form.

1. 3 pounds of meat for 4 people

2. 30 ounces in 24 glasses

3. $80 for 12 boards

4. 84 cents for 6 bars of soap

5. 300 miles on 15 gallons

6. 88 feet in 8 seconds

7. 20 children in 8 families

8. 48 leaves on 9 plants

9. 16 gallons in 2 hours

10. 25 ounces in 5 minutes

Objective B

Write as a unit rate.

11. 10 feet in 4 seconds

12. 816 miles in 6 days

13. $1300 earned in 4 weeks

14. $27,000 earned in 12 months

15. 1100 trees planted on 10 acres

16. 3750 words on 15 pages

17. $32.97 earned in 7 hours

18. $315.70 earned in 22 hours

19. 628.8 miles in 12 hours

20. 388.8 miles in 8 hours

21. 344.4 miles on 12.3 gallons of gasoline

22. 409.4 miles on 11.5 gallons of gasoline

23. $349.80 for 212 pounds

24. $11.05 for 3.4 pounds

Objective C   Application Problems

25. An automobile was driven 326.6 miles on 11.5 gallons of gas. Find the number of miles driven per gallon of gas.

26. You drive 246.6 miles in 4.5 hours. Find the number of miles driven per hour.

27. The Saturn-5 rocket uses 534,000 gallons of fuel in 2.5 minutes. How much fuel does the rocket use per minute?

28. An investor paid $116.75 a share for 500 shares of Citicorp and receives $1050 in yearly dividends. Find the dividend per share.

29. An investor purchased 280 shares of GTE for $11,830.
   a. Estimate the cost per share.
   b. Find the actual cost per share.

30. Enova investors have bought 116,572,000 shares of the corporation. In 1996 Enova paid $181,852,000 in dividends. Find the dividend per share paid by the corporation. Round to the nearest cent.

31. Assume that Apple Computer produced 5000 compact disks for $26,536.32. Of the disks made, 122 did not meet company standards.
   a. How many compact disks did meet company standards?
   b. What was the cost per disk for those disks that met company standards?

32. The Pierre family purchased a 250-pound side of beef for $365.75 and had it packaged. During the packaging, 75 pounds of beef were discarded as waste.
   a. How many pounds of beef were packaged?
   b. What was the cost per pound for the packaged beef?

33. The Bear Valley Fruit Stand purchased 250 boxes of strawberries for $162.50. All the strawberries were sold for $312.50. What was the profit per box of strawberries?

34. The table at the right shows the population and the area of four countries. Find the population density (people per square mile) for each country. Round to the nearest tenth.

Country	Population	Area (square miles)
Australia	18,322,000	2,968,000
Cambodia	10,561,000	70,000
India	936,546,000	1,269,000
United States	267,000,000	3,619,000

APPLYING THE CONCEPTS

35. You have a choice of receiving a wage of $34,000 per year, $2840 per month, $650 per week, or $18 per hour. Which pay choice would you take? Assume a 40-hour week with 52 weeks per year.

36. The price–earnings ratio of a company's stock is one measure used by stock market analysts to assess the financial well-being of the company. Explain the meaning of the price–earnings ratio.

4.3 Proportions

Objective A *To determine whether a proportion is true* ....

A **proportion** is an expression of the equality of two ratios or rates.

$$\frac{50 \text{ miles}}{4 \text{ gallons}} = \frac{25 \text{ miles}}{2 \text{ gallons}}$$

Note that the units of the numerators are the same and the units of the denominators are the same.

$$\frac{3}{6} = \frac{1}{2}$$

This is the equality of two ratios.

A proportion is **true** if the fractions are equal when written in lowest terms.

In any true proportion, the "cross products" are equal.

➡ Is $\frac{2}{3} = \frac{8}{12}$ a true proportion?

$$\frac{2}{3} \bowtie \frac{8}{12} \begin{array}{l} \rightarrow 3 \times 8 = 24 \\ \rightarrow 2 \times 12 = 24 \end{array}$$

The cross products *are* equal.
$\frac{2}{3} = \frac{8}{12}$ is a true proportion.

A proportion is **not true** if the fractions are not equal when reduced to lowest terms.

If the cross products are not equal, then the proportion is not true.

➡ Is $\frac{4}{5} = \frac{8}{9}$ a true proportion?

$$\frac{4}{5} \bowtie \frac{8}{9} \begin{array}{l} \rightarrow 5 \times 8 = 40 \\ \rightarrow 4 \times 9 = 36 \end{array}$$

The cross products *are not* equal.
$\frac{4}{5} = \frac{8}{9}$ is not a true proportion.

Example 1

Use cross products to determine whether $\frac{5}{8} = \frac{10}{16}$ is a true proportion.

Solution

$$\frac{5}{8} \bowtie \frac{10}{16} \begin{array}{l} \rightarrow 8 \times 10 = 80 \\ \rightarrow 5 \times 16 = 80 \end{array}$$

The proportion is true.

Example 2

Use cross products to determine whether $\frac{62 \text{ miles}}{4 \text{ gallons}} = \frac{33 \text{ miles}}{2 \text{ gallons}}$ is a true proportion.

Solution

$$\frac{62}{4} \bowtie \frac{33}{2} \begin{array}{l} \rightarrow 4 \times 33 = 132 \\ \rightarrow 62 \times 2 = 124 \end{array}$$

The proportion is not true.

You Try It 1

Use cross products to determine whether $\frac{6}{10} = \frac{9}{15}$ is a true proportion.

Your solution

You Try It 2

Use cross products to determine whether $\frac{\$32}{6 \text{ hours}} = \frac{\$90}{8 \text{ hours}}$ is a true proportion.

Your solution

Solutions on p. S11

Objective B   *To solve proportions* ....

Sometimes one of the numbers in a proportion is unknown. In this case, it is necessary to *solve* the proportion.

To **solve** a proportion, find a number to replace the unknown so that the proportion is true.

➡ Solve: $\dfrac{9}{6} = \dfrac{3}{n}$

$$\dfrac{9}{6} = \dfrac{3}{n}$$

$9 \times n = 6 \times 3$ • **Find the cross products.**

$9 \times n = 18$

$n = 18 \div 9$ • **Think of $9 \times n = 18$ as $9\overline{)18}$.**

$n = 2$

Check:

$\dfrac{9}{6} \diagdown \diagup \dfrac{3}{2}$ ➡ $6 \times 3 = 18$
  ➡ $9 \times 2 = 18$

Example 3

Solve $\dfrac{n}{12} = \dfrac{25}{60}$ and check.

Solution

$n \times 60 = 12 \times 25$
$n \times 60 = 300$
$\quad n = 300 \div 60$
$\quad n = 5$

Check:

$\dfrac{5}{12} \diagdown \diagup \dfrac{25}{60}$ ➡ $12 \times 25 = 300$
  ➡ $5 \times 60 = 300$

Example 4

Solve $\dfrac{4}{9} = \dfrac{n}{16}$. Write the answer to the nearest tenth.

Solution

$4 \times 16 = 9 \times n$
$\quad 64 = 9 \times n$
$64 \div 9 = n$
$\quad 7.1 \approx n$

Note: A rounded answer is an approximation. Therefore, the answer to a check will not be exact.

You Try It 3

Solve $\dfrac{n}{14} = \dfrac{3}{7}$ and check.

Your solution

You Try It 4

Solve $\dfrac{5}{8} = \dfrac{n}{20}$.

Your solution

Solutions on p. S11

Example 5

Solve $\frac{28}{52} = \frac{7}{n}$ and check.

Solution

$28 \times n = 52 \times 7$
$28 \times n = 364$
$\quad\ n = 364 \div 28$
$\quad\ n = 13$

Check:

$\frac{28}{52} \qquad \frac{7}{13}$ → $52 \times 7 = 364$
   → $28 \times 13 = 364$

You Try It 5

Solve $\frac{15}{20} = \frac{12}{n}$ and check.

Your solution

Example 6

Solve $\frac{15}{n} = \frac{8}{3}$. Write the answer to the nearest hundredth.

Solution

$15 \times 3 = n \times 8$
$\quad\ 45 = n \times 8$
$45 \div 8 = n$
$\quad 5.63 \approx n$

You Try It 6

Solve $\frac{12}{n} = \frac{7}{4}$. Write the answer to the nearest hundredth.

Your solution

Example 7

Solve $\frac{n}{9} = \frac{3}{1}$ and check.

Solution

$n \times 1 = 9 \times 3$
$n \times 1 = 27$
$\quad\ n = 27 \div 1$
$\quad\ n = 27$

Check:

$\frac{27}{9} \qquad \frac{3}{1}$ → $9 \times 3 = 27$
   → $27 \times 1 = 27$

You Try It 7

Solve $\frac{n}{12} = \frac{4}{1}$ and check.

Your solution

Example 8

Solve $\frac{5}{9} = \frac{15}{n}$ and check.

Solution

$5 \times n = 9 \times 15$
$5 \times n = 135$
$\quad\ n = 135 \div 5$
$\quad\ n = 27$

Check:

$\frac{5}{9} \qquad \frac{15}{27}$ → $9 \times 15 = 135$
  → $5 \times 27 = 135$

You Try It 8

Solve $\frac{3}{8} = \frac{12}{n}$ and check.

Your solution

Solutions on p. S11

Objective C   To solve application problems ....

Example 9
A mason determines that 9 cement blocks are required for a retaining wall 2 feet long. At this rate, how many cement blocks are required for a retaining wall that is 24 feet long?

Strategy
To find the number of cement blocks for a retaining wall 24 feet long, write and solve a proportion, using n to represent the number of blocks required.

Solution
$$\frac{9 \text{ cement blocks}}{2 \text{ feet}} = \frac{n \text{ cement blocks}}{24 \text{ feet}}$$

$$9 \times 24 = 2 \times n$$
$$216 = 2 \times n$$
$$216 \div 2 = n$$
$$108 = n$$

108 cement blocks are required for a 24-foot retaining wall.

You Try It 9
Twenty-four jars can be packed in 6 identical boxes. At this rate, how many jars can be packed in 15 boxes?

Your strategy

Your solution

Example 10
The dosage of a certain medication is 2 ounces for every 50 pounds of body weight. How many ounces of this medication are required for a person who weighs 175 pounds?

Strategy
To find the number of ounces of medication for a person weighing 175 pounds, write and solve a proportion, using n to represent the number of ounces of medication for a 175-pound person.

Solution
$$\frac{2 \text{ ounces}}{50 \text{ pounds}} = \frac{n \text{ ounces}}{175 \text{ pounds}}$$

$$2 \times 175 = 50 \times n$$
$$350 = 50 \times n$$
$$350 \div 50 = n$$
$$7 = n$$

7 ounces of medication are required for a 175-pound person.

You Try It 10
Three tablespoons of a liquid plant fertilizer are to be added to every 4 gallons of water. How many tablespoons of fertilizer are required for 10 gallons of water?

Your strategy

Your solution

Solutions on p. S11

4.3 Exercises

Objective A

Determine whether the proportion is true or not true.

1. $\dfrac{4}{8} = \dfrac{10}{20}$

2. $\dfrac{39}{48} = \dfrac{13}{16}$

3. $\dfrac{7}{8} = \dfrac{11}{12}$

4. $\dfrac{15}{7} = \dfrac{17}{8}$

5. $\dfrac{27}{8} = \dfrac{9}{4}$

6. $\dfrac{3}{18} = \dfrac{4}{19}$

7. $\dfrac{45}{135} = \dfrac{3}{9}$

8. $\dfrac{3}{4} = \dfrac{54}{72}$

9. $\dfrac{16}{3} = \dfrac{48}{9}$

10. $\dfrac{15}{5} = \dfrac{3}{1}$

11. $\dfrac{7}{40} = \dfrac{7}{8}$

12. $\dfrac{9}{7} = \dfrac{6}{5}$

13. $\dfrac{50 \text{ miles}}{2 \text{ gallons}} = \dfrac{25 \text{ miles}}{1 \text{ gallon}}$

14. $\dfrac{16 \text{ feet}}{10 \text{ seconds}} = \dfrac{24 \text{ feet}}{15 \text{ seconds}}$

15. $\dfrac{6 \text{ minutes}}{5 \text{ cents}} = \dfrac{30 \text{ minutes}}{25 \text{ cents}}$

16. $\dfrac{16 \text{ pounds}}{12 \text{ days}} = \dfrac{20 \text{ pounds}}{14 \text{ days}}$

17. $\dfrac{\$15}{4 \text{ pounds}} = \dfrac{\$45}{12 \text{ pounds}}$

18. $\dfrac{270 \text{ trees}}{6 \text{ acres}} = \dfrac{90 \text{ trees}}{2 \text{ acres}}$

19. $\dfrac{300 \text{ feet}}{4 \text{ rolls}} = \dfrac{450 \text{ feet}}{7 \text{ rolls}}$

20. $\dfrac{1 \text{ gallon}}{4 \text{ quarts}} = \dfrac{7 \text{ gallons}}{28 \text{ quarts}}$

21. $\dfrac{\$65}{5 \text{ days}} = \dfrac{\$26}{2 \text{ days}}$

22. $\dfrac{80 \text{ miles}}{2 \text{ hours}} = \dfrac{110 \text{ miles}}{3 \text{ hours}}$

23. $\dfrac{7 \text{ tiles}}{4 \text{ feet}} = \dfrac{42 \text{ tiles}}{20 \text{ feet}}$

24. $\dfrac{15 \text{ feet}}{3 \text{ yards}} = \dfrac{90 \text{ feet}}{18 \text{ yards}}$

Objective B

Solve. Round to the nearest hundredth.

25. $\dfrac{n}{4} = \dfrac{6}{8}$ **26.** $\dfrac{n}{7} = \dfrac{9}{21}$ **27.** $\dfrac{12}{18} = \dfrac{n}{9}$ **28.** $\dfrac{7}{21} = \dfrac{35}{n}$

29. $\dfrac{6}{n} = \dfrac{24}{36}$ **30.** $\dfrac{3}{n} = \dfrac{15}{10}$ **31.** $\dfrac{n}{45} = \dfrac{17}{135}$ **32.** $\dfrac{9}{4} = \dfrac{18}{n}$

33. $\dfrac{n}{6} = \dfrac{2}{3}$ **34.** $\dfrac{5}{12} = \dfrac{n}{144}$ **35.** $\dfrac{n}{5} = \dfrac{7}{8}$ **36.** $\dfrac{4}{n} = \dfrac{9}{5}$

37. $\dfrac{n}{11} = \dfrac{32}{4}$ **38.** $\dfrac{3}{4} = \dfrac{8}{n}$ **39.** $\dfrac{5}{12} = \dfrac{n}{8}$ **40.** $\dfrac{36}{20} = \dfrac{12}{n}$

41. $\dfrac{n}{15} = \dfrac{21}{12}$ **42.** $\dfrac{40}{n} = \dfrac{15}{8}$ **43.** $\dfrac{32}{n} = \dfrac{1}{3}$ **44.** $\dfrac{5}{8} = \dfrac{42}{n}$

45. $\dfrac{18}{11} = \dfrac{16}{n}$ **46.** $\dfrac{25}{4} = \dfrac{n}{12}$ **47.** $\dfrac{28}{8} = \dfrac{12}{n}$ **48.** $\dfrac{n}{30} = \dfrac{65}{120}$

49. $\dfrac{0.3}{5.6} = \dfrac{n}{25}$ **50.** $\dfrac{1.3}{16} = \dfrac{n}{30}$ **51.** $\dfrac{0.7}{9.8} = \dfrac{3.6}{n}$ **52.** $\dfrac{1.9}{7} = \dfrac{13}{n}$

Objective C   *Application Problems*

Solve. Round to the nearest hundredth.

53. A 6-ounce package of Puffed Wheat contains 600 calories. How many calories are in a 0.5-ounce serving of the cereal?

54. A car travels 70.5 miles on 3 gallons of gas. Find the distance that the car can travel on 14 gallons of gas.

55. Ron Stokes uses 2 pounds of fertilizer for every 100 square feet of lawn for landscape maintenance. At this rate, how many pounds of fertilizer did he use on a lawn that measures 2500 square feet?

Solve. Round to the nearest hundredth.

56. A nursery provides a liquid plant food by adding 1 gallon of water for each 2 ounces of plant food. At this rate, how many ounces of plant food are required for 25 gallons of water?

57. A manufacturer of baseball equipment makes 4 aluminum bats for every 15 bats made from wood. On a day when 100 aluminum bats are made, how many wooden bats are produced?

58. A brick wall 20 feet in length contains 1040 bricks. At the same rate, how many bricks would it take to build a wall 48 feet in length?

59. The scale on the map at the right is 1.25 inches equals 10 miles. Find the distance between Carlsbad and Del Mar, which are 2 inches apart on the map.

60. The scale on the plans for a new house is 1 inch equals 3 feet. Find the length and width of a room that measures 5 inches by 8 inches on the drawing.

61. The dosage for a medication is $\frac{1}{3}$ ounce for every 40 pounds of body weight. At this rate, how many ounces of medication should a physician prescribe for a patient who weighs 150 pounds?

62. A bank requires a monthly payment of $33.45 on a $2500 loan. At the same rate, find the monthly payment on a $10,000 loan.

63. A pre-election survey showed that 2 out of every 3 eligible voters would cast ballots in the county election. At this rate, how many people in a county of 240,000 eligible voters would vote in the election?

64. A paint manufacturer suggests using 1 gallon of paint for every 400 square feet of a wall. At this rate, how many gallons of paint would be required for a room that has 1400 square feet of wall?

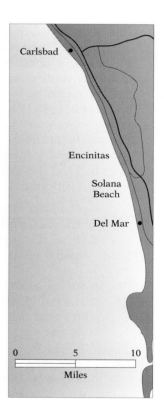

65. A 60-year-old male can obtain $10,000 of life insurance for $35.35 per month. At this rate, what is the monthly cost of $50,000 of life insurance?

66. Suppose a computer chip manufacturer knows from experience that in an average production run of 2000 circuit boards, 60 will be defective. What number of defective circuit boards can be expected from a run of 25,000 circuit boards?

67. You own 240 shares of a computer stock. The company declares a stock split of 5 shares for every 3 owned. How many shares of stock will you own after the stock split?

Solve. Round to the nearest hundredth.

68. Carlos Capasso owns 50 shares of Texas Utilities that pay dividends of $153. At this rate, what dividend would Carlos receive after buying 300 additional shares of Texas Utilities?

69. The director of data processing at a college estimates that the ratio of student time to administrative time on a certain computer is 3:2. During a month in which the computer was used 200 hours for administration, how many hours was it used by students?

APPLYING THE CONCEPTS

The table at the right shows how each dollar of projected spending by the federal government for 1998 is distributed. Social security, interest payments, Medicare, Medicaid, and other entitlements are fixed expenditures. Nondefense discretionary and defense are the only discretionary spending by the federal government. The projected budget for 1998 is $1,687.5 billion.

How Your Federal Tax Dollar Is Spent	
Social Security	23 cents
Interest payments	15 cents
Medicare	12 cents
Medicaid	6 cents
Other entitlements	12 cents
Non-defense discretionary	17 cents
Defense	15 cents

Source: Office of Management and Budget

70. a. Is at least one-half of federal spending discretionary spending?
   b. Find the ratio of the fixed expenditures to the discretionary spending.
   c. Find the amount of the 1998 budget to be spent on fixed expenditures.
   d. Find the amount of the 1998 budget to be spent on social security.

71. A survey of voters in a city claimed that 2 people of every 5 who voted cast a ballot in favor of city amendment A and that 3 people of every 4 who voted cast a ballot against amendment A. Is this possible? Explain your answer.

72. The ratio of weight on the moon to weight on Earth is 1:6. If a bowling ball weighs 16 pounds on Earth, what would it weigh on the moon?

73. When engineers design a new car, they first build a model of the car. The ratio of the size of a part on the model to the actual size of the part is 2:5. If a door is 1.3 feet long on the model, what is the length of the door on the car?

74. Write a word problem that requires solving a proportion to find the answer.

75. Choose a local pizza restaurant and a particular type of pizza. Determine the size and cost of a medium pizza and of a large pizza. (Use regular prices and no special discounts.) Is $\frac{\text{cost of medium}}{\text{size of medium}}$ approximately equal to $\frac{\text{cost of large}}{\text{size of large}}$? Explain your answer.

Focus on Problem Solving

A very useful problem-solving strategy is looking for a pattern.

Problem   A legend says that a peasant invented the game of chess and gave it to a very rich king as a present. The king so enjoyed the game that he gave the peasant the choice of anything in the kingdom. The peasant's request was simple: "Place one grain of wheat on the first square, 2 grains on the second square, 4 grains on the third square, 8 on the fourth square, and continue doubling the number of grains until the last square of the chessboard is reached." How many grains of wheat must the king give the peasant?

Solution   A chessboard consists of 64 squares. To find the total number of grains of wheat on the 64 squares, we begin by looking at the amount of wheat on the first few squares.

Square 1	Square 2	Square 3	Square 4	Square 5	Square 6	Square 7	Square 8
1	2	4	8	16	32	64	128
1	3	7	15	31	63	127	255

The bottom row of numbers represents the sum of the number of grains of wheat up to and including that square. For instance, the number of grains of wheat on the first 7 squares is $1 + 2 + 4 + 8 + 16 + 32 + 64 = 127$.

One pattern to observe is that the number of grains of wheat on a square can be expressed as a power of 2.

The number of grains on square $n = 2^{n-1}$.

For example, the number of grains on square $7 = 2^{7-1} = 2^6 = 64$.

A second pattern of interest is that the number *below* a square (the total number of grains up to and including that square) is one less than the number of grains of wheat *on* the next square. For example, the number *below* square 7 is one less than the number *on* square 8 ($128 - 1 = 127$). From this observation, the number of grains of wheat on the first 8 squares is the number on square 8 (128) plus one less than the number on square 8 (127): The total number of grains of wheat on the first 8 squares is $128 + 127 = 255$.

From this observation,

$$\begin{array}{ll} \text{Number of grains of} \\ \text{wheat on the chessboard} \end{array} = \begin{array}{l} \text{number of grains} \\ \text{on square 64} \end{array} + \begin{array}{l} \text{one less than the number} \\ \text{of grains on square 64} \end{array}$$

$$= 2^{64-1} + (2^{64-1} - 1)$$

$$= 2^{63} + 2^{63} - 1 \approx 18{,}000{,}000{,}000{,}000{,}000{,}000$$

To give you an idea of the magnitude of this number, this is more wheat than has been produced in the world since chess was invented.

The same king decided to have a banquet in the long banquet room of the palace to celebrate the invention of chess. The king had 50 square tables, and each table could seat only one person on each side. The king pushed the tables together to form one long banquet table. How many people can sit at this table? *Hint:* Try constructing a pattern by using 2 tables, 3 tables, and 4 tables.

Projects and Group Activities

The Golden Ratio   There are certain designs that have been repeated over and over in both art and architecture. One of these involves the **golden rectangle.**

A golden rectangle is drawn at the right. Begin with a square that measures, say, 2 inches on a side. Now measure the distance from *A* to *B*. Place this length along the bottom of the square, starting at *A*. The resulting rectangle is a golden rectangle.

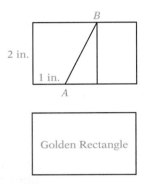

The **golden ratio** is the ratio of the length of the golden rectangle to its width. If you have drawn the rectangle following the procedure above, you will find that the golden ratio is approximately 1.6.

The golden ratio appears in many different situations. Some historians claim that some of the great pyramids of Egypt are based on the golden ratio. The drawing at the right shows the Pyramid of Gizeh, which dates from approximately 2600 B.C. The ratio of the height to a side of the base is approximately 1.6.

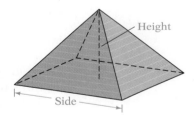

1. The canvas of the Mona Lisa painted by Leonardo da Vinci is a golden rectangle. However, there are other instances of the golden rectangle in the painting itself. Do some research on this painting and write a few paragraphs summarizing your findings.
2. What do 3 × 5 and 5 × 8 index cards have to do with the golden rectangle?
3. When was the United Nations building in New York built? What does the front of that building have to do with a golden rectangle?
4. When was the Parthenon in Athens, Greece, built? What does the front of that building have to do with a golden rectangle?

Drawing the Floor Plans for a Building

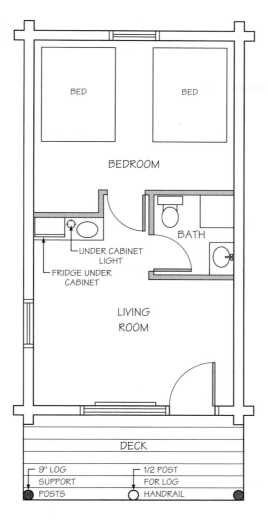

BED

BED

BEDROOM

BATH

UNDER CABINET LIGHT

FRIDGE UNDER CABINET

LIVING ROOM

DECK

9" LOG SUPPORT POSTS

1/2 POST FOR LOG HANDRAIL

The drawing at the left is a sketch of the floor plan for a cabin at a resort in the mountains of Utah. The measurements are missing. Assume that you are the architect and will finish the drawing. You will have to decide the size of the rooms and put in the measurements to scale.

Design a cabin that you would like to own. Select a scale and draw all the rooms to scale.

If you are interested in architecture, visit an architect who is using CAD (computer-aided design). Computer technology has revolutionized the field of architectural design.

Small Business and Gross Income

Do some research on your local newspaper. What is the newspaper's circulation? How often is an edition of the newspaper published? What is the cost per issue?

Is there a special rate for subscribers to the newspaper? What is the cost of a subscription? What is the length of time for which a subscription is paid? How many of the newspaper's readers are subscribers? Use this figure and the newspaper's total circulation to determine the number of copies sold at newsstands.

Use the figures you have gathered in answering the questions above to determine the total annual income, or **gross income,** derived from sales of the newspaper.

Search the World Wide Web

The framers of the Constitution decided to use a ratio to determine the number of representatives from each state. It was determined that each state would have one representative for every 30,000 citizens, with a minimum of one representative. Congress has changed this ratio over the years, so that we now have 435 representatives.

Find the number of representatives from your state. Determine the ratio of citizens to representatives. Also do this for the most populous state and for the least populous state.

The following Web sites will give you information on the number of representatives for each state and the population of each state.

For representatives: http://clerkweb.house.gov/mbrcmtee/members/mbrsstate/ UnOfLMbr.htm

For population: http://wwwnt.state.id.us/dfm/pop9601.txt

Chapter Summary

Key Words

Quantities such as 8 feet and 60 miles are number quantities written with *units*.

A *ratio* is a comparison of two quantities that have the same units.

A ratio is in *simplest form* when the two numbers that form the ratio have no common factors.

A *rate* is a comparison of two quantities that have different units.

A rate is in *simplest form* when the numbers that form the rate have no common factors.

A *unit rate* is a rate in which the number in the denominator is 1.

A *proportion* is an expression of the equality of two ratios or rates.

Essential Rules

To Find Unit Rates
To find unit rates, divide the number in the numerator of the rate by the number in the denominator of the rate.

To Solve a Proportion
One of the numbers in a proportion may be unknown. To solve a proportion, find a number to replace the unknown so that the proportion is true.

Ways to Express a Ratio
A ratio can be written three different ways:

1. As a fraction
2. As two numbers separated by a colon (:)
3. As two numbers separated by the word *to*

Chapter Review

1. Determine whether the proportion is true or not true.
   $$\frac{2}{9} = \frac{10}{45}$$

2. Write the comparison 32 dollars to 80 dollars as a ratio in simplest form using a fraction, a colon (:), and the word *to*.

3. Write "250 miles in 4 hours" as a unit rate.

4. Determine whether the proportion is true or not true.
   $$\frac{8}{15} = \frac{32}{60}$$

5. Solve the proportion.
   $$\frac{16}{n} = \frac{4}{17}$$

6. Write "$300 earned in 40 hours" as a unit rate.

7. Write "$8.75 for 5 pounds" as a unit rate.

8. Write the comparison 8 feet to 28 feet as a ratio in simplest form using a fraction, a colon (:), and the word *to*.

9. Solve the proportion.
   $$\frac{n}{8} = \frac{9}{2}$$

10. Solve the proportion. Round to hundredths.
    $$\frac{18}{35} = \frac{10}{n}$$

11. Write the comparison 6 inches to 15 inches as a ratio in simplest form using a fraction, a colon (:), and the word *to*.

12. Determine whether the proportion is true or not true.
    $$\frac{3}{8} = \frac{10}{24}$$

13. Write "$15 in 4 hours" as a rate in simplest form.

14. Write "326.4 miles on 12 gallons" as a unit rate.

15. Write the comparison 12 days to 12 days as a ratio in simplest form using a fraction, a colon (:), and the word *to*.

16. Determine whether the proportion is true or not true.
    $$\frac{5}{7} = \frac{25}{35}$$

17. Solve the proportion. Round to hundredths.
$$\frac{24}{11} = \frac{n}{30}$$

18. Write "100 miles in 3 hours" as a rate in simplest form.

19. In 5 years, the price of a calculator went from $40 to $24. What is the ratio of the decrease in price to the original price?

20. The property tax on a $45,000 home is $900. At the same rate, what is the property tax on a home valued at $120,000?

21. The high temperature during a 24-hour period was 84 degrees, and the low temperature was 42 degrees. Write the ratio of the high temperature to the low temperature for the 24-hour period.

22. The total cost of manufacturing 1000 radios was $36,600. Of the radios made, 24 did not pass inspection. Find the cost per radio of the radios that did pass inspection.

23. A brick wall 40 feet in length contains 448 concrete blocks. At the same rate, how many blocks would it take to build a wall that is 120 feet in length?

24. A retail computer store spends $30,000 a year on TV advertising and $12,000 on newspaper advertising. Find the ratio of TV advertising to newspaper advertising.

25. A 15-pound turkey costs $10.20. What is the cost per pound?

26. Mahesh drove 198.8 miles in 3.5 hours. Find the average number of miles he drove per hour.

27. An insurance policy costs $3.87 for every $1000 of insurance. At this rate, what is the cost of $50,000 of insurance?

28. Pascal Hollis purchased 80 shares of stock for $3580. What is the cost per share?

29. Monique used 1.5 pounds of fertilizer for every 200 square feet of lawn. How many pounds of fertilizer will she have to use on a lawn that measures 3000 square feet?

30. A house had an original value of $80,000 but increased in value to $120,000 in 2 years. Find the ratio of the increase to the original value.

Chapter Test

1. Write "$22,036.80 earned in 12 months" as a unit rate.

2. Write the comparison 40 miles to 240 miles as a ratio in simplest form using a fraction, a colon (:), and the word *to*.

3. Write "18 supports for every 8 feet" as a rate in simplest form.

4. Determine whether the proportion is true or not true.
$$\frac{40}{125} = \frac{5}{25}$$

5. Write the comparison 12 days to 8 days as a ratio in simplest form using a fraction, a colon (:), and the word *to*.

6. Solve the proportion.
$$\frac{5}{12} = \frac{60}{n}$$

7. Write "256.2 miles on 8.4 gallons of gas" as a unit rate.

8. Write the comparison 27 dollars to 81 dollars as a ratio in simplest form using a fraction, a colon (:), and the word *to*.

9. Determine whether the proportion is true or not true.
$$\frac{5}{14} = \frac{25}{70}$$

10. Solve the proportion.
$$\frac{n}{18} = \frac{9}{4}$$

11. Write "$81 for 12 boards" as a rate in simplest form.

12. Write the comparison 18 feet to 30 feet as a ratio in simplest form using a fraction, a colon (:), and the word *to*.

13. Fifty shares of a utility stock pay a dividend of $62.50. At the same rate, find the dividend paid on 500 shares of the utility stock.

14. The average summer temperature in a California desert is 112 degrees. In a city 100 miles away, the average temperature is 86 degrees. Write the ratio of the average city temperature to the average desert temperature.

15. A plane travels 2421 miles in 4.5 hours. Find the plane's speed in miles per hour.

16. A research scientist estimates that the human body contains 88 pounds of water for every 100 pounds of body weight. At this rate, estimate the number of pounds of water in a college student who weighs 150 pounds.

17. If 40 feet of lumber costs $69.20, what is the per-foot cost of the lumber?

18. The dosage of a medicine is $\frac{1}{4}$ ounce for every 50 pounds of body weight. How many ounces of this medication are required for a person who weighs 175 pounds?

19. An automobile sales company spends $25,000 each month for television advertising and $40,000 each month for radio advertising. Find, as a fraction in simplest form, the ratio of the cost of radio advertising to the total cost of advertising.

20. The property tax on a house valued at $175,000 is $1500. At the same rate, find the property tax on a house valued at $140,000.

Cumulative Review

1. Subtract: 20,095
                  − 10,937

2. Write $2 \cdot 2 \cdot 2 \cdot 2 \cdot 3 \cdot 3 \cdot 3$ in exponential notation.

3. Simplify: $4 - (5 - 2)^2 \div 3 + 2$

4. Find the prime factorization of 160.

5. Find the LCM of 9, 12, and 18.

6. Find the GCF of 28 and 42.

7. Reduce $\frac{40}{64}$ to simplest form.

8. Find $4\frac{7}{15}$ more than $3\frac{5}{6}$.

9. What is $4\frac{5}{9}$ less than $10\frac{1}{6}$?

10. Multiply: $\frac{11}{12} \times 3\frac{1}{11}$

11. Find the quotient of $3\frac{1}{3}$ and $\frac{5}{7}$.

12. Simplify: $\left(\frac{2}{5} + \frac{3}{4}\right) \div \frac{3}{2}$

13. Write 4.0709 in words.

14. Round 2.09762 to the nearest hundredth.

15. Divide: $8.09\overline{)16.0976}$
    Round to the nearest thousandth.

16. Convert $0.06\frac{2}{3}$ to a fraction.

17. Write the comparison 25 miles to 200 miles as a ratio in simplest form.

18. Write "87 cents for 6 bars of soap" as a rate in simplest form.

19. Write "250.5 miles on 7.5 gallons of gas" as a unit rate.

20. Solve $\frac{40}{n} = \frac{160}{17}$. Round to the nearest hundredth.

21. A car traveled 457.6 miles in 8 hours. Find the car's speed in miles per hour.

22. Solve the proportion.
$$\frac{12}{5} = \frac{n}{15}$$

23. You had $1024 in your checking account. You then wrote checks for $192 and $88. What is your new checking account balance?

24. Malek Khatri buys a tractor for $22,760. A down payment of $5000 is required. The balance remaining is paid in 48 equal monthly installments. What is the monthly payment?

25. Yuko is assigned to read a book containing 175 pages. She reads $\frac{2}{5}$ of the book during Thanksgiving vacation. How many pages of the assignment remain to be read?

26. A building contractor bought $2\frac{1}{3}$ acres of land for $84,000. What was the cost of each acre?

27. Benjamin Eli bought a shirt for $21.79 and a tie for $8.59. He used a $50 bill to pay for the purchases. Find the amount of change.

28. A college baseball player had 42 hits in 155 at-bats. Find the baseball player's batting average. Round to the nearest thousandth.

29. A soil conservationist estimates that a river bank is eroding at the rate of 3 inches every 6 months. At this rate, how many inches will be eroded in 50 months?

30. The dosage of a medicine is $\frac{1}{2}$ ounce for every 50 pounds of body weight. How many ounces of this medication are required for a person who weighs 160 pounds?

5

Percents

A water treatment technologist monitors and controls the quality of a city's drinking water. To ensure the quality of the water, the technicians must measure the percent of various quantities, including chlorine, bacteria, and sediment.

Objectives

Section 5.1
To write a percent as a fraction or a decimal
To write a fraction or a decimal as a percent

Section 5.2
To find the amount when the percent and the base are given
To solve application problems

Section 5.3
To find the percent when the base and amount are given
To solve application problems

Section 5.4
To find the base when the percent and amount are given
To solve application problems

Section 5.5
To solve percent problems using proportions
To solve application problems

Percent Symbol

The idea of using percent dates back many hundreds of years. Percents are used in business for all types of consumer and business loans, in chemistry to measure the percent concentration of an acid, in economics to measure the increases or decreases in the consumer price index (CPI), and in many other areas that affect our daily lives.

The word *percent* comes from the Latin phrase *per centum,* which means "by the hundred." The symbol that is used today for percent is %, but this was not always the symbol.

The present symbol apparently is a result of abbreviations for the word "percent."

One abbreviation was p. cent; later, p. 100 and p. $\overset{o}{c}$ were used. From p. $\overset{o}{c}$, the abbreviation changed to p$\frac{o}{o}$ around the 17th century. This probably was a result of continual writing of p. $\overset{o}{c}$ and the eventual closing of the "c" to make an "o." By the 19th century, the "p" in front of the symbol p$\frac{o}{o}$ was no longer written. The bar that separated the o's became a slash, and the modern symbol % became widely used.

5.1 Introduction to Percents

Objective A *To write a percent as a fraction or a decimal* ....

Percent means "parts of 100." In the figure at the right, there are 100 parts. Because 13 of the 100 parts are shaded, 13% of the figure is shaded.

In most applied problems involving percents, it is necessary either to rewrite a percent as a fraction or a decimal or to rewrite a fraction or a decimal as a percent.

To write a percent as a fraction, remove the percent sign and multiply by $\frac{1}{100}$.

$$13\% = 13 \times \frac{1}{100} = \frac{13}{100}$$

To rewrite a percent as a decimal, remove the percent sign and multiply by 0.01.

$$13\% \quad = \quad 13 \times 0.01 \quad = \quad 0.13$$

> Move the decimal point two places to the left. Then remove the percent sign.

Example 1 Write 120% as a fraction and as a decimal.

Solution
$$120\% = 120 \times \frac{1}{100} = \frac{120}{100}$$
$$= 1\frac{1}{5}$$

$$120\% = 120 \times 0.01 = 1.2$$

Note that percents larger than 100 are greater than 1.

You Try It 1 Write 125% as a fraction and as a decimal.

Your solution

Example 2 Write $16\frac{2}{3}\%$ as a fraction.

Solution
$$16\frac{2}{3}\% = 16\frac{2}{3} \times \frac{1}{100}$$
$$= \frac{50}{3} \times \frac{1}{100} = \frac{50}{300} = \frac{1}{6}$$

You Try It 2 Write $33\frac{1}{3}\%$ as a fraction.

Your solution

Example 3 Write 0.5% as a decimal.

Solution $0.5\% = 0.5 \times 0.01 = 0.005$

You Try It 3 Write 0.25% as a decimal.

Your solution

Solutions on p. S12

Objective B   *To write a fraction or a decimal as a percent* ....

A fraction or a decimal can be written as a percent by multiplying by 100%.

➡ Write $\frac{3}{8}$ as a percent.

$$\frac{3}{8} = \frac{3}{8} \times 100\% = \frac{3}{8} \times \frac{100}{1}\% = \frac{300}{8}\% = 37\frac{1}{2}\% \text{ or } 37.5\%$$

➡ Write 0.37 as a percent.

$$0.37 \quad = \quad 0.37 \times 100\% \quad = \quad 37\%$$

> Move the decimal point two places to the right. Then write the percent sign.

Example 4   Write 0.015 as a percent.

Solution $0.015 = 0.015 \times 100\%$
$\qquad = 1.5\%$

You Try It 4   Write 0.048 as a percent.

Your solution

Example 5   Write 2.15 as a percent.

Solution $2.15 = 2.15 \times 100\% = 215\%$

You Try It 5   Write 3.67 as a percent.

Your solution

Example 6   Write $0.33\frac{1}{3}$ as a percent.

Solution $0.33\frac{1}{3} = 0.33\frac{1}{3} \times 100\%$
$\qquad = 33\frac{1}{3}\%$

You Try It 6   Write $0.62\frac{1}{2}$ as a percent.

Your solution

Example 7   Write $\frac{2}{3}$ as a percent.
Write the remainder in fractional form.

Solution $\frac{2}{3} = \frac{2}{3} \times 100\% = \frac{200}{3}\%$
$\qquad = 66\frac{2}{3}\%$

You Try It 7   Write $\frac{5}{6}$ as a percent.
Write the remainder in fractional form.

Your solution

Example 8   Write $2\frac{2}{7}$ as a percent.
Round to the nearest tenth.

Solution $2\frac{2}{7} = \frac{16}{7} = \frac{16}{7} \times 100\%$
$\qquad = \frac{1600}{7}\% \approx 228.6\%$

You Try It 8   Write $1\frac{4}{9}$ as a percent.
Round to the nearest tenth.

Your solution

Solutions on p. S12

5.1 Exercises

Objective A

Write as a fraction and as a decimal.

1.	25%	**2.**	40%	**3.**	130%	**4.**	150%
5.	100%	**6.**	87%	**7.**	73%	**8.**	45%
9.	383%	**10.**	425%	**11.**	70%	**12.**	55%
13.	88%	**14.**	64%	**15.**	32%	**16.**	18%

Write as a fraction.

17. $66\frac{2}{3}\%$ **18.** $12\frac{1}{2}\%$ **19.** $83\frac{1}{3}\%$ **20.** $3\frac{1}{8}\%$ **21.** $11\frac{1}{9}\%$ **22.** $\frac{3}{8}\%$

23. $45\frac{5}{11}\%$ **24.** $15\frac{3}{8}\%$ **25.** $4\frac{2}{7}\%$ **26.** $5\frac{3}{4}\%$ **27.** $6\frac{2}{3}\%$ **28.** $8\frac{2}{3}\%$

Write as a decimal.

29.	6.5%	**30.**	12.3%	**31.**	0.55%	**32.**	2%
33.	8.25%	**34.**	5.05%	**35.**	6.75%	**36.**	3.08%
37.	0.45%	**38.**	6.4%	**39.**	80.4%	**40.**	16.7%

Objective B

Write as a percent.

41.	0.16	**42.**	0.73	**43.**	0.05	**44.**	0.13	**45.**	0.01	**46.**	0.95
47.	0.70	**48.**	1.07	**49.**	1.24	**50.**	2.07	**51.**	0.004	**52.**	0.37
53.	0.006	**54.**	1.012	**55.**	3.106	**56.**	0.12				

Write as a percent. Round to the nearest tenth.

57. $\frac{27}{50}$ **58.** $\frac{37}{100}$ **59.** $\frac{1}{3}$ **60.** $\frac{2}{5}$

61. $\dfrac{5}{8}$ **62.** $\dfrac{1}{8}$ **63.** $\dfrac{1}{6}$ **64.** $1\dfrac{1}{2}$

65. $\dfrac{7}{40}$ **66.** $1\dfrac{2}{3}$ **67.** $1\dfrac{7}{9}$ **68.** $\dfrac{7}{8}$

Write as a percent. Write the remainder in fractional form.

69. $\dfrac{15}{50}$ **70.** $\dfrac{12}{25}$ **71.** $\dfrac{7}{30}$ **72.** $\dfrac{1}{3}$

73. $2\dfrac{3}{8}$ **74.** $1\dfrac{2}{3}$ **75.** $2\dfrac{1}{6}$ **76.** $\dfrac{7}{8}$

APPLYING THE CONCEPTS

77. Determine whether the statement is true or false. If the statement is false, give an example to show that the statement is false.
  a. Multiplying a number by a percent always decreases the number.
  b. Dividing by a percent always increases the number.
  c. The word *percent* means "per hundred."
  d. A percent is always less than one.

78. Write the part of the square that is shaded as a fraction, as a decimal, and as a percent. Write the part of the square that is not shaded as a fraction, as a decimal, and as a percent.

79. Explain in your own words how to change a percent to a decimal and a decimal to a percent.

80. A sale on computers advertised $\dfrac{1}{3}$ off the regular price. What percent of the regular price does this represent?

81. A suit was priced at 50% off the regular price. What fraction of the regular price does this represent?

82. If $\dfrac{2}{5}$ of the population voted in an election, what percent of the population did not vote?

83. Is $\dfrac{1}{2}\%$ the same as 0.5? If not, what is the difference between 0.5 and the decimal equivalent of $\dfrac{1}{2}\%$?

84. Is 9.4% the same as $9\dfrac{2}{5}\%$? If not, what is the difference between the decimal equivalent of 9.4% and $9\dfrac{2}{5}\%$?

85. How can you recognize a fraction that represents a number that is less than 1%?

5.2 Percent Equations: Part I

Objective A **To find the amount when the percent and the base are given** ....

A real estate broker receives a payment that is 4% of an $85,000 sale. To find the amount the broker receives requires answering the question "4% of $85,000 is what?"

This sentence can be written using mathematical symbols and then solved for the unknown number.

4%	of	$85,000	is	what?
↓	↓	↓	↓	↓

$$\boxed{\text{percent } 4\%} \times \boxed{\text{base } \$85{,}000} = \boxed{\text{amount } n}$$

of is written as × (times)
is is written as = (equals)
what is written as n (the unknown number)

$$0.04 \times \$85{,}000 = n$$
$$\$3400 = n$$

Note that the percent is written as a decimal.

The broker receives a payment of $3400.

The solution was found by solving the basic percent equation for amount.

The Basic Percent Equation

$$\boxed{\text{Percent}} \times \boxed{\text{base}} = \boxed{\text{amount}}$$

In most cases, the percent is written as a decimal before the basic percent equation is solved. However, some percents are more easily written as a fraction than as a decimal. For example,

$$33\frac{1}{3}\% = \frac{1}{3} \qquad 66\frac{2}{3}\% = \frac{2}{3} \qquad 16\frac{2}{3}\% = \frac{1}{6} \qquad 83\frac{1}{3}\% = \frac{5}{6}$$

Example 1 Find 5.7% of 160.

Solution $n = 0.057 \times 160$
$n = 9.12$

Note that the words "what is" are missing from the problem but are implied by the word "Find."

You Try It 1 Find 6.3% of 150.

Your solution

Example 2 What is $33\frac{1}{3}\%$ of 90?

Solution $n = \dfrac{1}{3} \times 90$
$n = 30$

You Try It 2 What is $16\frac{2}{3}\%$ of 66?

Your solution

Solutions on p. S12

Objective B   *To solve application problems* ....

Solving percent problems requires identifying the three elements of the basic percent equation. Recall that these three parts are *percent, base,* and *amount.* Usually the base follows the phrase "percent of."

The table at the right shows the approximate value of various cars in 1997. According to *Money* magazine (March 1997), in 5 years, the value of the Firebird will decrease from its 1997 value by 45%, the value of the Camaro will decrease by 43% of its 1997 value, and the value of the Mustang will decrease by 40% of its 1997 value.

Car	1997 Retail Value
Pontiac Firebird	$22,884
Chevrolet Camaro	$23,170
Ford Mustang	$23,985

⇒ Using the estimates from *Money* magazine, find the resale value of the Ford Mustang in 5 years.

Strategy   To find the resale value in 5 years,

- Write and solve the basic percent equation to find the amount of the decrease. The percent is 40%; the base is $23,985.
- Subtract the decrease in value from the 1997 retail value.

Solution   Percent × base = amount

$$40\% \times 23{,}985 = n$$
$$0.40 \times 23{,}985 = n$$
$$9594 = n \quad \bullet \text{ Amount of decrease}$$
$$23{,}985 - 9594 = 14{,}391$$

The value of the Mustang in 5 years will be $14,391.

Example 3
A quality control inspector found that 1.2% of 2500 telephones inspected were defective. How many telephones inspected were not defective?

Strategy
To find the number of nondefective phones:

- Find the number of defective phones. Write and solve a basic percent equation, using n to represent the number of defective phones (amount). The percent is 1.2% and the base is 2500.
- Subtract the number of defective phones from the number of phones inspected (2500).

Solution $1.2\% \times 2500 = n$
$0.012 \times 2500 = n$
$30 = n$ defective phones
$2500 - 30 = 2470$

2470 telephones were not defective.

You Try It 3
An electrician's hourly wage was $13.50 before an 8% raise. What is the new hourly wage?

Your strategy

Your solution

Solution on p. S12

5.2 Exercises

· ·

Objective A

Solve.

1. 8% of 100 is what?

2. 16% of 50 is what?

3. 27% of 40 is what?

4. 52% of 95 is what?

5. 0.05% of 150 is what?

6. 0.075% of 625 is what?

7. 125% of 64 is what?

8. 210% of 12 is what?

9. Find 10.7% of 485.

10. Find 12.8% of 625.

11. What is 0.25% of 3000?

12. What is 0.06% of 250?

13. 80% of 16.25 is what?

14. 26% of 19.5 is what?

15. What is $1\frac{1}{2}$% of 250?

16. What is $5\frac{3}{4}$% of 65?

17. $16\frac{2}{3}$% of 120 is what?

18. $83\frac{1}{3}$% of 246 is what?

19. What is $33\frac{1}{3}$% of 630?

20. What is $66\frac{2}{3}$% of 891?

21. Which is larger: 5% of 95, or 75% of 6?

22. Which is larger: 112% of 5, or 0.45% of 800?

23. Which is smaller: 79% of 16, or 20% of 65?

24. Which is smaller: 15% of 80, or 95% of 15?

25. Which is smaller: 2% of 1500, or 72% of 40?

26. Which is larger: 22% of 120, or 84% of 32?

27. Find 31.294% of 82,460.

28. Find 123.94% of 275,976.

Objective B   *Application Problems*

29. A car dealer's sticker price for a Ford Ranger V-6 is $17,820. The dealer's invoice cost for this truck is 94% of the sticker price. Find the dealer's invoice cost for this truck.

30. Natasha Gomez receives a salary of $2240 per month. Of this amount, 18% is deducted for income tax. Find the amount deducted for income tax.

31. In 1987, the average number of worker hours required to build a car was 36. In 1997, the average number of worker hours required to build a car was approximately 70% of the average number of worker hours required in 1987. What was the average number of worker hours required to build a car in 1997?

32. In one year, Blockbuster Video customers rented 24% of the approximately 3,700,000,000 videos rented that year. How many videos did Blockbuster Video rent that year?

33. A General Motors buyers' incentive program offered a 3.5% rebate on the sale price of a new car. What rebate would a customer receive who purchased a $23,500 car under this program?

34. A farmer is given an income tax credit of 10% of the cost of farm machinery. What tax credit would the farmer receive on farm equipment that cost $85,000?

35. A sales tax of 6% of the cost of a car was added to the purchase price of $9500.
    a. How much was the sales tax?
    b. What is the total cost of the car including sales tax?

36. During the packaging process for oranges, spoiled oranges are discarded by an inspector. In one day an inspector found that 4.8% of the 20,000 pounds of oranges inspected were spoiled.
    a. How many pounds of oranges were spoiled?
    b. How many pounds of oranges were not spoiled?

37. Funtimes Amusement Park has 550 employees and must hire an additional 22% for the vacation season. What is the total number of employees needed for the vacation season?

38. Between two model years, Chrysler Corporation increased the average price of its new models by 2.6%. If the average price of a car was $7856, what was the amount of the increase?

APPLYING THE CONCEPTS

The two circle graphs at the right show how surveyed employees actually spend their time and the way they would prefer to spend their time. (WSJ Supplement, Work & Family 3/31/97; from *Families and Work Institute*.) Assuming that employees have 112 hours a week of time that is not spent sleeping, answer Exercises 39 to 42. Round to the nearest tenth of an hour.

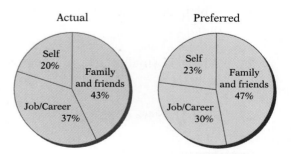

39. What is the actual number of hours per week that employees spend with family and friends?

40. What is the number of hours per week that employees would prefer to spend on job/career?

41. What is the difference between the number of hours an employee preferred to spend on self and the actual amount of time the employee spent on self?

42. What is the difference between the number of hours an employee preferred to spend on family and friends and the actual amount of time the employee spent on family and friends?

5.3   Percent Equations: Part II

Objective A   **To find the percent when the base and amount are given** ....

A recent promotional game at a grocery store listed the probability of winning a prize as "1 chance in 2." A percent can be used to describe the chance of winning. This requires answering the question "What percent of 2 is 1?"

The chance of winning can be found by solving the basic percent equation for *percent*.

What percent of   2   is   1?
$\downarrow$ $\downarrow$ $\downarrow$ $\downarrow$ $\downarrow$

$$\boxed{\begin{array}{c}\text{percent}\\n\end{array}} \times \boxed{\begin{array}{c}\text{base}\\2\end{array}} = \boxed{\begin{array}{c}\text{amount}\\1\end{array}}$$

$$n \times 2 = 1$$
$$n = 1 \div 2$$
$$n = 0.5$$
$$n = 50\%$$

• The solution must be written as a percent to answer the question.

There is a 50% chance of winning a prize.

Example 1   What percent of 40 is 30?

Solution
$$n \times 40 = 30$$
$$n = 30 \div 40$$
$$n = 0.75$$
$$n = 75\%$$

You Try It 1   What percent of 32 is 16?

Your solution

Example 2   What percent of 12 is 27?

Solution
$$n \times 12 = 27$$
$$n = 27 \div 12$$
$$n = 2.25$$
$$n = 225\%$$

You Try It 2   What percent of 15 is 48?

Your solution

Example 3   25 is what percent of 75?

Solution
$$25 = n \times 75$$
$$25 \div 75 = n$$
$$\frac{1}{3} = n$$
$$33\frac{1}{3}\% = n$$

You Try It 3   30 is what percent of 45?

Your solution

Solutions on p. S12

Objective B **To solve application problems** ....

To solve percent problems, remember that it is necessary to identify the percent, base, and amount. Usually the base follows the phrase "percent of."

Example 4

The monthly house payment for the Kaminski family is $787.50. What percent of the Kaminskis' monthly income of $3750 is the house payment?

Strategy

To find what percent of the income the house payment is, write and solve the basic percent equation, using n to represent the percent. The base is $3750 and the amount is $787.50.

Solution

$n \times \$3750 = \$787.50$
$n = \$787.50 \div \$3750$
$n = 0.21 = 21\%$

The house payment is 21% of the monthly income.

You Try It 4

Tomo Nagata had an income of $33,500 and paid $5025 in income tax. What percent of the income is the income tax?

Your strategy

Your solution

Example 5

On one Thursday night, 33.4 million of the approximately 64.5 million people watching television on the four major networks were not watching *Seinfeld*. What percent of these viewers were watching *Seinfeld*? Round the answer to the nearest percent.

Strategy

To find the percent of viewers watching *Seinfeld*:

- Subtract to find the number of people who were watching *Seinfeld* (64.5 million − 33.4 million).
- Write and solve the basic percent equation, using n to represent the percent. The base is 64.5 and the amount is the number of people watching *Seinfeld*.

Solution

64.5 million − 33.4 million = 31.1 million people were watching *Seinfeld*.

$n \times 64.5 = 31.1$
$n = 31.1 \div 64.5$
$n \approx 0.482$

Approximately 48% of the viewers were watching *Seinfeld*.

You Try It 5

Of the approximately 1,300,000 enlisted women and men in the U.S. military, 416,000 are over the age of 30. What percent of the enlisted people are under the age of 30?

Your strategy

Your solution

Solutions on pp. S12–S13

5.3 Exercises

Objective A

Solve.

1. What percent of 75 is 24?

2. What percent of 80 is 20?

3. 15 is what percent of 90?

4. 24 is what percent of 60?

5. What percent of 12 is 24?

6. What percent of 6 is 9?

7. What percent of 16 is 6?

8. What percent of 24 is 18?

9. 18 is what percent of 100?

10. 54 is what percent of 100?

11. 5 is what percent of 2000?

12. 8 is what percent of 2500?

13. What percent of 6 is 1.2?

14. What percent of 2.4 is 0.6?

15. 16.4 is what percent of 4.1?

16. 5.3 is what percent of 50?

17. 1 is what percent of 40?

18. 0.3 is what percent of 20?

19. What percent of 48 is 18?

20. What percent of 11 is 88?

21. What percent of 2800 is 7?

22. What percent of 400 is 12?

23. 4.2 is what percent of 175?

24. 41.79 is what percent of 99.5?

25. What percent of 86.5 is 8.304?

26. What percent of 1282.5 is 2.565?

Objective B   *Application Problems*

27. In 1997, total revenues for all U.S. software companies were approximately $18 billion. Of this amount, Microsoft Corporation accounted for approximately $9.5 billion. What percent of total revenues were Microsoft's revenues? Round to the nearest tenth of a percent.

28. Of the approximately 27 million personal and portable computers sold in 1997, Compaq Corporation sold an estimated 3.5 million. What percent of the total did Compaq sell? Round to the nearest tenth of a percent.

29. According to the U.S. Department of Agriculture, of the 63 billion pounds of vegetables produced in the United States in one year, 16 billion pounds were wasted. What percent of vegetables produced were wasted? Round to the nearest tenth of a percent.

30. In 1997, a Porsche 911 Carrera Targa sold for approximately $70,500. According to *Money* magazine (March 1997), the estimated value of the car 5 years later will be approximately $42,300. What percent of the 1997 selling price is the value 5 years later?

31. The total number of people in the United States Army is approximately 570,000 (*USA Today*, 3/11/97), of whom 78,000 are women and 492,000 are men. What percent of the total number of people in the Army are women? Round to the nearest tenth of a percent.

32. Of $2,100,000,000 in revenues that major league baseball anticipates for the 162 games of the 1997 season, $1,200,000,000 will go to the players.
a. How much will go to the owners?
b. What percent of the total revenues will go to the owners? Round to the nearest tenth of a percent.

33. The speed of a PowerPC computer chip in one year was 225 megahertz. Two years later, the speed of the chip increased to 300 megahertz.
a. Find the increase in speed.
b. What percent of the 225-megahertz speed is the increase?

34. To receive a license to sell insurance, an insurance account executive must answer correctly 70% of the 250 questions on a test. Nicholas Mosley answered 177 questions correctly. Did he pass the test?

35. In a test of the breaking strength of concrete slabs for freeway construction, 3 of the 200 slabs tested did not meet safety requirements. What percent of the slabs did meet safety requirements?

APPLYING THE CONCEPTS

The graph at the right shows several categories of average lifetime costs of dog ownership. Use this graph for Exercises 36 and 37. Round answers to the nearest tenth of a percent.

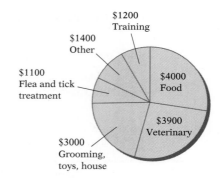

36. What percent of the total amount is spent on food?

37. What percent of the total is spent on veterinary care?

38. Public utility companies will provide consumers with an analysis of their energy bills. For one customer, it was determined that $76 of a total bill of $134 was spent for home heating. What percent (to the nearest tenth) of the total bill was for home heating?

39. The original cost of the Statue of Liberty was approximately $24,000. The cost to refurbish the statue today is approximately $780,000. What percent of the original cost is the cost to refurbish the Statue of Liberty?

40. Write a paragraph on some of the uses of percent that you find in newspapers or magazines.

41. The Fun in the Sun organization claims to have taken a survey of 350 people, asking them to give their favorite outdoor temperature for hiking. The results are given in the table at the right. Explain why these results are not possible.

Favorite Temperature	Percent
Greater than 90	5%
80–89	28%
70–79	35%
60–69	32%
Below 60	13%

5.4 Percent Equations: Part III

Objective A *To find the base when the percent and amount are given* ....

In 1997, the average salary of a major league baseball player was $1,320,000. This was 60% of the average salary of a professional basketball player in the NBA. To find the average salary of a NBA basketball player, you must answer the question "60% of what salary is $1,320,000?"

The average salary of a NBA basketball player can be found by solving the basic percent equation for the base.

$$60\% \quad \text{of} \quad \text{what} \quad \text{is} \quad 1{,}320{,}000?$$

$$\boxed{\begin{array}{c}\text{percent}\\60\%\end{array}} \times \boxed{\begin{array}{c}\text{base}\\n\end{array}} = \boxed{\begin{array}{c}\text{amount}\\1{,}320{,}000\end{array}}$$

$$0.60 \times n = 1{,}320{,}000$$
$$n = 1{,}320{,}000 \div 0.60$$
$$n = 2{,}200{,}000$$

The average salary of a NBA basketball player was $2,200,000.

Example 1 18% of what is 900?

Solution $0.18 \times n = 900$
$n = 900 \div 0.18$
$n = 5000$

You Try It 1 86% of what is 215?

Your solution

Example 2 30 is 1.5% of what?

Solution $0.015 \times n = 30$
$n = 30 \div 0.015$
$n = 2000$

You Try It 2 15 is 2.5% of what?

Your solution

Example 3 $33\frac{1}{3}\%$ of what is 7?

Solution $\frac{1}{3} \times n = 7$
$n = 7 \div \frac{1}{3}$
$n = 21$

• Note that the percent is written as a fraction.

You Try It 3 $16\frac{2}{3}\%$ of what is 5?

Your solution

Solutions on p. S13

Objective B *To solve application problems* ....

To solve percent problems, it is necessary to identify the percent, base, and amount. Usually the base follows the phrase "percent of."

Example 4

A business office bought a used copy machine for $450, which was 75% of the original cost. What was the original cost of the copier?

Strategy

To find the original cost of the copier, write and solve the basic percent equation, using n to represent the original cost (base). The percent is 75% and the amount is $450.

Solution

$75\% \times n = \$450$
$0.75 \times n = \$450$
$\qquad n = \$450 \div 0.75$
$\qquad n = \$600$

The original cost of the copier was $600.

You Try It 4

A used car has a value of $5229, which is 42% of the car's original value. What was the car's original value?

Your strategy

Your solution

Example 5

A carpenter's wage this year is $19.80 per hour, which is 110% of last year's wage. What was the increase in the hourly wage over last year?

Strategy

To find the increase in the hourly wage over last year:

• Find last year's wage. Write and solve the basic percent equation, using n to represent last year's wage (base). The percent is 110% and the amount is $19.80.
• Subtract last year's wage from this year's wage ($19.80).

Solution

$110\% \times n = \$19.80$
$\;1.10 \times n = \$19.80$
$\qquad n = \$19.80 \div 1.10$
$\qquad n = \$18.00$ • **Last year's wage**
$\$19.80 - \$18.00 = \$1.80$

The increase in the hourly wage was $1.80.

You Try It 5

Chang's Sporting Goods has a tennis racket on sale for $44.80, which is 80% of the original price. What is the difference between the original price and the sale price?

Your strategy

Your solution

Solutions on p. S13

5.4 Exercises

Objective A

Solve. Round to the nearest hundredth.

1. 12% of what is 9?

2. 38% of what is 171?

3. 8 is 16% of what?

4. 54 is 90% of what?

5. 10 is 10% of what?

6. 37 is 37% of what?

7. 30% of what is 25.5?

8. 25% of what is 21.5?

9. 2.5% of what is 30?

10. 10.4% of what is 52?

11. 125% of what is 24?

12. 180% of what is 21.6?

13. 18 is 240% of what?

14. 24 is 320% of what?

15. 4.8 is 15% of what?

16. 87.5 is 50% of what?

17. 25.6 is 12.8% of what?

18. 45.014 is 63.4% of what?

19. 0.7% of what is 0.56?

20. 0.25% of what is 1?

21. 30% of what is 2.7?

22. 78% of what is 3.9?

23. 84 is $16\frac{2}{3}$% of what?

24. 120 is $33\frac{1}{3}$% of what?

25. $66\frac{2}{3}$% of what is 72?

26. $83\frac{1}{3}$% of what is 13.5?

27. 6.59% of what is 469.35?

28. 182.3% of what is 46,253?

Objective B   *Application Problems*

29. The average size of a house in 1997 was 2100 square feet. This is approximately 125% of the average size of a house in 1977. What was the average size of a house in 1977?

30. A used Chevrolet Blazer was purchased for $22,400. This was 70% of the cost of the Blazer when new. What was the cost of the Blazer when it was new?

31. A salesperson received a commission of $820 for selling a car. This was 5% of the selling price of the car. What was the selling price of the car?

32. The per capita personal income in the United States for 1997 was approximately $25,289. This was 104.5% of the per capita personal income in 1996. What was the per capita income in 1996?

33. According to John Dvorak of *PC Magazine* (July 1997), the amount of money spent on Internet products in 1996 was approximately $19,000,000,000. This is approximately 20.6% of the projected amount that will be spent in the year 2000. What amount, to the nearest billion, is expected to be spent in 2000?

34. The Internal Revenue Service says that the average deduction for medical expenses for taxpayers in the $40,000–$50,000 bracket is $4500. This is 26% of the medical expenses claimed by taxpayers in the over $200,000 bracket. How much is the average deduction for medical expense claimed by taxpayers in the over $200,000 bracket? Round to the nearest $1000.

35. During a quality control test, Micronics found that 24 computer boards were defective. This amount was 0.8% of the computer boards tested.
    a. How many computer boards were tested?
    b. How many computer boards tested were not defective?

36. Of the calls a directory assistance operator received, 441 were requests for telephone numbers listed in the current directory. This accounted for 98% of the calls for assistance that the operator received.
    a. How many calls did the operator receive?
    b. How many telephone numbers requested were not listed in the current directory?

APPLYING THE CONCEPTS

The table at the right contains nutrition information about a breakfast cereal. Solve Exercises 37 and 38 using information from this table.

37. The recommended daily amount of thiamin for an adult is 1.5 milligrams. Find the amount of thiamin in one serving of cereal with skim milk.

38. The recommended daily amount of copper for an adult is 2 milligrams. Find the amount of copper in one serving of cereal with skim milk.

39. Increase a number by 10%. Now decrease the number by 10%. Is the result the original number? Explain.

40. When a company goes bankrupt, another company may offer to purchase the assets of the company for "25 cents on the dollar." What percent of the value of the company is the buyer paying?

NUTRITION INFORMATION
SERVING SIZE: 1.4 OZ WHEAT FLAKES WITH
0.4 OZ. RAISINS: 39.4 g. ABOUT 1/2 CUP
SERVINGS PER PACKAGE: 14

	CEREAL & RAISINS	WITH 1/2 CUP VITAMINS A & D SKIM MILK

PERCENTAGE OF U.S. RECOMMENDED DAILY ALLOWANCES (U.S. RDA)

	CEREAL & RAISINS	WITH 1/2 CUP VITAMINS A & D SKIM MILK
PROTEIN	4	15
VITAMIN A	15	20
VITAMIN C	**	2
THIAMIN	25	30
RIBOFLAVIN	25	35
NIACIN	25	35
CALCIUM	**	15
IRON	100	100
VITAMIN D	10	25
VITAMIN B$_6$	25	25
FOLIC ACID...............	25	25
VITAMIN B$_{12}$	25	30
PHOSPHOROUS.........	10	15
MAGNESIUM	10	20
ZINC	25	30
COPPER...................	2	4

* 2% MILK SUPPLIES AN ADDITIONAL 20 CALORIES.
  2 g FAT, AND 10 mg CHOLESTEROL.
** CONTAINS LESS THAN 2% OF THE U.S. RDA OF
   THIS NUTRIENT

5.5 Percent Problems: Proportion Method

Objective A   *To solve percent problems using proportions* ....

Problems that can be solved using the basic percent equation can also be solved using proportions.

The proportion method is based on writing two ratios. One ratio is the percent ratio, written as $\frac{percent}{100}$. The second ratio is the amount-to-base ratio, written as $\frac{amount}{base}$. These two ratios form the proportion:

$$\frac{percent}{100} = \frac{amount}{base}$$

To use the proportion method, first identify the percent, the amount, and the base (the base usually follows the phrase "percent of").

What is 23% of 45?	What percent of 25 is 4?	12 is 60% of what number?
$\frac{23}{100} = \frac{n}{45}$	$\frac{n}{100} = \frac{4}{25}$	$\frac{60}{100} = \frac{12}{n}$
$23 \times 45 = 100 \times n$	$n \times 25 = 100 \times 4$	$60 \times n = 100 \times 12$
$1035 = 100 \times n$	$n \times 25 = 400$	$60 \times n = 1200$
$1035 \div 100 = n$	$n = 400 \div 25$	$n = 1200 \div 60$
$10.35 = n$	$n = 16\%$	$n = 20$

Example 1   15% of what is 7? Round to the nearest hundredth.

Solution
$$\frac{15}{100} = \frac{7}{n}$$
$$15 \times n = 100 \times 7$$
$$15 \times n = 700$$
$$n = 700 \div 15$$
$$n \approx 46.67$$

You Try It 1   26% of what is 22? Round to the nearest hundredth.

Your solution

Example 2   30% of 63 is what?

Solution
$$\frac{30}{100} = \frac{n}{63}$$
$$30 \times 63 = n \times 100$$
$$1890 = n \times 100$$
$$1890 \div 100 = n$$
$$18.90 = n$$

You Try It 2   16% of 132 is what?

Your solution

Solutions on p. S13

Objective B   *To solve application problems* ··· (9) CT

Example 3

An antiques dealer found that 86% of the 250 items that were sold during one month sold for under $1000. How many items sold for under $1000?

Strategy

To find the number of items that sold for under $1000, write and solve a proportion, using n to represent the number of items sold (amount) for less than $1000. The percent is 86% and the base is 250.

Solution

$$\frac{86}{100} = \frac{n}{250}$$
$$86 \times 250 = 100 \times n$$
$$21{,}500 = 100 \times n$$
$$21{,}500 \div 100 = n$$
$$215 = n$$

215 items sold for under $1000.

You Try It 3

Last year it snowed 64% of the 150 days of the ski season at a resort. How many days did it snow?

Your strategy

Your solution

Example 4

In a test of the strength of nylon rope, 5 pieces of the 25 pieces tested did not meet the test standards. What percent of the nylon ropes tested did meet the standards?

Strategy

To find the percent of ropes tested that met the standards:

- Find the number of ropes that met the test standards (25 − 5).
- Write and solve a proportion, using n to represent the percent of ropes that met the test standards. The base is 25. The amount is the number of ropes that met the standards.

Solution

$25 - 5 = 20$ ropes met test standards

$$\frac{n}{100} = \frac{20}{25}$$
$$n \times 25 = 100 \times 20$$
$$n \times 25 = 2000$$
$$n = 2000 \div 25$$
$$n = 80\%$$

80% of the ropes tested did meet the test standards.

You Try It 4

Five ballpoint pens in a box of 200 were found to be defective. What percent of the pens were not defective?

Your strategy

Your solution

Solutions on pp. S13–S14

Focus on Problem Solving

Using a Calculator as a Problem-Solving Tool

A calculator is an important tool for problem solving. Here are a few problems to solve with a calculator. You may need to research some of the questions to find information you do not know.

1. Choose any single-digit positive number. Multiply the number by 1507 and 7373. What is the answer? Choose another positive single-digit number and again multiply by 1507 and 7373. What is the answer? What pattern do you see? Why does this work?

2. The gross domestic product in 1997 was $7,200,000,000,000. Is this more or less than the amount of money that would be placed on the last square of a standard checkerboard if 1 cent is placed on the first square, 2 cents is placed on the second square, 4 cents is placed on the third square, 8 cents is placed on the fourth square, and so on until the 64th square is reached?

3. Which of the reciprocals of the first 16 natural numbers have a terminating decimal representation and which have a repeating decimal representation?

4. What is the largest natural number n for which $4^n > 1 \cdot 2 \cdot 3 \cdot 4 \cdot 5 \cdot \cdots \cdot n$?

5. If $1000 bills are stacked one on top of another, is the height of $1 billion less than or more than the height of the Washington Monument?

6. What is the value of $1 + \cfrac{1}{1 + \cfrac{1}{1 + \cfrac{1}{1 + \cfrac{1}{1 + 1}}}}$?

7. Calculate 15^2, 35^2, 65^2, and 85^2. Study the results. Make a conjecture about a relationship between a number ending in 5 and its square. Use your conjecture to find 75^2 and 95^2. Does your conjecture work for 125^2?

8. Find the sum of the first 1000 natural numbers. (*Hint:* You could just start adding $1 + 2 + 3 + \cdots$, but even if you performed one operation every 3 seconds, it would take you an hour to find the sum. Instead, try pairing the numbers and then adding the number of pairs. Pair 1 and 1000, 2 and 999, 3 and 998, and so on. What is the sum of each pair? How many pairs are there? Use this information to answer the original question.)

9. To qualify for a home loan, a bank requires that the monthly mortgage payment be less than 25% of a borrower's monthly take-home income. A laboratory technician has deductions for taxes, insurance, and retirement that amount to 25% of the technician's monthly gross income. What minimum monthly income must this technician earn to receive a bank loan that has a $1200 per month mortgage payment?

Projects and Group Activities

Consumer Price Index   The Consumer Price Index (CPI) is a percent that is written without the percent sign. For instance, a CPI of 160.1 means 160.1%. This number means that an item that cost $100 between 1982 and 1984 (the base years) would cost $160.10 today. Determining the cost is an application of the basic percent equation.

$$\text{Percent} \times \text{base} = \text{amount}$$
$$\text{CPI} \times \text{cost in base year} = \text{cost today}$$
$$1.601 \times 100 = 160.1 \qquad \bullet \; \mathbf{160.1\% = 1.601}$$

The table below gives the CPI for various products in July of 1997. If you have Internet access, you can obtain current data for the items below plus other items not on this list. Visit http://stats.bls.gov.

Product	CPI
All items	160.1
Food	156.6
Housing	155.8
Clothes	136.1
New cars	145.2
Used cars	154.3
Medical care	233.8
Tobacco	243.2
School books	235.8
Entertainment	162.2

1. Of the items listed, are there any items that in 1997 cost more than twice as much as they cost during the base year? If so, which ones?

2. Of the items listed, are there any items that in 1997 cost more than one-and-one-half times as much as they cost during the base years but less than twice as much as they cost during the base years? If so, which ones?

3. If the cost for textbooks for one semester was $120 in the base years, how much did similar textbooks cost in 1997?

4. If a new car cost $16,000 in 1997, what would a comparable new car have cost during the base years?

5. If a movie ticket cost $7.50 in 1997, what would a comparable movie ticket have cost during the base years?

6. The base year for the CPI was 1967 before the change to 1982–1984. If 1967 were still used as the base year, the CPI for all items (not just those listed above) would be 479.7.
   a. Using the base year of 1967, explain the meaning of a CPI of 479.7.
   b. Using the base year of 1967 and a CPI of 479.7, if textbooks cost $75 for one semester in 1967, how much did similar textbooks cost in 1997?
   c. Using the base year of 1967 and a CPI of 479.7, if a new car cost $16,000 in 1997, what would a comparable new car have cost in 1967?

Economic Inflation   A government report that indicates that the *inflation* rate last year was 4% means that the CPI for all items has *increased* by 4% from the previous year. That is, on average, all the items you are buying today (food, clothes, gas, rent, movie tickets, detergent, insurance, etc.) cost 4% more than they cost last year.

This does not mean that the inflation rate is the same for all items. For instance, from 1996 to 1997, the inflation rate for food and beverages was 2.2%. However, gasoline *decreased* in price by 7%. The *deflation* rate for gasoline was 7%.

The amount of inflation (increase in price) or deflation (decrease in price) can be calculated by using the basic percent equation.

$$\text{Percent} \times \text{base} = \text{amount}$$
$$\text{Inflation rate} \times \text{cost last year} = \text{increase over last year's cost}$$
$$\text{Deflation rate} \times \text{cost last year} = \text{decrease over last year's cost}$$

For instance, the inflation rate for clothing was 4.4% and the cost of a pair of jeans last year was $33. To calculate the increase over last year's cost, use the basic percent equation.

$$\text{Inflation rate} \times \text{cost last year} = \text{increase over last year's cost}$$
$$0.044 \times 33 = 1.452$$

The increase over last year's cost was $1.45. Since this amount is the *increase* over last year and last year's cost was $33, *add* this amount to the cost last year to obtain the cost this year.

$$\text{Cost last year} + \text{increase in cost} = \text{cost this year}$$
$$33 + 1.45 = 34.45$$

The cost of the jeans this year is $34.45.

Here is an example of what happens to a commodity that experiences deflation. Suppose the cost of a gallon of gas last year was $1.31 and the deflation rate for gasoline was 7.7%. To calculate the decrease over last year's cost, use the basic percent equation.

$$\text{Deflation rate} \times \text{cost last year} = \text{decrease over last year's cost}$$
$$0.077 \times 1.31 = 0.10087$$

The decrease over last year's cost was $.10. Since this amount is the *decrease* over last year and last year's cost was $1.31, *subtract* this amount from the cost last year to obtain the cost this year.

$$\text{Cost last year} - \text{decrease in cost} = \text{cost this year}$$
$$1.31 - 0.10 = 1.21$$

The cost of one gallon of gas this year is $1.21.

The inflation rate and deflation rate are also calculated by using the basic percent equation.

Suppose that a set of four steel-belted radial tires cost $220 last year, and this year those same tires cost $235. Since the price has *increased*, we will calculate an *inflation* rate by letting n represent the unknown inflation rate.

1. Calculate the increase in cost.

$$\text{Cost this year} - \text{cost last year} = \text{increase in cost}$$
$$235 - 220 = 15$$

2. Calculate the inflation rate.

$$\text{Inflation rate} \times \text{cost last year} = \text{increase in cost}$$
$$n \times 220 = 15$$
$$n = 15 \div 220 \approx 0.068$$

The inflation rate was 6.8%.

The deflation rate is calculated in a similar manner.

Suppose that during the last year the price of the average pair of running shoes decreased from $53 to $49. Since the price has decreased, we will calculate the *deflation* rate by letting *n* represent the unknown deflation rate.

1. Calculate the decrease in cost.

   Cost last year − cost this year = decrease in cost
   $$53 - 49 = 4$$

2. Calculate the deflation rate.

   Deflation rate × cost last year = decrease in cost
   $$n \times 53 = 4$$
   $$n = 4 \div 53 \cong 0.075$$

The deflation rate was 7.5%.

1. The inflation rate for meats, poultry, fish, and eggs was 4.6% between 1996 and 1997. If a dozen eggs cost $.89 in 1996, how much did the cost of a dozen eggs increase in 1997? Round to the nearest cent.

2. The deflation rate for energy products was 6.2% between 1996 and 1997. If a gallon of heating oil cost $.74 in 1996, how much did the cost of a gallon of heating oil decrease in 1997? Round to the nearest cent.

3. The deflation rate for the typical used car was 2.3% between 1996 and 1997. If a typical used car cost $4500 in 1996, what is the cost of a comparable used car in 1997? Round to the nearest cent.

4. The inflation rate for tobacco products was 4.6% between 1996 and 1997. If a typical pack of cigarettes cost $2.30 in 1996, what is the cost of a typical pack of cigarettes in 1997? Round to the nearest cent.

5. Suppose the inflation rate for airline tickets was 4% between 1996 and 1997 and 3% between 1997 and 1998. What is the cost at the end of 1998 of an airline ticket that cost $250 at the beginning of 1996?

Chapter Summary

Key Words   *Percent* means "parts of 100."

Essential Rules   *To Write a Percent as a Fraction*
To write a percent as a fraction, remove the percent sign and multiply by $\frac{1}{100}$.

To Write a Percent as a Decimal
To write a percent as a decimal, remove the percent sign and multiply by 0.01.

To Write a Decimal as a Percent
To write a decimal as a percent, multiply by 100%.

To Write a Fraction as a Percent
To write a fraction as a percent, multiply by 100%.

Basic Percent Equation   Percent × base = amount

Proportion Method to Solve Percent Equations $\dfrac{\text{Percent}}{100} = \dfrac{\text{amount}}{\text{base}}$

Chapter Review

1. What is 30% of 200?

2. 16 is what percent of 80?

3. Write $1\frac{3}{4}$ as a percent.

4. 20% of what is 15?

5. Write 12% as a fraction.

6. Find 22% of 88.

7. What percent of 20 is 30?

8. $16\frac{2}{3}\%$ of what is 84?

9. Write 42% as a decimal.

10. What is 7.5% of 72?

11. $66\frac{2}{3}\%$ of what is 105?

12. Write 7.6% as a decimal.

13. Find 125% of 62.

14. Write $16\frac{2}{3}\%$ as a fraction.

15. Use the proportion method to find what percent of 25 is 40.

16. 20% of what number is 15? Use the proportion method.

17. Write 0.38 as a percent.

18. 78% of what is 8.5? Round to the nearest tenth.

19. What percent of 30 is 2.2? Round to the nearest tenth of a percent.

20. What percent of 15 is 92? Round to the nearest tenth of a percent.

21. Trent missed 9 out of 60 questions on a history exam. What percent of the questions did he answer correctly? Use the proportion method.

22. A company used 7.5% of its $60,000 advertising budget for TV advertising. How much of the advertising budget was spent for TV advertising?

23. In a two-year period, the population of Houston, Texas, increased from 1,600,000 to 1,700,000. What percent increase does this represent?

24. Joshua purchased a video camera for $980 and paid a sales tax of 6.25% of the cost. What was the total cost of the video camera?

25. In a survey of 350 women and 420 men, 275 of the women and 300 of the men reported that they wore sunscreen often. To the nearest tenth of a percent, what percent of the women wore sunscreen often?

26. It is estimated that the world's population will be 6,100,000,000 by the year 2000. This is 105% of the population in 1997. What was the world's population in 1997? Round to the nearest hundred million.

27. A computer system can be purchased for $1800. This is 60% of what the computer cost 4 years ago. What was the cost of the computer 4 years ago? Use the proportion method.

28. In a recent basketball game, Michael Jordan scored 33 points. This was 30% of all the points scored by the team. How many points were scored by the team?

Chapter Test

1. Write 97.3% as a decimal.

2. Write $83\frac{1}{3}\%$ as a fraction.

3. Write 0.3 as a percent.

4. Write 1.63 as a percent.

5. Write $\frac{3}{2}$ as a percent.

6. Write $\frac{2}{3}$ as a percent.

7. What is 77% of 65?

8. 47.2% of 130 is what?

9. Which is larger:
7% of 120, or 76% of 13?

10. Which is smaller:
13% of 200, or 212% of 12?

11. A fast-food company uses 6% of its $75,000 budget for advertising. What amount of the budget is spent on advertising?

12. During the packaging process for vegetables, spoiled vegetables are discarded by an inspector. In one day an inspector found that 6.4% of the 1250 pounds of vegetables were spoiled. How many pounds of vegetables were not spoiled?

The table at the right contains nutrition information about a breakfast cereal. Solve Exercises 13 and 14 with information taken from this table.

13. The recommended amount of potassium per day for an adult is 3000 milligrams (mg). What percent, to the nearest tenth of a percent, of the daily recommended amount of potassium is provided by one serving of cereal with skim milk?

14. The daily recommended number of calories for a 190-pound man is 2200 calories. What percent, to the nearest tenth of a percent, of the daily recommended number of calories is provided by one serving of cereal with 2% milk?

NUTRITION INFORMATION

SERVING SIZE: 1.4 OZ WHEAT FLAKES WITH
0.4 OZ. RAISINS: 39.4 g. ABOUT 1/2 CUP
SERVINGS PER PACKAGE: 14

	CEREAL & RAISINS	WITH 1/2 CUP VITAMINS A & D SKIM MILK
CALORIES	120	180
PROTEIN, g	3	7
CARBOHYDRATE, g	28	34
FAT, TOTAL, g	1	1*
UNSATURATED, g 1		
SATURATED, g 0		
CHOLESTEROL, mg	0	0*
SODIUM, mg	125	190
POTASSIUM, mg	240	440

* 2% MILK SUPPLIES AN ADDITIONAL 20 CALORIES.
  2 g FAT, AND 10 mg CHOLESTEROL.
** CONTAINS LESS THAN 2% OF THE U.S. RDA OF
  THIS NUTRIENT

15. The Urban Center Department Store has 125 permanent employees and must hire an additional 20 temporary employees for the holiday season. What percent of the permanent employees is the number hired as temporary employees for the holiday season?

16. Conchita missed 7 out of 80 questions on a math exam. What percent of the questions did she answer correctly? (Round to the nearest tenth of a percent.)

17. 12 is 15% of what?

18. 42.5 is 150% of what? Round to the nearest tenth.

19. A manufacturer of transistors found 384 defective transistors during a quality control study. This amount was 1.2% of the transistors tested. Find the number of transistors tested.

20. A new house was bought for $95,000; 5 years later the house sold for $152,000. The increase was what percent of the original price?

21. 123 is 86% of what number? Round to the nearest tenth.

22. What percent of 12 is 120?

23. A secretary receives a wage of $9.52 per hour. This amount is 112% of last year's salary. What is the dollar increase in the hourly wage over last year?

24. A city has a population of 71,500; 10 years ago the population was 32,500. The population now is what percent of what the population was 10 years ago?

25. The annual license fee on a car is 1.4% of the value of the car. If the license fee during a year was $91.00, what is the value of the car?

Cumulative Review

1. Simplify $18 \div (7 - 4)^2 + 2$.

2. Find the LCM of 16, 24, and 30.

3. Find the sum of $2\frac{1}{3}$, $3\frac{1}{2}$, and $4\frac{5}{8}$.

4. Subtract: $27\frac{5}{12} - 14\frac{9}{16}$

5. Multiply: $7\frac{1}{3} \times 1\frac{5}{7}$

6. What is $\frac{14}{27}$ divided by $1\frac{7}{9}$?

7. Simplify: $\left(\frac{3}{4}\right)^3 \cdot \left(\frac{8}{9}\right)^2$

8. Simplify: $\left(\frac{2}{3}\right)^2 - \left(\frac{3}{8} - \frac{1}{3}\right) \div \frac{1}{2}$

9. Round 3.07973 to the nearest hundredth.

10. Subtract: 3.0902
       $-$ 1.9706

11. Divide: $0.032\overline{)1.097}$
    Round to the nearest ten-thousandth.

12. Convert $3\frac{5}{8}$ to a decimal.

13. Convert 1.75 to a fraction.

14. Place the correct symbol, $<$ or $>$, between the two numbers.
    $\frac{3}{8}$ 0.87

15. Solve the proportion $\frac{3}{8} = \frac{20}{n}$. Round to the nearest tenth.

16. Write "$76.80 earned in 8 hours" as a unit rate.

17.   Write $18\frac{1}{3}\%$ as a fraction.

18.   Write $\frac{5}{6}$ as a percent.

19.   16.3% of 120 is what? Round to the nearest hundredth.

20.   24 is what percent of 18?

21.   12.4 is 125% of what?

22.   What percent of 35 is 120? Round to the nearest tenth.

23.   Sergio has an income of $740 per week. One-fifth of his income is deducted for income tax payments. Find his take-home pay.

24.   Eunice bought a car for $4321, with a down payment of $1000. The balance was paid in 36 equal monthly payments. Find the monthly payment.

25.   The gasoline tax is $.19 a gallon. Find the number of gallons of gasoline used during a month in which $79.80 was paid in gasoline taxes.

26.   The real estate tax on a $72,000 home is $1440. At the same rate, find the real estate tax on a home valued at $150,000.

27.   Ken purchased a stereo set for $490 and paid $29.40 in sales tax. What percent of the purchase price was the sales tax?

28.   A survey of 300 people showed that 165 people favored a certain candidate for mayor. What percent of the people surveyed did not favor this candidate?

29.   The value of a home in the northern part of the United States was $62,000 in 1985. The same home in 1990 had a value of $155,000. What percent of the 1985 value is the 1990 value?

30.   The Environmental Protection Agency found that 990 out of 5500 children tested had levels of lead in their blood exceeding federal guidelines. What percent of the children tested had levels of lead in the blood that exceeded federal standards?

6

Applications for Business and Consumers

Buyers are employed by department stores to purchase the various items that are sold in the stores. Sometimes buyers are very specialized and buy only, for instance, clothes. Once the clothes have been purchased, the buyer must determine a price at which to sell the clothes so that the store will make a profit. Markup, one of the topics of this chapter, is the amount that must be added to the cost of the clothes so that the department store makes a profit.

Objectives

Section 6.1
To find unit cost
To find the most economical purchase
To find total cost

Section 6.2
To find percent increase
To apply percent increase to business—markup
To find percent decrease
To apply percent decrease to business—discount

Section 6.3
To calculate simple interest
To calculate compound interest

Section 6.4
To calculate the initial expenses of buying a home
To calculate ongoing expenses of owning a home

Section 6.5
To calculate the initial expenses of buying a car
To calculate ongoing expenses of owning a car

Section 6.6
To calculate commissions, total hourly wages,
    and salaries

Section 6.7
To calculate checkbook balances
To balance a checkbook

A Penny a Day

A fictitious job offer in a newspaper claimed that the salary would be 1¢ on the first of the month and 2¢ on the second of the month. Next month, the salary would be 4¢ on the first and 8¢ on the second of the month. The next month and each succeeding month for 12 months, the procedure of doubling the previous payment would be continued the same way. Do you think you would want the job under these conditions?

This problem is an example of compounding interest, which is similar to the type of interest that is earned on bank or savings deposits. The big difference is that banks and savings and loans do not compound the interest as quickly as in the given example.

The table below shows the amount of salary you would earn for each month and then gives the total annual salary.

January	$.01 + $.02 = $.03
February	$.04 + $.08 = $.12
March	$.16 + $.32 = $.48
April	$.64 + $1.28 = $1.92
May	$2.56 + $5.12 = $7.68
June	$10.24 + $20.48 = $30.72
July	$40.96 + $81.92 = $122.88
August	$163.84 + $327.68 = $491.52
September	$655.36 + $1310.72 = $1966.08
October	$2621.44 + $5242.88 = $7864.32
November	$10,485.76 + $20,971.52 = $31,457.28
December	$41,943.04 + $83,886.08 = $125,829.12

Total Annual Salary = $167,772.15

Not a bad annual salary!

6.1 Applications to Purchasing

Objective A   To find unit cost ....

Frequently stores advertise items for purchase as, say, 2 Red Baron Deep Dish Pizzas for $5.50 or 5 cans of StarKist tuna for $4.25.

The **unit cost** is the cost of one Red Baron Deep Dish Pizza or for one can of StarKist tuna. To find the unit cost, divide the total cost by the number of units.

2 pizzas for $5.50   5 cans for $4.25

$5.50 \div 2 = 2.75$ $4.25 \div 5 = 0.85$

$2.75 is the cost of one pizza. $.85 is the cost of one can.

Unit cost: $2.75 per pizza   Unit cost: $.85 per can

Example 1

Find the unit cost. Round to the nearest tenth of a cent.
a. 3 gallons of mint chip ice cream for $10
b. 4 ounces of Crest toothpaste for $3.29

Strategy
To find the unit cost, divide the total cost by the number of units.

Solution
a. $10 \div 3 \approx 3.3333$
    $3.333 per gallon
b. $3.29 \div 4 = 0.8225$
    $.823 per ounce

You Try It 1

Find the unit cost. Round to the nearest tenth of a cent.
a. 8 size AA Energizer batteries for $5.59
b. 15 ounces of Revlon shampoo for $1.89

Your strategy

Your solution

Solution on p. S14

Objective B   To find the most economical purchase ....

Comparison shoppers often find the most economical buy by comparing unit costs.

One store is selling 6 twelve-ounce cans of ginger ale for $1.79, and a second store is selling 24 twelve-ounce cans of ginger ale for $7.47. To find the better buy, compare the unit costs.

$1.79 \div 6 \approx 0.298$ $7.47 \div 24 \approx 0.311$

Unit cost: $.298 per can   Unit cost: $.311 per can

Because $.298 < $.311, the better buy is 6 cans for $1.79.

Example 2
Find the more economical purchase:
5 pounds of nails for $3.25, or 4 pounds of nails for $2.58.

Strategy
To find the more economical purchase, compare the unit costs.

Solution
$3.25 \div 5 = 0.65$
$2.58 \div 4 = 0.645$
$\$.645 < \$.65$

The more economical purchase is 4 pounds for $2.58.

You Try It 2
Find the more economical purchase:
6 cans of fruit for $2.52, or 4 cans of fruit for $1.66.

Your strategy

Your solution

Solution on p. S14

Objective C   *To find total cost* ....

An installer of floor tile found the unit cost of identical floor tiles at three stores.

Store 1	Store 2	Store 3
$1.22 per tile	$1.18 per tile	$1.28 per tile

By comparing the unit costs, the installer determined that store 2 would provide the most economical purchase.

The installer also uses the unit cost to find the total cost of purchasing 300 floor tiles at store 2. The **total cost** is found by multiplying the unit cost by the number of units purchased.

Unit cost	$\times$	number of units	$=$	total cost

$$1.18 \quad \times \quad 300 \quad = \quad 354$$

The total cost is $354.

Example 3
Clear redwood lumber costs $2.43 per foot. How much would 25 feet of clear redwood cost?

Strategy
To find the total cost, multiply the unit cost ($2.43) by the number of units (25).

Solution

Unit cost	$\times$	number of units	$=$	total cost

$$2.43 \quad \times \quad 25 \quad = \quad 60.75$$

The total cost is $60.75.

You Try It 3
Pine saplings cost $4.96 each. How much would 7 pine saplings cost?

Your strategy

Your solution

Solution on p. S14

6.1 Exercises

. . . .

Objective A   *Application Problems*

Find the unit cost. Round to the nearest tenth of a cent.

1. Chris & Pitts Bar·B·Q sauce, 23 ounces for $1.69

2. Birds-eye maple, 6 feet for $18.75

3. Diamond walnuts, $2.99 for 8 ounces

4. A&W root beer, 6 cans for $1.69

5. Ibuprofen, 50 tablets for $1.99

6. Visine eye drops, 0.5 ounce for $2.69

7. Adjustable wood clamps, 2 for $9.95

8. Corn, 6 ears for $1.25

9. Cheerios cereal, 15 ounces for $2.29

10. Doritos tortilla chips, 14.5 ounces for $3.79

11. Sheet metal screws, 8 for $.95

12. MJB coffee, 39 ounces for $9.99

Objective B   *Application Problems*

Suppose your local supermarket offers the following products at the given prices. Find the more economical purchase.

13. Sutter Home pasta sauce, 25.5 ounces for $3.29, or Muir Glen Organic pasta sauce, 26 ounces for $3.79

14. Kraft mayonnaise, 40 ounces for $2.98, or Springfield mayonnaise, 32 ounces for $2.39

15. Ortega salsa, 20 ounces for $2.59, or La Victoria salsa, 16 ounces for $1.98

16. L'Oreal shampoo, 13 ounces for $3.99, or Cortexx shampoo, 12 ounces for $3.69

17. Golden Sun vitamin E, 200 tablets for $7.39 or 400 tablets for $12.99

18. Ultra Mr. Clean, 20 ounces for $2.67, or Ultra Spic and Span, 14 ounces for $2.19

19. 16 ounces Kraft cheddar cheese, $4.37, or 9 ounces of Land to Lake cheddar cheese, $2.29

20. Bertolli olive oil, 34 ounces for $9.49, or Pompeian olive oil, 8 ounces for $2.39

21. Maxwell House coffee, 7 ounces for $5.59, or Sanka coffee, 2 ounces for $2.37

22. Wagner's vanilla extract, $3.29 for 1.5 ounces, or Durkee vanilla extract, 1 ounce for $2.74

23. Purina Cat Chow, $4.19 for 56 ounces, or Friskies Chef's Blend, $3.37 for 50.4 ounces

24. Kleenex tissues, $1.73 for 250 tissues or Puffs tissues, $1.23 for 175 tissues

---

Objective C   *Application Problems*

25. If Hormel sliced bacon costs $2.59 per pound, find the total cost of 3 pounds.

26. Used red brick costs $.98 per brick. Find the total cost of 75 bricks.

27. Kiwi fruit cost $.23 each. Find the total cost of 8 kiwi.

28. Boneless chicken filets cost $4.69 per pound. Find the cost of 3.6 pounds. Round to the nearest cent.

29. Herbal tea costs $.98 per ounce. Find the total cost of 6.5 ounces.

30. If Stella Swiss Lorraine cheese costs $5.99 per pound, find the total cost of 0.65 pound. Round to the nearest cent.

31. Red Delicious apples cost $.59 per pound. Find the total cost of 2.1 pounds. Round to the nearest cent.

32. Choice rib eye steak costs $4.49 per pound. Find the total cost of 2.8 pounds. Round to the nearest cent.

33. If Godiva chocolate costs $7.95 per pound, find the total cost of $\frac{3}{4}$ pound. Round to the nearest cent.

34. Color photocopying costs $.11 per page. Find the total cost for photocopying 120 pages.

APPLYING THE CONCEPTS

35. Explain in your own words the meaning of unit pricing.

36. What is the UPC (Universal Product Code) and how is it used?

6.2 Percent Increase and Percent Decrease

Objective A   *To find percent increase* ···

Percent increase is used to show how much a quantity has increased over its original value. The statements "food prices increased by 2.3% last year" and "city council members received a 4% pay increase" are examples of percent increase.

➡ According to the Energy Information Administration, the number of alternative-fuel vehicles will increase from 251,000 to 386,000 in 5 years. Find the percent increase in alternative-fuel vehicles.

New value	−	original value	=	amount of increase

386,000 − 251,000 = 135,000

Now solve the basic percent equation for percent.

Percent × base = amount

Percent increase	×	original value	=	amount of increase

n × 251,000 = 135,000
   n = 135,000 ÷ 251,000
   n ≈ 0.5378

The number of alternative-fuel vehicles will increase by approximately 54%.

Amount of Increase (135,000)

Original Value (251,000)

New Value (386,000)

Example 1

The average wholesale price of coffee increased from $2 per pound to $3 per pound in one year. What was the percent increase in the price of one pound of coffee?

Strategy

To find the percent increase:

- Find the amount of the increase.
- Solve the basic percent equation for *amount*.

Solution

New value	−	original value	=	amount of increase

3 − 2 = 1

Percent × base = amount
 n × 2 = 1
   $n = 1 ÷ 2$
   $n = 0.50 = 50\%$

The percent increase was 50%.

You Try It 1

According to the U.S. Geological Society, the number of earthquakes measuring 6.0 to 6.9 on the Richter scale increased from 80 to 150 over a 7-year period. What percent increase in the number of these earthquakes does this represent?

Your strategy

Your solution

Solution on p. S14

Example 2
Chris Carley was earning $6.50 an hour as a nursing assistant before receiving a 10% increase in pay. What is Chris's new hourly pay?

You Try It 2
Yolanda Liyama was making a wage of $8.50 an hour as a baker before receiving a 14% increase in hourly pay. What is Yolanda's new hourly wage?

Strategy
To find the new hourly wage:

* Solve the basic percent equation for *amount*.
* Add the amount of the increase to the original wage.

Your strategy

Solution
Percent × base = amount
  0.10   × 6.50 =   *n*
            0.65 = *n*

The amount of the increase was $.65.

6.50 + 0.65 = 7.15

The new hourly wage is $7.15.

Your solution

Solution on p. S14

Objective B   *To apply percent increase to business—markup* ....

Some of the expenses involved in operating a business are salaries, rent, equipment, and utilities. To pay these expenses and earn a profit, a business must sell a product at a higher price than it paid for the product.

Cost is the price a business pays for a product, and **selling price** is the price at which a business sells a product to a customer. The difference between selling price and cost is called **markup**.

| Selling price | − | cost | = | markup |

or

| Cost | + | markup | = | selling price |

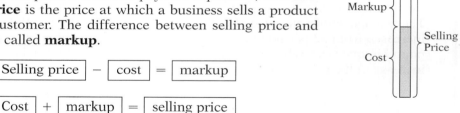

Markup is frequently expressed as a percent of a product's cost. This percent is called the **markup rate.**

| Markup rate | × | cost | = | markup |

⟹ Suppose Bicycles Galore purchases an AMP Research B4 bicycle for $2119.20 and sells it for $2649. What markup rate does Bicycles Galore use?

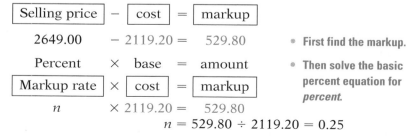

| Selling price | − | cost | = | markup |
| 2649.00 | − | 2119.20 = | | 529.80 |

* First find the markup.

Percent   ×   base   =   amount

| Markup rate | × | cost | = | markup |
| *n* | × | 2119.20 = | | 529.80 |

* Then solve the basic percent equation for *percent*.

n = 529.80 ÷ 2119.20 = 0.25

The markup rate is 25%.

Example 3

The manager of a sporting goods store determines that a markup rate of 36% is necessary to make a profit. What is the markup on a pair of Olin skis that costs the store $225?

Strategy

To find the markup, solve the basic percent equation for *amount*.

Solution

Percent $\times$ base = amount

Markup rate	$\times$	cost	=	markup

$$0.36 \quad \times \quad 225 \quad = \quad n$$
$$81 \quad = n$$

The markup is $81.

You Try It 3

A bookstore manager determines that a markup rate of 20% is necessary to make a profit. What is the markup on a book that costs the bookstore $8?

Your strategy

Your solution

Example 4

A plant nursery bought a citrus tree for $4.50 and used a markup rate of 46%. What is the selling price?

Strategy

To find the selling price:

- Find the markup by solving the basic percent equation for *amount*.
- Add the markup to the cost.

Solution

Percent $\times$ base = amount

Markup rate	$\times$	cost	=	markup

$$0.46 \quad \times \quad 4.50 \quad = \quad n$$
$$2.07 = n$$

Cost	+	markup	=	selling price

$$4.50 \quad + \quad 2.07 \quad = \quad 6.57$$

The selling price is $6.57.

You Try It 4

A clothing store bought a suit for $72 and used a markup rate of 55%. What is the selling price?

Your strategy

Your solution

Solutions on p. S14

Objective C **To find percent decrease** ....

Percent decrease is used to show how much a quantity has decreased over its original value. The statements "the number of family farms decreased by 2% last year" and "there has been a 50% decrease in the cost of a Pentium chip" are examples of percent decrease.

➡ Between 1996 and 1997, the average cost of gasoline in Washington, D.C., decreased from $1.34 per gallon to $1.20 per gallon. Find the percent decrease in the cost of gasoline for Washington, D.C.

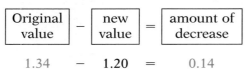

$$1.34 \quad - \quad 1.20 \quad = \quad 0.14$$

Now solve the basic percent equation for percent.

$$\text{Percent} \quad \times \quad \text{base} \quad = \quad \text{amount}$$

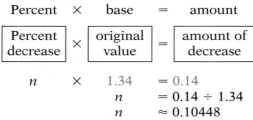

$$n \quad \times \quad 1.34 \quad = 0.14$$
$$n \quad = 0.14 \div 1.34$$
$$n \quad \approx 0.10448$$

The price of one gallon of gasoline decreased by approximately 10.4%.

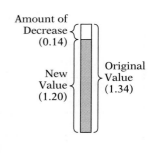

Example 5

Arms control treaties between the United States and Russia specify that the United States will decrease the number of its strategic nuclear weapons from 7150 to 3500 by the end of 2003. Find the percent decrease in strategic nuclear weapons for the U.S. Round to the nearest percent.

Strategy

To find the percent decrease:

• Find the amount of the decrease.
• Solve the basic percent equation for *percent*.

Solution

$$\boxed{\text{Original value}} - \boxed{\text{new value}} = \boxed{\text{amount of decrease}}$$

$$7150 \quad - \quad 3500 \quad = \quad 3650$$

$$\text{Percent} \times \text{base} = \text{amount}$$
$$n \quad \times 7150 = \quad 3650$$
$$n = 3650 \div 7150$$
$$n \approx 0.51049$$

The percent decrease is approximately 51%.

You Try It 5

Arms control treaties between the United States and Russia specify that Russia will decrease the number of its strategic nuclear weapons from 6670 to 3100 by the end of 2003. Find the percent decrease in strategic nuclear weapons for Russia. Round to the nearest percent.

Your strategy

Your solution

Solution on p. S14

Example 6

The total sales for December for a stationery store were $16,000. For January, total sales showed an 8% decrease from December's sales. What were the total sales for January?

Strategy

To find the total sales for January:

- Find the amount of decrease by solving the basic percent equation for *amount.*
- Subtract the amount of decrease from the December sales.

Solution

Percent × base = amount

$$0.08 \times 16,000 = n$$
$$1280 = n$$

The decrease in sales was $1280.

$$16,000 - 1280 = 14,720$$

The total sales for January were $14,720.

You Try It 6

Fog decreased the normal 5-mile visibility at an airport by 40%. What was the visibility in the fog?

Your strategy

Your solution

Solution on p. S15

Objective D   To apply percent decrease to business—discount ....

To promote sales, a store may reduce the regular price of some of its products temporarily. The reduced price is called the **sale price.** The difference between the regular price and the sale price is called the **discount**.

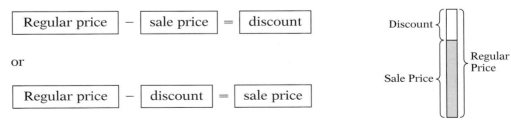

Discount is frequently stated as a percent of a product's regular price. This percent is called the **discount rate.**

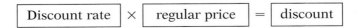

Example 7

A GE 25-inch stereo television that regularly sells for $299 is on sale for $250. Find the discount rate. Round to the nearest tenth of a percent.

Strategy

To find the discount rate:

- Find the discount.
- Solve the basic percent equation for *percent*.

Solution

Regular price	−	sale price	=	discount
299	−	250	=	49

Percent	×	base	=	amount
Discount rate	×	regular price	=	discount
n	×	299	=	49

$$n = 49 \div 299$$
$$n \approx 0.1639$$

The discount rate is 16.4%.

Example 8

A Scotts 6-horsepower lawn mower is on sale for 25% off the regular price of $325. Find the sale price.

Strategy

To find the sale price:

- Find the discount by solving the basic percent equation for *amount*.
- Subtract to find the sale price.

Solution

Percent	×	base	=	amount
Discount rate	×	regular price	=	discount
0.25	×	325	=	n

$$81.25 = n$$

Regular price	−	discount	=	sale price
325	−	81.25	=	243.75

The sale price is $243.75.

You Try It 7

A star jasmine shrub that regularly sells for $4.25 is on sale for $3.75. Find the discount rate. Round to the nearest tenth of a percent.

Your strategy

Your solution

You Try It 8

A hardware store is selling a Newport security door for 15% off the regular price of $110. Find the sale price.

Your strategy

Your solution

Solutions on p. S15

6.2 Exercises

· ·

Objective A   *Application Problems*

1. Porsche Cars Canada, Ltd. increased its sales from 310 cars in one year to 372 cars the next year. What percent increase does this represent?

2. Wolfgang Lutz of the International Institute for Applied Systems Analysis estimates that the world's population will increase from approximately 5,800,000,000 in 1997 to 10,600,000,000 in 2077. What percent increase does this represent? Round to the nearest tenth of a percent.

3. The number of students donating $1000 or more to presidential campaigns increased from 1078 in 1992 to 1746 in the 1996 presidential election. What percent increase does this represent? Round to the nearest tenth of a percent.

4. In 1991, there were 13.6 million new cars sold in the United States. From 1991 to 1997, there was approximately a 17% increase in the number of new car sales. How many new cars were sold in the United States in 1997? Round to the nearest hundred thousand.

5. On average, only 25 of every 1000 people worldwide owned a computer in 1991. Four years later, 45 of every 1000 people owned a computer. What percent increase does this represent?

6. The average size of airplanes in the world fleet is estimated to grow from 197 seats in 1996 to 219 seats by 2016. What percent increase does this represent? Round to the nearest tenth of a percent.

7. During one year, the number of people subscribing to direct broadcast satellite systems increased from 2.3 million to 4.3 million. What percent increase does this represent? Round to the nearest tenth of a percent.

8. In 1996 IBM was issued 1876 patents. This was approximately a 35% increase from the previous year. How many patents were issued to IBM in the previous year? Round to the nearest whole number.

9. The National Hockey League increased its spending on newspaper, magazine, television, and billboard advertising from $42,000 to $2,500,000 in 3 years. What percent increase does this represent? Round to the nearest percent.

10. The Nike shoe company employed approximately 14,200 people in 1995. By the end of 1996, it had increased the number of employees by 21%. How many employees did Nike have at the end of 1996?

Objective B   *Application Problems*

11. A window air conditioner cost AirRite Air Conditioning Systems $285. Find the markup on the air conditioner if the markup rate is 25% of the cost.

12. The owner of Kerr Electronics purchased 300 Craig portable CD players at a cost of $85 each. If the owner uses a markup rate of 42%, what is the markup on each of the Craig CD players?

13. A ski and tennis store uses a markup rate of 30% on tennis rackets and 40% on skis. What is the markup on a Prince tennis racket that cost $80?

14. The manager of Brass Antiques has determined that a markup rate of 38% is necessary for a profit to be made. What is the markup on a brass doorknob that costs $45?

15. Computer Inc. uses a markup of $975 on an AMD K6 computer system that costs $3250. What is the markup rate on this system?

16. Saizon Pen & Office Supply uses a markup of $12 on a calculator that costs $20. What markup rate does this amount represent?

17. Giant Photo Service uses a markup rate of 48% on its Model ZA cameras, which cost the shop $162.
   a. What is the markup?
   b. What is the selling price?

18. The Circle R golf pro shop uses a markup rate of 45% on a set of Tour Pro golf clubs that costs the shop $210.
   a. What is the markup?
   b. What is the selling price?

19. According to *Managing a Small Business* from Liraz Publishing Co., goods in a store are often marked up 50 to 100% of the cost. This allows a business to make a profit of 5% to 10%. Suppose a store purchases a pair of jeans for $32 and uses a markup rate of 80% of the cost.
   a. What is the markup?
   b. What is the selling price?

20. Harvest Time Produce Inc. uses a 55% markup rate and pays $.60 for a box of strawberries.
   a. What is the markup?
   b. What is the selling price of a box of strawberries?

21. Resner Builders' Hardware uses a markup rate of 42% for a table saw that costs $160. What is the selling price of the table saw?

22. Brad Burt's Magic Shop uses a markup rate of 48%. What is the selling price of a telescoping sword that costs $50?

Objective C   Application Problems

23. A new bridge reduced the normal 45-minute travel time between two cities by 18 minutes. What percent decrease does this represent?

24. According to *Popular Mechanics* (7/97), the Honda EV Plus electric weighs approximately 3600 pounds. The General Motors EV1 weighs 630 pounds less. What percent decrease in weight does this represent?

25. By installing energy-saving equipment, the Pala Rey Youth Camp reduced its normal $800-per-month utility bill by $320. What percent decrease does this amount represent?

26. During the last 40 years, the consumption of eggs in the United States has dropped from 400 eggs per person per year to 260 eggs per person per year. What percent decrease does this amount represent?

27. It is estimated that the value of a new car is reduced 30% after 1 year of ownership. Using this estimate, find how much value a $11,200 new car loses after 1 year.

28. A department store employs 1200 people during the holiday. At the end of the holiday season, the store reduces the number of employees by 45%. What is the decrease in the number of employees?

29. Because of a decrease in demand for super-8 video cameras, Kit's Cameras reduced the orders for these models from 20 per month to 8 per month.
   a. What is the amount of the decrease?
   b. What percent decrease does this amount represent?

30. A new computer system reduced the time for printing the payroll from 52 minutes to 39 minutes.
   a. What is the amount of the decrease?
   b. What percent decrease does this amount represent?

31. Juanita's average expense for gasoline was $76. After joining a car pool, she was able to reduce the expense by 20%.
   a. What was the amount of the decrease?
   b. What is the average monthly gasoline bill now?

32. An oil company paid a dividend of $1.60 per share. After a reorganization, the company reduced the dividend by 37.5%.
   a. What was the amount of the decrease?
   b. What is the new dividend?

33. Because of an improved traffic pattern at a sports stadium, the average amount of time a fan waits to park decreased from 3.5 minutes to 2.8 minutes. What percent decrease does this amount represent?

Objective D   *Application Problems*

34. The Austin College Bookstore is giving a discount of $8 on calculators that normally sell for $24. What is the discount rate?

35. A discount clothing store is selling a $72 sport jacket for $24 off the regular price. What is the discount rate?

36. A disk player that regularly sells for $340 is selling for 20% off the regular price. What is the discount?

37. Dacor Appliances is selling its $450 washing machine for 15% off the regular price. What is the discount?

38. An electric grill that regularly sells for $140 is selling for $42 off the regular price. What is the discount rate?

39. You bought a computer system for $350 off the regular price of $1400. What is the discount rate?

40. Quick Service Gas Station has its regularly priced $45 tune-up on sale for 16% off the regular price.
   a. What is the discount?
   b. What is the sale price?

41. Turkey that regularly sells for $.85 per pound is on sale for 20% off the regular price.
   a. What is the discount?
   b. What is the sale price?

42. An outdoor supply store has regularly priced $160 sleeping bags on sale for $120.
    a. What is the discount?
    b. What is the discount rate?

43. Standard Brands paint that regularly sells for $16 per gallon is on sale for $12 per gallon.
    a. What is the discount?
    b. What is the discount rate?

APPLYING THE CONCEPTS

44. In many instances, there are three numbers that a consumer might expect to see or hear at a new car dealership: sticker price, invoice cost, and selling price. According to CarBargains, a not-for-profit new-car shopping service, the selling price of a car for a consumer may be more or less than the dealer's invoice cost, but rarely is the sticker price. Here are some CarBargains estimates for 1997 full-size trucks.

Model	Best price	Amount to add or subtract from dealer's invoice	Dealer's invoice cost	Sticker price
Ford F-150 XL Styleside	$21,751	Subtract $100	$21,851	$25,060
GMC 1500 Sierra Wideside	$21,180	Add $300	$20,880	$23,821
Chevy C/K 1500 Fleetside	$21,427	Add $500	$20,927	$23,868
Dodge Ram BR1500 Sweptline	$22,011	Add $500	$21,511	$24,535

   a. For the GMC, what is the discount rate from the sticker price to the dealer's invoice cost?
   b. For the Ford F-150, what is the percent decrease from the dealer's invoice cost to the best price?
   c. For which of the trucks listed is the percent increase from the dealer's invoice cost to the best price the greatest?

45. A welder earning $12 per hour is given a 10% raise. To find the new wage, we can multiply $12 by 0.10 and add the product to $12. Can the new wage be found by multiplying $12 by 1.10? Try both methods and compare your answers.

46. Grocers, florists, bakers, and other businesses must consider spoilage when deciding the markup of a product. For instance, suppose a florist purchased 200 roses at a cost of $.86 per rose. The florist wants a markup rate of 50% of the total cost of all the roses and expects 7% of the roses to wilt and therefore not be salable. Find the selling price per rose by answering each of the following questions.
    a. What is the florist's total cost for the 200 roses?
    b. Find the total selling price without spoilage.
    c. Find the number of roses the florist expects to sell. *Hint:* The number of roses the florist expects to sell is:
          % of salable roses × number of roses purchased
    d. To find the selling price per rose, divide the total selling price without spoilage by the number of roses the florist expects to sell. Round to the nearest cent.

47. A promotional sale at a department store offers 25% off the sale price. The sale price is 25% off the regular price. Is this the same as a sale that offers 50% off the regular price? If not, which sale gives the better price? Explain your answer.

48. In your own words, explain how to find a percent increase or a percent decrease.

6.3 Interest

Objective A   To calculate simple interest ....

When money is deposited in a bank account, the bank pays the depositor for the privilege of using that money. The amount paid to the depositor is called **interest.** When money is borrowed from a bank, the borrower pays for the privilege of using that money. The amount paid to the bank is also called interest.

The original amount deposited or borrowed is called the **principal.** The amount of interest is a percent of the principal. The percent used to determine the amount of interest is the **interest rate.** Interest rates are given for specific periods of time, usually months or years.

Interest computed on the original principal is called **simple interest.** To calculate simple interest, multiply the principal by the interest rate per period by the number of time periods.

→ Calculate the simple interest on $1500 deposited for 2 years at an annual interest rate of 7.5%.

Principal × annual interest rate × time (in years) = interest
  1500   ×   0.075   ×   2   =   225

The interest earned is $225.

Example 1
Kamal borrowed $500 from a savings and loan association for 6 months at an annual interest rate of 7%. What is the simple interest due on the loan?

Strategy
To find the simple interest, multiply:

Principal × $\dfrac{\text{annual}}{\text{interest rate}}$ × $\dfrac{\text{time}}{\text{(in years)}}$

Solution $500 \times 0.07 \times 0.5 = 17.5$

The interest due is $17.50.

You Try It 1
A company borrowed $15,000 from a bank for 18 months at an annual interest rate of 8%. What is the simple interest due on the loan?

Your strategy

Your solution

Example 2
A credit card company charges a customer 1.5% per month on the unpaid balance of charges on the credit card. What is the interest due in a month when the customer has an unpaid balance of $54?

Strategy
To find the interest due, multiply the principal by the monthly interest rate by the time (in months).

Solution $54 \times 0.015 \times 1 = 0.81$

The interest charge is $.81.

You Try It 2
A bank offers short-term loans at a simple interest rate of 1.2% per month. What is the interest due on a short-term loan of $400 for 2 months?

Your strategy

Your solution

Solutions on p. S15

Objective B   *To calculate compound interest* ....

Usually the interest paid on money deposited or borrowed is compound interest. **Compound interest** is computed not only on the original principal but also on interest already earned. Compound interest is usually compounded annually (once a year), semiannually (twice a year), quarterly (four times a year), or daily.

$100 is invested for 3 years at an annual interest rate of 9% compounded annually. The interest earned over the 3 years is calculated by first finding the interest earned each year.

1st year	Interest earned:	$0.09 \times \$100.00 = \$9.00$
	New principal:	$\$100.00 + \$9.00 = \underline{\$109.00}$
2nd year	Interest earned:	$0.09 \times \$109.00 = \$9.81$
	New principal:	$\$109.00 + \$9.81 = \underline{\$118.81}$
3rd year	Interest earned:	$0.09 \times \$118.81 \approx \$10.69$
	New principal:	$\$118.81 + \$10.69 = \$129.50$

To find the interest earned, subtract the original principal from the new principal.

New principal	−	original principal	=	interest earned
129.50	−	100	=	29.5

The interest earned is $29.50.

Note that the compound interest earned is $29.50. The simple interest earned on the investment would have been only $100 \times 0.09 \times 3 = \$27$.

Calculating compound interest can be very tedious, so there are tables that can be used to simplify these calculations. A portion of a compound interest table is given in the Appendix.

Example 3
An investment of $650 pays 8% annual interest compounded semiannually. What is the interest earned in 5 years?

Strategy
To find the interest earned:

- Find the new principal by multiplying the original principal by the factor (1.48024) found in the compound interest table.
- Subtract the original principal from the new principal.

Solution
$650 \times 1.48024 \approx 962.16$

The new principal is $962.16.

$962.16 − 650 = 312.16$

The interest earned is $312.16.

You Try It 3
An investment of $1000 pays 6% annual interest compounded quarterly. What is the interest earned in 20 years?

Your strategy

Your solution

Solution on p. S15

6.3 Exercises

· ·

Objective A   *Application Problems*

1. To finance the purchase of 15 new cars, the Tropical Car Rental Agency borrowed $100,000 for 9 months at an annual interest rate of 9%. What is the simple interest due on the loan?

2. A home builder obtained a pre-construction loan of $50,000 for 8 months at an annual interest rate of 9.5%. What is the simple interest due on the loan? Round to the nearest cent.

3. The Mission Valley Credit Union charges its customers an interest rate of 2% per month on money that is transferred into an account that is overdrawn. Find the interest owed to the credit union for 1 month when $800 is transferred into an overdrawn account.

4. The Visa card that Francesca has charges her 1.6% per month on her unpaid balance. Find the interest owed to Visa when her unpaid balance for the month is $1250.

5. Action Machining Company purchased a robot-controlled lathe for $225,000 and financed the full amount at 8% simple annual interest for 4 years.
   a. Find the interest on the loan.
   b. Find the monthly payment.

   $$\left(\text{Monthly payment} = \frac{\text{loan amount + interest}}{\text{number of months}}\right)$$

6. For the purchase of an entertainment center, an $1800 loan is obtained for 2 years at a simple interest rate of 9.4%.
   a. Find the interest due on the loan.
   b. Find the monthly payment using the formula in part b of Exercise 5.

7. To attract new customers, Heller Ford is offering car loans at a simple interest rate of 4.5%.
   a. Find the interest charged to a customer who finances a car loan of $12,000 for 2 years.
   b. Find the monthly payment using the formula in part b of Exercise 5.

8. Cimarron Homes Inc. purchased a small plane for $57,000 and financed the full amount for 5 years at a simple annual interest rate of 9%.
   a. Find the interest due on the loan.
   b. Find the monthly payment using the formula in part b of Exercise 5.

9. Dennis Pappas decided to build onto an existing structure instead of buying a new home. He borrowed $42,000 for $3\frac{1}{2}$ years at a simple interest rate of 9.5%. Find the monthly payment.

Objective B   *Application Problems*

Solve. Use the table in the Appendix. Round to the nearest cent.

10. North Island Federal Credit Union pays 4% annual interest, compounded daily, on time savings deposits. Find the value of $750 deposited in this account after 1 year.

Solve. Use the table in the Appendix. Round to the nearest cent.

11. What is the value after 5 years of $1000 invested at 7% annual interest compounded quarterly?

12. An investment club invested $50,000 in a certificate of deposit that pays 5% annual interest compounded quarterly. Find the value of this investment after 10 years.

13. Tanya invested $2500 in a tax-sheltered annuity that pays 8% annual interest compounded daily. Find the value of her investment after 20 years.

14. Sal Trovato invested $3000 in a corporate retirement account that pays 6% annual interest compounded semiannually. Find the value of his investment after 15 years.

15. To replace equipment, a farmer invested $20,000 in an account that pays 7% annual interest compounded semiannually. What is the value of the investment after 5 years?

16. Green River Lodge invests $75,000 in a trust account that pays 8% interest compounded quarterly.
   a. What will the value of the investment be in 5 years?
   b. How much interest will be earned in the 5 years?

17. To save for retirement, a couple deposited $3000 in an account that pays 7% annual interest compounded daily.
   a. What will the value of the investment be in 10 years?
   b. How much interest will be earned in the 10 years?

18. To save for a child's college education, the Petersens deposited $2500 into an account that pays 6% annual interest compounded daily. Find the amount of interest earned in this account over a 20-year period.

APPLYING THE CONCEPTS

19. Explain the fundamental difference between simple interest and compound interest.

20. Visit a brokerage business and obtain information on annuities. Explain the advantages of an annuity as a retirement option.

21. Suppose you have a savings account that earns interest at the rate of 6% per year compounded monthly. On January 1, you open this account with a deposit of $100.
   a. On February 1, you deposit an additional $100 into the account. What is the value of the account after the deposit?
   b. On March 1, you deposit an additional $100 into the account. What is the value of the account after the deposit? *Note:* This type of savings plan, wherein equal amounts ($100) are saved at equal time intervals (every month), is called an annuity.

6.4   Real Estate Expenses

Objective A   *To calculate the initial expenses of buying a home* .... CT

One of the largest investments most people ever make is the purchase of a home. The major initial expense in the purchase is the down payment. The amount of the down payment is normally a percent of the purchase price. This percent varies among banks, but it usually ranges from 5% to 25%.

The **mortgage** is the amount that is borrowed to buy real estate. The mortgage amount is the difference between the purchase price and the down payment.

➡ A home is purchased for $140,000, and a down payment of $21,000 is made. Find the mortgage.

Purchase price	−	down payment	=	mortgage
140,000	−	21,000	=	119,000

The mortgage is $119,000.

Another large initial expense in buying a home is the loan origination fee, which is a fee that the bank charges for processing the mortgage papers. The loan origination fee is usually a percent of the mortgage and is expressed in **points,** which is the term banks use to mean percent. For example, "5 points" means "5 percent."

Points	×	mortgage	=	loan origination fee

Example 1

A house is purchased for $125,000, and a down payment, which is 20% of the purchase price, is made. Find the mortgage.

Strategy

To find the mortgage:

- Find the down payment by solving the basic percent equation for *amount*.
- Subtract the down payment from the purchase price.

Solution

Percent × base = amount

Percent	×	purchase price	=	down payment
0.20	×	125,000	=	n
		25,000	=	n

Purchase price	−	down payment	=	mortgage
125,000	−	25,000	=	100,000

The mortgage is $100,000.

You Try It 1

An office building is purchased for $216,000, and a down payment, which is 25% of the purchase price, is made. Find the mortgage.

Your strategy

Your solution

Solution on p. S15

Example 2

A home is purchased with a mortgage of $65,000. The buyer pays a loan origination fee of $3\frac{1}{2}$ points. How much is the loan origination fee?

Strategy

To find the loan origination fee, solve the basic percent equation for *amount*.

Solution

$$\text{Percent} \times \text{base} = \text{amount}$$

$$\boxed{\text{Points}} \times \boxed{\text{mortgage}} = \boxed{\text{fee}}$$

$$0.035 \times 65,000 = n$$

$$2275 = n$$

The loan origination fee is $2275.

You Try It 2

The mortgage on a real estate investment is $80,000. The buyer paid a loan origination fee of $4\frac{1}{2}$ points. How much was the loan origination fee?

Your strategy

Your solution

Solution on p. S15

Objective B   ***To calculate ongoing expenses of owning a home*** ....

Besides the initial expenses of buying a home, there are continuing monthly expenses involved in owning a home. The monthly mortgage payment, utilities, insurance, and taxes are some of these ongoing expenses. Of these expenses, the largest one is normally the monthly mortgage payment.

For a fixed-rate mortgage, the monthly mortgage payment remains the same throughout the life of the loan. The calculation of the monthly mortgage payment is based on the amount of the loan, the interest rate on the loan, and the number of years required to pay back the loan. Calculating the monthly mortgage payment is fairly difficult, so tables such as the one in the Appendix are used to simplify these calculations.

$\Rightarrow$ Find the monthly mortgage payment on a 30-year $60,000 mortgage at an interest rate of 9%. Use the monthly payment table in the Appendix.

$$60,000 \times \underset{\substack{\downarrow \\ \text{from the} \\ \text{table}}}{0.0080462} \approx 482.77$$

The monthly mortgage payment is $482.77.

The monthly mortgage payment includes the payment of both principal and interest on the mortgage. The interest charged during any one month is charged on the unpaid balance of the loan. Therefore, during the early years of the mortgage, when the unpaid balance is high, most of the monthly mortgage payment is interest charged on the loan. During the last few years of a mortgage, when the unpaid balance is low, most of the monthly mortgage payment goes toward paying off the loan.

➡ Find the interest paid on a mortgage during a month when the monthly mortgage payment is $186.26 and $58.08 of that amount goes toward paying off the principal.

Monthly mortgage payment	−	principal	=	interest
186.26	−	58.08	=	128.18

The interest paid on the mortgage is $128.18.

Property tax is another ongoing expense of owning a house. Property tax is normally an annual expense that may be paid on a monthly basis. The monthly property tax, which is determined by dividing the annual property tax by 12, is usually added to the monthly mortgage payment.

➡ A homeowner must pay $534 in property tax annually. Find the property tax that must be added each month to the homeowner's monthly mortgage payment.

$$534 \div 12 = 44.5$$

Each month, $44.50 must be added to the monthly mortgage payment for property tax.

Example 3

Serge purchased some land for $120,000 and made a down payment of $25,000. The savings and loan association charges an annual interest rate of 8% on Serge's 25-year mortgage. Find the monthly mortgage payment.

Strategy

To find the monthly mortgage payment:

- Subtract the down payment from the purchase price to find the mortgage.
- Multiply the mortgage by the factor found in the monthly payment table in the Appendix.

Solution

Purchase price	−	down payment	=	mortgage
120,000	−	25,000	=	95,000

$$95,000 \quad \times \quad 0.0077182 \quad \approx \quad 733.23$$
$$\uparrow$$
from the table

The monthly mortgage payment is $733.23.

You Try It 3

A new condominium project is selling townhouses for $75,000. A down payment of $15,000 is required, and a 20-year mortgage at an annual interest rate of 9% is available. Find the monthly mortgage payment.

Your strategy

Your solution

Solution on p. S15

Example 4

A home has a mortgage of $134,000 for 25 years at an annual interest rate of 7%. During a month when $375.88 of the monthly mortgage payment is principal, how much of the payment is interest?

Strategy

To find the interest:

- Multiply the mortgage by the factor found in the monthly payment table in the Appendix to find the monthly mortgage payment.
- Subtract the principal from the monthly mortgage payment.

Solution

$134,000 \times 0.0070678 \approx 947.09$

$\qquad\quad\uparrow\qquad\qquad\quad\uparrow$

$\qquad$ from the $\qquad$ monthly mortgage
$\qquad$ table $\qquad\qquad$ payment

$$\boxed{\begin{array}{c}\text{Monthly}\\\text{mortgage}\\\text{payment}\end{array}} - \boxed{\begin{array}{c}\text{prin-}\\\text{cipal}\end{array}} = \boxed{\text{interest}}$$

$\quad$ 947.09 $\quad - \quad$ 375.88 $\quad = \quad$ 571.21

$571.21 is interest on the mortgage.

You Try It 4

An office building has a mortgage of $125,000 for 25 years at an annual interest rate of 9%. During a month when $492.65 of the monthly mortgage payment is principal, how much of the payment is interest?

Your strategy

Your solution

Example 5

The monthly mortgage payment for a home is $598.75. The annual property tax is $900. Find the total monthly payment for the mortgage and property tax.

Strategy

To find the monthly payment:

- Divide the annual property tax by 12 to find the monthly property tax.
- Add the monthly property tax to the monthly mortgage payment.

Solution

$900 \div 12 = 75$ $\quad$ (monthly property tax)

$598.75 + 75 = 673.75$

The total monthly payment is $673.75.

You Try It 5

The monthly mortgage payment for a home is $415.20. The annual property tax is $744. Find the total monthly payment for the mortgage and property tax.

Your strategy

Your solution

Solutions on p. S16

6.4 Exercises

· ·

Objective A *Application Problems*

---

1. A condominium at Mt. Baldy Ski Resort was purchased for $97,000, and a down payment of $14,550 was made. Find the mortgage.

2. An insurance business was purchased for $173,000, and a down payment of $34,600 was made. Find the mortgage.

3. A building lot was purchased for $25,000. The lender requires a down payment of 30% of the purchase price. Find the down payment.

4. Ian Goldman purchased a new home for $88,500. The lender requires a down payment of 20% of the purchase price. Find the down payment.

5. Brian Stedman made a down payment of 25% of the $850,000 purchase price of an apartment building. How much was the down payment?

6. A clothing store was purchased for $125,000, and a down payment that was 25% of the purchase price was made. How much was the down payment?

7. A loan of $150,000 is obtained to purchase a home. The loan origination fee is $2\frac{1}{2}$ points. Find the amount of the loan origination fee.

8. Security Savings & Loan requires a borrower to pay $3\frac{1}{2}$ points for a loan. Find the amount of the loan origination fee for a loan of $90,000.

9. Baja Construction Inc. is selling homes for $150,000. A down payment of 5% is required.
   a. Find the down payment.
   b. Find the mortgage.

10. A cattle rancher purchased some land for $240,000. The bank requires a down payment of 15% of the purchase price.
   a. Find the down payment.
   b. Find the mortgage.

11. Vivian Tom purchased a home for $210,000. Find the mortgage if the down payment Vivian made is 10% of the purchase price.

12. A mortgage lender requires a down payment of 5% of the $80,000 purchase price of a condominium. How much is the mortgage?

Objective B *Application Problems*

Solve. Use the monthly payment table in the Appendix. Round to the nearest cent.

13. An investor obtained a loan of $150,000 to buy a car wash business. The monthly mortgage payment was based on 25 years at 8%. Find the monthly mortgage payment.

14. A beautician obtained a 20-year mortgage of $90,000 to expand the business. The credit union charges an annual interest rate of 9%. Find the monthly mortgage payment.

Solve. Use the monthly payment table in the Appendix. Round to the nearest cent.

15. A couple interested in buying a home determines that they can afford a monthly mortgage payment of $800. Can they afford to buy a home with a 30-year $110,000 mortgage at 8% interest?

16. A lawyer is considering purchasing a new office building with a 20-year $400,000 mortgage at 9% interest. The lawyer can afford a monthly mortgage payment of $4000. Can the lawyer afford the monthly mortgage payment on the new office building?

17. The county tax assessor has determined that the annual property tax on a $125,000 house is $1348.20. Find the monthly property tax.

18. The annual property tax on a $155,000 home is $1992. Find the monthly property tax.

19. Abacus Imports Inc. has a warehouse with a 25-year mortgage of $200,000 at an annual interest rate of 9%.
    a. Find the monthly mortgage payment.
    b. During a month when $941.72 of the monthly mortgage payment is principal, how much of the payment is interest?

20. A vacation home has a mortgage of $135,000 for 30 years at an annual interest rate of 7%.
    a. Find the monthly mortgage payment.
    b. During a month when $392.47 of the monthly mortgage payment is principal, how much of the payment is interest?

21. The annual mortgage payment on a duplex is $10,844.40. The owner must pay an annual property tax of $948. Find the total monthly payment for the mortgage and property tax.

22. The monthly mortgage payment on a home is $716.40, and the homeowner pays an annual property tax of $792. Find the total monthly payment for the mortgage and property tax.

23. Maria Hernandez purchased a home for $210,000 and made a down payment of $15,000. The balance was financed for 30 years at an annual interest rate of 8%. Find the monthly mortgage payment.

24. A customer of a savings and loan purchased a $185,000 home and made a down payment of $20,000. The savings and loan charges its customers an annual interest rate of 7% for 30 years for a home mortgage. Find the monthly mortgage payment.

APPLYING THE CONCEPTS

25. A couple considering a mortgage of $100,000 have a choice of loans. One loan is an 8% loan for 20 years, and the other loan is at 8% for 30 years. Find the amount of interest that can be saved by choosing the 20-year loan.

26. Find out what an adjustable-rate mortgage is. What is the difference between this type of loan and a fixed-rate mortgage? List some of the advantages and disadvantages of each.

6.5   Car Expenses

Objective A   *To calculate the initial expenses of buying a car* ·································· 🖫 CT

The initial expenses in the purchase of a car usually include the down payment, the license fees, and the sales tax. The down payment may be very small or as much as 25% or 30% of the purchase price of the car, depending on the lending institution. License fees and sales tax are regulated by each state, so these expenses vary from state to state.

Example 1

A car is purchased for $8500, and the lender requires a down payment of 15% of the purchase price. Find the amount financed.

Strategy

To find the amount financed:

- Find the down payment by solving the basic percent equation for *amount*.
- Subtract the down payment from the purchase price.

Solution

Percent   ×   base   =   amount

Percent	×	purchase price	=	down payment
0.15	×	8500	=	n

$$1275 = n$$

$$8500 - 1275 = 7225$$

The amount financed is $7225.

You Try It 1

A down payment of 20% of the $9200 purchase price of a new car is made. Find the amount financed.

Your strategy

Your solution

Example 2

A sales clerk purchases a car for $6500 and pays a sales tax that is 5% of the purchase price. How much is the sales tax?

Strategy

To find the sales tax, solve the basic percent equation for *amount*.

Solution

Percent   ×   base   = amount

Percent	×	purchase price	=	sales tax
0.05	×	6500	=	n

$$325 = n$$

The sales tax is $325.

You Try It 2

A car is purchased for $7350. The car license fee is 1.5% of the purchase price. How much is the license fee?

Your strategy

Your solution

Solutions on p. S16

Objective B   *To calculate ongoing expenses of owning a car* ....

Besides the initial expenses of buying a car, there are continuing expenses involved in owning a car. These ongoing expenses include car insurance, gas and oil, general maintenance, and the monthly car payment. The monthly car payment is calculated in the same manner as monthly mortgage payments on a home loan. A monthly payment table, such as the one in the Appendix, is used to simplify the calculation of monthly car payments.

Example 3
At a cost of $.27 per mile, how much does it cost to operate a car during a year in which the car is driven 15,000 miles?

Strategy
To find the cost, multiply the cost per mile by the number of miles driven.

Solution
$15,000 \times 0.27 = 4050$ The cost is $4050.

You Try It 3
At a cost of $.22 per mile, how much does it cost to operate a car during a year in which the car is driven 23,000 miles?

Your strategy

Your solution

Example 4
During 1 month the total gasoline bill was $84 and the car was driven 1200 miles. What was the cost per mile for gasoline?

Strategy
To find the cost per mile for gasoline, divide the cost for gasoline by the number of miles driven.

Solution $84 \div 1200 = 0.07$

The cost per mile was $.07.

You Try It 4
In a year in which the total car insurance bill was $360 and the car was driven 15,000 miles, what was the cost per mile for car insurance?

Your strategy

Your solution

Example 5
A car is purchased for $8500 with a down payment of $1700. The balance is financed for 3 years at an annual interest rate of 9%. Find the monthly car payment.

Strategy
To find the monthly payment:

• Subtract the down payment from the purchase price to find the amount financed.
• Multiply the amount financed by the factor found in the monthly payment table in the Appendix.

Solution
$8500 - 1700 = 6800$

The amount financed is $6800.

$6800 \times 0.0317997 \approx 216.24$

The monthly payment is $216.24.

You Try It 5
A truck is purchased for $15,900 with a down payment of $3975. The balance is financed for 4 years at an annual interest rate of 8%. Find the monthly payment.

Your strategy

Your solution

Solutions on p. S16

6.5 Exercises

- -

Objective A   *Application Problems*

1. Amanda has saved $780 to make a down payment on a car. The car dealer requires a down payment of 12% of the purchase price. Does she have enough money to make the down payment on a used minivan that costs $7100?

2. A Ford Ranger was purchased for $23,500. A down payment of 15% of the purchase price was required. How much was the down payment?

3. A drapery installer bought a minivan to carry drapery samples. The purchase price of the van was $16,500, and a 4.5% sales tax was paid. How much was the sales tax?

4. A delivery truck for the Dixieline Lumber Company was purchased for $18,500. A sales tax of 4% of the purchase price was paid. Find the sales tax.

5. A license fee of 2% of the purchase price of a truck is to be paid on a pickup truck costing $12,500. How much is the license fee for the truck?

6. Your state charges a license fee of 1.5% on the purchase price of a car. How much is the license fee for a car that costs $6998?

7. An electrician bought a $12,000 flatbed truck. A state license fee of $175 and a sales tax of 3.5% of the purchase price are required.
   a. Find the sales tax.
   b. Find the total cost of the sales tax and the license fee.

8. A physical therapist bought a car for $9375 and made a down payment of $1875. The sales tax is 5% of the purchase price.
   a. Find the sales tax.
   b. Find the total cost of the sales tax and the down payment.

9. Martin bought a motorcycle for $2200 and made a down payment that is 25% of the purchase price.
   a. Find the down payment.
   b. Find the amount financed.

10. A carpenter bought a utility van for $14,900 and made a down payment that is 15% of the purchase price.
    a. Find the down payment.
    b. Find the amount financed.

11. An author bought a sports car for $35,000 and made a down payment of 20% of the purchase price. Find the amount financed.

12. Tania purchased a new car for $13,500 and made a down payment of 25% of the cost. Find the amount financed.

Objective B   *Application Problems*

Solve. Use the monthly payment table in the Appendix. Round to the nearest cent.

13. A rancher financed $14,000 for the purchase of a truck through a credit union at 9% interest for 4 years. Find the monthly truck payment.

14. A car loan of $8000 is financed for 3 years at an annual interest rate of 10%. Find the monthly car payment.

15. An estimate of the cost of owning a compact car is $.32 per mile. Using this estimate, how much does it cost to operate a car during a year in which the car is driven 16,000 miles?

16. An estimate of the cost of care and maintenance of automobile tires is $.015 per mile. Using this estimate, how much would it cost for care and maintenance of tires during a year in which the car is driven 14,000 miles?

17. A family spent $1600 on gas, oil, and car insurance during a period in which the car was driven 14,000 miles. Find the cost per mile for gas, oil, and car insurance.

18. Last year you spent $1050 for gasoline for your car. The car was driven 15,000 miles. What was your cost per mile for gasoline?

19. Elena's monthly car payment is $143.50. During a month in which $68.75 of the monthly payment is principal, how much of the payment is interest?

20. The cost for a pizza delivery truck for the year included $1870 in truck payments, $1200 for gasoline, and $675 for insurance. Find the total cost for truck payments, gasoline, and insurance for the year.

21. The city of Colton purchased a fire truck for $82,000 and made a down payment of $5400. The balance is financed for 5 years at an annual rate of 9%.
a. Find the amount financed.
b. Find the monthly truck payment.

22. A used car is purchased for $4995, and a down payment of $995 is made. The balance is financed for 3 years at an interest rate of 8%.
a. Find the amount financed.
b. Find the monthly car payment.

23. An artist purchased a new car costing $27,500 and made a down payment of $5500. The balance is financed for 3 years at an annual interest rate of 10%. Find the monthly car payment.

24. A half-ton truck with a camper is purchased for $19,500, and a down payment of $2500 is made. The balance is financed for 4 years at an annual interest rate of 9%. Find the monthly payment.

APPLYING THE CONCEPTS

25. One bank offers a 4-year car loan at an annual interest rate of 10% plus a loan application fee of $45. A second bank offers 4-year car loans at an annual interest rate of 11% but charges no loan application fee. If you need to borrow $5800 to purchase a car, which of the two bank loans has the lesser loan costs? Assume you keep the car for 4 years.

26. How much interest is paid on a 5-year car loan of $9000 if the interest rate is 9%?

6.6 Wages

Objective A **To calculate commissions, total hourly wages, and salaries** ....

Commissions, hourly wage, and salary are three ways to receive payment for doing work.

Commissions are usually paid to salespersons and are calculated as a percent of total sales.

➡ As a real estate broker, Emma Smith receives a commission of 4.5% of the selling price of a house. Find the commission she earned for selling a home for $75,000.

To find the commission Emma earned, solve the basic percent equation for *amount*.

Percent	×	base	=	amount
Commission rate	×	total sales	=	commission
0.045	×	75,000	=	3375

The commission is $3375.

An employee who receives an **hourly wage** is paid a certain amount for each hour worked.

➡ A plumber receives an hourly wage of $13.25. Find the plumber's total wages for working 37 hours.

To find the plumber's total wages, multiply the hourly wage by the number of hours worked.

Hourly wage	×	number of hours worked	=	total wages
13.25	×	37	=	490.25

The plumber's total wages for working 37 hours are $490.25.

An employee who is paid a **salary** receives payment based on a weekly, biweekly (every other week), monthly, or annual time schedule. Unlike the employee who receives an hourly wage, the salaried worker does not receive additional pay for working more than the regularly scheduled workday.

➡ Ravi Basar is a computer operator who receives a weekly salary of $395. Find his salary for 1 month (4 weeks).

To find Ravi's salary for 1 month, multiply the salary per pay period by the number of pay periods.

Salary per pay period	×	number of pay periods	=	total salary
395	×	4	=	1580

Ravi's total salary for 1 month is $1580.

Example 1

A pharmacist's hourly wage is $28. On Saturday, the pharmacist earns time and a half ($1\frac{1}{2}$ times the regular hourly wage). How much does the pharmacist earn for working 6 hours on Saturday?

Strategy

To find the pharmacist's earnings:

- Find the hourly wage for working on Saturday by multiplying the hourly wage by $1\frac{1}{2}$.
- Multiply the hourly wage by the number of hours worked.

Solution

$28 \times 1.5 = 42 \qquad 42 \times 6 = 252$

The pharmacist earns $252.

Example 2

An efficiency expert received a contract for $3000. The consultant spent 75 hours on the project. Find the consultant's hourly wage.

Strategy

To find the hourly wage, divide the total earnings by the number of hours worked.

Solution

$3000 \div 75 = 40$

The hourly wage was $40.

Example 3

Dani Greene earns $18,500 per year plus a $5\frac{1}{2}\%$ commission on sales over $100,000.

During one year, Dani sold $150,000 worth of computers. Find Dani's total earnings for the year.

Strategy

To find the total earnings:

- Find the commission earned by multiplying the commission rate by sales over $100,000.
- Add the commission to the annual pay.

Solution

$150,000 - 100,000 = 50,000$
$50,000 \times 0.055 = 2750$ commission
$18,500 + 2750 = 21,250$

Dani earned $21,250.

You Try It 1

A construction worker's hourly wage is $8.50. The worker earns double time (2 times the regular hourly wage) for working overtime. How much does the worker earn for working 8 hours of overtime?

Your strategy

Your solution

You Try It 2

A contractor for a bridge project receives an annual salary of $28,224. What is the contractor's salary per month?

Your strategy

Your solution

You Try It 3

An insurance agent earns $12,000 per year plus a $9\frac{1}{2}\%$ commission on sales over $50,000. During one year, the agent's sales totaled $175,000. Find the agent's total earnings for the year.

Your strategy

Your solution

Solutions on pp. S16–S17

6.6 Exercises

. . . .

Objective A *Application Problems*

1. Lewis works in a clothing store and earns $7.50 per hour. How much does he earn in a 40-hour week?

2. Sasha pays a gardener an hourly wage of $9. How much does she pay the gardener for working 25 hours?

3. A real estate agent receives a 3% commission for selling a house. Find the commission that the agent earned for selling a house for $131,000.

4. Ron Caruso works as an insurance agent and receives a commission of 40% of the first year's premium. Find Ron's commission for selling a life insurance policy with a first-year premium of $1050.

5. A stockbroker receives a commission of 1.5% of the price of stock that is bought or sold. Find the commission on 100 shares of stock that were bought for $5600.

6. The owner of the Carousel Art Gallery receives a commission of 20% on paintings that are sold on consignment. Find the commission on a painting that sold for $22,500.

7. Keisha Brown receives an annual salary of $38,928 as an Italian Language teacher. How much does Keisha receive each month?

8. An apprentice plumber receives an annual salary of $27,900. How much does the plumber receive per month?

9. An electrician's hourly wage is $15.80. For working overtime, the electrician earns double time. What is the electrician's hourly wage for working overtime?

10. Carlos receives a commission of 12% of his weekly sales as a sales representative for a medical supply company. Find the commission he earned during a week in which sales were $4500.

11. A golf pro receives a commission of 25% for selling a golf set. Find what commission the pro earned for selling a golf set costing $450.

12. Steven receives $1.75 per square yard to install carpet. How much does he receive for installing 160 square yards of carpet?

13. A typist charges $1.75 per page for typing technical material. How much does the typist earn for typing a 225-page book?

14. A nuclear chemist received $15,000 in consulting fees while working on a nuclear power plant. The chemist worked 120 hours on the project. Find the consultant's hourly wage.

15. Maxine received $3400 for working on a project as a computer consultant for 40 hours. Find her hourly wage.

16. Gil Stratton's hourly wage is $10.78. For working overtime, he receives double time.
   a. What is Gil's hourly wage for working overtime?
   b. How much does he earn for working 16 hours of overtime?

17. Mark is a lathe operator and receives an hourly wage of $12.90. When working on Saturday, he receives time and a half.
   a. What is Mark's hourly wage on Saturday?
   b. How much does he earn for working 8 hours on Saturday?

18. A stock clerk at a supermarket earns $8.20 an hour. For working the night shift, the clerk's wage increases by 15%.
   a. What is the increase in hourly pay for working the night shift?
   b. What is the clerk's hourly wage for working the night shift?

19. A nurse earns $16.50 an hour. For working the night shift, the nurse receives a 10% increase in pay.
   a. What is the increase in hourly pay for working the night shift?
   b. What is the hourly pay for working the night shift?

20. Tony's hourly wage as a service station attendant is $6.40. For working the night shift, his wage is increased 25%. What is Tony's hourly wage for working the night shift?

21. Nicole Tobin, a door-to-door salesperson, receives a salary of $150 per week plus a commission of 15% on all sales over $1500. Find her earnings during a week in which sales totaled $3000.

APPLYING THE CONCEPTS

The table at the right shows the top five and bottom five starting salaries for recent college graduates (Source: Michigan State University). Use this table for Exercises 22 to 25.

22. What was the amount of increase in the starting salary for journalism majors from the previous year? Round to the nearest dollar.

23. What was the starting salary in the previous year for an industrial engineer? Round to the nearest dollar.

24. Between electrical engineers and computer scientists, which received the larger amount of increase in starting salary from the previous year?

25. Between telecommunications majors and liberal arts majors, which received the smaller amount of increase in starting salary from the previous year?

Estimated Starting Salaries
Top 5

Bachelor's Degree	Estimated Starting Salary	% Change from Previous Year
Chemical Engineering	$42,758	4.3%
Mechanical Engineering	$39,852	4.5%
Electrical Engineering	$38,811	4.0%
Industrial Engineering	$37,732	4.0%
Computer Science	$36,964	4.5%

Bottom 5

Liberal Arts	$24,102	3.5%
Natural Resources	$22,950	3.5%
Home Economics	$22,916	3.5%
Telecommunications	$22,447	4.0%
Journalism	$22,102	4.0%

6.7 Bank Statements

Objective A **To calculate checkbook balances** ....

A checking account can be opened at most banks or savings and loan associations by depositing an amount of money in the bank. A checkbook contains checks and deposit slips and a checkbook register in which to record checks written and amounts deposited in the checking account. Each time a check is written, the amount of the check is subtracted from the amount in the account. When a deposit is made, the amount deposited is added to the amount in the account.

A portion of a checkbook register is shown below. The account holder had a balance of $587.93 before writing two checks, one for $286.87 and the other for $102.38, and making one deposit of $345.00.

| | | RECORD ALL CHARGES OR CREDITS THAT AFFECT YOUR ACCOUNT | | | | | BALANCE | |
NUMBER	DATE	DESCRIPTION OF TRANSACTION	PAYMENT/DEBIT (-)	√ T	FEE (IF ANY) (-)	DEPOSIT/CREDIT (+)	$ 587	93
108	8/4	*Plumber*	$ 286 87		$	$	301	06
109	8/10	*Car Payment*	102 38				198	68
	8/14	*Deposit*				345 00	543	68

To find the current checking account balance, subtract the amount of each check from the previous balance. Then add the amount of the deposit.

The current checking account balance is $543.68.

Example 1 A mail carrier had a checking account balance of $485.93 before writing two checks, one for $18.98 and another for $35.72, and making a deposit of $250. Find the current checking account balance.

You Try It 1 A cement mason had a checking account balance of $302.46 before writing a check for $20.59 and making two deposits, one in the amount of $176.86 and another in the amount of $94.73. Find the current checking account balance.

Strategy To find the current balance:
- Subtract the amount of each check from the old balance.
- Add the amount of the deposit.

Your strategy

Solution

```
  485.93
−  18.98   first check
  466.95
−  35.72   second check
  431.23
+ 250.00   deposit
  681.23
```

The current checking account balance is $681.23.

Your solution

Solution on p. S17

Objective B   To balance a checkbook ....

Each month a bank statement is sent to the account holder. The bank statement shows the checks that the bank has paid, the deposits received, and the current bank balance.

A bank statement and checkbook register are shown on the next page.

Balancing the checkbook, or determining if the checking account balance is accurate, requires a number of steps.

   1. In the checkbook register, put a check mark (✓) by each check paid by the bank and each deposit recorded by the bank.

		RECORD ALL CHARGES OR CREDITS THAT AFFECT YOUR ACCOUNT							BALANCE	
NUMBER	DATE	DESCRIPTION OF TRANSACTION	PAYMENT/DEBIT (-)		√ T	FEE (IF ANY) (-)	DEPOSIT/CREDIT (+)		$ 840	27
263	5/20	Dentist	$ 25	00	√	$	$		815	27
264	5/22	Meat Market	33	61	√				781	66
265	5/22	Gas Company	67	14					714	52
	5/29	Deposit			√		192	00	906	52
266	5/29	Pharmacy	18	95	√				887	57
267	5/30	Telephone	43	85					843	72
268	6/2	Groceries	43	19	√				800	53
	6/3	Deposit			√		215	00	1015	53
269	6/7	Insurance	103	00	√				912	53
	6/10	Deposit					225	00	1137	53
270	6/15	Clothing Store	16	63	√				1120	90
271	6/18	Newspaper	7	00					1113	90

CHECKING ACCOUNT Monthly Statement   Account Number: 924-297-8

Date	Transaction	Amount	Balance
5/20	OPENING BALANCE		840.27
5/21	CHECK	25.00	815.27
5/23	CHECK	33.61	781.66
5/29	DEPOSIT	192.00	973.66
6/1	CHECK	18.95	954.71
6/1	INTEREST	4.47	959.18
6/3	CHECK	43.19	915.99
6/3	DEPOSIT	215.00	1130.99
6/9	CHECK	103.00	1027.99
6/16	CHECK	16.63	1011.36
6/20	SERVICE CHARGE	3.00	1008.36
6/20	CLOSING BALANCE		1008.36

2. Add to the current checkbook balance all checks that have been written but have not yet been paid by the bank and any interest paid on the account.

3. Subtract any service charges and any deposits not yet recorded by the bank. This is the checkbook balance.

4. Compare the balance with the bank balance listed on the bank statement. If the two numbers are equal, the bank statement and checkbook balance.

Current checkbook balance:	1113.90
Checks: 265	67.14
267	43.85
271	7.00
Interest:	+ 4.47
	1236.36
Service charge:	− 3.00
	1233.36
Deposit:	− 225.00
Checkbook balance:	1008.36

Checking bank balance from bank statement   Checkbook balance

$1008.36   = $1008.36

The bank statement and checkbook balance.

		RECORD ALL CHARGES OR CREDITS THAT AFFECT YOUR ACCOUNT									
NUMBER	DATE	DESCRIPTION OF TRANSACTION	PAYMENT/DEBIT (-)		√ T	FEE (IF ANY) (-)	DEPOSIT/CREDIT (+)		$	BALANCE 1620	42
413	3/2	Car Payment	$ 132	15	√	$	$			1488	27
414	3/2	Utility	67	14	√					1421	13
415	3/5	Restaurant - Dinner for 4	78	14						1342	99
	3/8	Deposit			√		1842	66		3185	65
416	3/10	House Payment	672	14	√					2513	51
417	3/14	Insurance	177	10						2336	41

CHECKING ACCOUNT Monthly Statement		Account Number: 924-297-8	
Date	Transaction	Amount	Balance
3/1	OPENING BALANCE		1620.42
3/4	CHECK	132.15	1488.27
3/5	CHECK	67.14	1421.13
3/8	DEPOSIT	1842.66	3263.79
3/10	INTEREST	6.77	3270.56
3/12	CHECK	672.14	2598.42
3/25	SERVICE CHARGE	2.00	2596.42
3/30	CLOSING BALANCE		2596.42

Balance the bank statement shown above.

1. In the checkbook register, put a check mark (√) by each check paid by the bank and each deposit recorded by the bank.

2. Add to the current checkbook balance all checks that have been written but have not yet been paid by the bank and any interest paid on the account.

3. Subtract any service charges and any deposits not yet recorded by the bank. This is the checkbook balance.

4. Compare the balance with the bank balance listed on the bank statement. If the two numbers are equal, the bank statement and checkbook balance.

Current checkbook balance:	2336.41
Checks: 415	78.14
417	177.10
Interest:	+ 6.77
	2598.42
Service charge:	− 2.00
Checkbook balance:	2596.42

Closing bank balance from bank statement	Checkbook balance
$2596.42	= $2596.42

The bank statement and checkbook balance.

RECORD ALL CHARGES OR CREDITS THAT AFFECT YOUR ACCOUNT

NUMBER	DATE	DESCRIPTION OF TRANSACTION	PAYMENT/DEBIT (-)	√ T	FEE (IF ANY) (-)	DEPOSIT/CREDIT (+)	BALANCE $ 412 64	
345	1/14	Phone Bill	$ 34 75	√	$	$	377	89
346	1/19	Magazine	8 98	√			368	91
347	1/23	Theatre Tickets	45 00				323	91
	1/31	Deposit		√		947 00	1270	91
348	2/5	Cash	250 00	√			1020	91
349	2/12	Rent	440 00				580	91

CHECKING ACCOUNT Monthly Statement			Account Number: 924-297-8
Date	Transaction	Amount	Balance
1/10	OPENING BALANCE		412.64
1/18	CHECK	34.75	377.89
1/23	CHECK	8.98	368.91
1/31	DEPOSIT	947.00	1315.91
2/1	INTEREST	4.52	1320.43
2/10	CHECK	250.00	1070.43
2/10	CLOSING BALANCE		1070.43

Example 2

Balance the bank statement shown above.

Solution

Current checkbook balance:	580.91
Checks: 347	45.00
349	440.00
Interest:	+ 4.52
	1070.43
Service charge:	− 0.00
	1070.43
Deposit:	− 0.00
Checkbook balance:	1070.43

Closing bank balance from bank statement: $1070.43

Checkbook balance: $1070.43

The bank statement and checkbook balance.

RECORD ALL CHARGES OR CREDITS THAT AFFECT YOUR ACCOUNT

NUMBER	DATE	DESCRIPTION OF TRANSACTION	PAYMENT/DEBIT (-)		√ T	FEE (IF ANY) (-)	DEPOSIT/CREDIT (+)		BALANCE $ 603	17
	2/15	*Deposit*	$			$	$ 523	84	1127	01
234	2/20	*Mortgage*	473	21					653	80
235	2/27	*Cash*	200	00					453	80
	3/1	*Deposit*					523	84	977	64
236	3/12	*Insurance*	275	50					702	14
237	3/12	*Telephone*	48	73					653	41

CHECKING ACCOUNT Monthly Statement   Account Number: 314-271-4

Date	Transaction	Amount	Balance
2/14	OPENING BALANCE		603.17
2/15	DEPOSIT	523.84	1127.01
2/21	CHECK	473.21	653.80
2/28	CHECK	200.00	453.80
3/1	INTEREST	2.11	455.91
3/14	CHECK	275.50	180.41
3/14	CLOSING BALANCE		180.41

You Try It 2

Balance the bank statement shown above.

Your solution

Solution on p. S17

6.7 Exercises

Objective A   *Application Problems*

1. You had a checking account balance of $342.51 before making a deposit of $143.81. What is your new checking account balance?

2. Carmen had a checking account balance of $493.26 before writing a check for $48.39. What is the current checking account balance?

3. A real estate firm had a balance of $2431.76 in its rental property checking account. What is the balance in this account after a check for $1209.29 has been written?

4. The business checking account for R and R Tires showed a balance of $1536.97. What is the balance in this account after a deposit of $439.21 has been made?

5. A nutritionist had a checking account balance of $1204.63 before writing one check for $119.27 and another check for $260.09. Find the current checkbook balance.

6. Sam had a checking account balance of $3046.93 before writing a check for $1027.33 and making a deposit of $150.00. Find the current checkbook balance.

7. The business checking account for Rachael's Dry Cleaning had a balance of $3476.85 before a deposit of $1048.53 was made. The store manager then wrote checks, one for $848.37 and another for $676.19. Find the current checkbook balance.

8. Joel had a checking account balance of $427.38 before a deposit of $127.29 was made. Joel then wrote two checks, one for $43.52 and one for $249.78. Find the current checkbook balance.

9. A carpenter had a checkbook balance of $104.96 before making a deposit of $350 and writing a check for $71.29. Is there enough money in the account to purchase a refrigerator for $375?

10. A taxi driver had a checkbook balance of $149.85 before making a deposit of $245 and writing a check for $387.68. Is there enough money in the account for the bank to pay the check?

11. A sporting goods store has the opportunity to buy downhill skis and cross-country skis at a manufacturer's closeout sale. The downhill skis will cost $3500, and the cross-country skis will cost $2050. There is currently $5625.42 in the sporting goods store's checking account. Is there enough money in the account to make both purchases by check?

12. A lathe operator's current checkbook balance is $643.42. The operator wants to purchase a utility trailer for $225 and a used piano for $450. Is there enough money in the account to make the two purchases?

Objective B   *Application Problems*

13. Balance the checkbook.

RECORD ALL CHARGES OR CREDITS THAT AFFECT YOUR ACCOUNT

NUMBER	DATE	DESCRIPTION OF TRANSACTION	PAYMENT/DEBIT (-)	√ T	FEE (IF ANY) (-)	DEPOSIT/CREDIT (+)	BALANCE $ 466 79
223	3/2	Groceries	$ 67 32		$	$	399 47
	3/5	Deposit				560 70	960 17
224	3/5	Rent	460 00				500 17
225	3/7	Gas & Electric	42 35				457 82
226	3/7	Cash	100 00				357 82
227	3/7	Insurance	118 44				239 38
228	3/7	Credit Card	119 32				120 06
229	3/12	Dentist	42 00				78 06
230	3/13	Drug Store	17 03				61 03
	3/19	Deposit				560 70	621 73
231	3/22	Car Payment	141 35				480 38
232	3/25	Cash	100 00				380 38
233	3/25	Oil Company	66 40				313 98
234	3/28	Plumber	55 73				258 25
235	3/29	Department Store	88 39				169 86

CHECKING ACCOUNT Monthly Statement   Account Number: 122-345-1

Date	Transaction	Amount	Balance
3/1	OPENING BALANCE		466.79
3/5	DEPOSIT	560.70	1027.49
3/7	CHECK	67.32	960.17
3/8	CHECK	460.00	500.17
3/8	CHECK	100.00	400.17
3/9	CHECK	42.35	357.82
3/12	CHECK	118.44	239.38
3/14	CHECK	42.00	197.38
3/18	CHECK	17.03	180.35
3/19	DEPOSIT	560.70	741.05
3/25	CHECK	141.35	599.70
3/27	CHECK	100.00	499.70
3/29	CHECK	55.73	443.97
3/30	INTEREST	13.22	457.19
4/1	CLOSING BALANCE		457.19

14. Balance the checkbook.

RECORD ALL CHARGES OR CREDITS THAT AFFECT YOUR ACCOUNT

NUMBER	DATE	DESCRIPTION OF TRANSACTION	PAYMENT/DEBIT (-)		√ T	FEE (IF ANY) (-)	DEPOSIT/CREDIT (+)		BALANCE $ 219	43
	5/1	Deposit	$			$	$ 219	14	438	57
515	5/2	Electric Bill	22	35					416	22
516	5/2	Groceries	55	14					361	08
517	5/4	Insurance	122	17					238	91
518	5/5	Theatre Tickets	24	50					214	41
	5/8	Deposit					219	14	433	55
519	5/10	Telephone	17	39					416	16
520	5/12	Newspaper	12	50					403	66
	5/15	Interest					7	82	411	48
	5/15	Deposit					219	14	630	62
521	5/20	Hotel	172	90					457	72
522	5/21	Credit Card	113	44					344	28
523	5/22	Eye Exam	42	00					302	28
524	5/24	Groceries	77	14					225	14
525	5/24	Deposit					219	14	444	28
526	5/25	Oil Company	44	16					400	12
527	5/30	Car Payment	88	62					311	50
528	5/30	Doctor	37	42					274	08

CHECKING ACCOUNT Monthly Statement   Account Number: 122-345-1

Date	Transaction	Amount	Balance
5/1	OPENING BALANCE		219.43
5/1	DEPOSIT	219.14	438.57
5/3	CHECK	55.14	383.43
5/4	CHECK	22.35	361.08
5/6	CHECK	24.50	336.58
5/8	CHECK	122.17	214.41
5/8	DEPOSIT	219.14	433.55
5/15	INTEREST	7.82	441.37
5/15	CHECK	17.39	423.98
5/15	DEPOSIT	219.14	643.12
5/23	CHECK	42.00	601.12
5/23	CHECK	172.90	428.22
5/24	CHECK	77.14	351.08
5/24	DEPOSIT	219.14	570.22
5/30	CHECK	88.62	481.60
6/1	CLOSING BALANCE		481.60

15.   Balance the checkbook.

		RECORD ALL CHARGES OR CREDITS THAT AFFECT YOUR ACCOUNT					BALANCE	
NUMBER	DATE	DESCRIPTION OF TRANSACTION	PAYMENT/DEBIT (-)	√ T	FEE (IF ANY) (-)	DEPOSIT/CREDIT (+)	$ 1035	18
218	7/2	*Mortgage*	$ 284 60	$	$		750	58
219	7/4	*Telephone*	23 36				727	22
220	7/7	*Cash*	200 00				527	22
	7/12	*Deposit*				792 60	1319	82
221	7/15	*Insurance*	192 30				1127	52
222	7/18	*Investment*	100 00				1027	52
223	7/20	*Credit Card*	214 83				812	69
	7/26	*Deposit*				792 60	1605	29
224	7/27	*Department Store*	113 37				1491	92

CHECKING ACCOUNT Monthly Statement   Account Number: 122-345-1

Date	Transaction	Amount	Balance
7/1	OPENING BALANCE		1035.18
7/1	INTEREST	5.15	1040.33
7/4	CHECK	284.60	755.73
7/6	CHECK	23.36	732.37
7/12	DEPOSIT	792.60	1524.97
7/20	CHECK	192.30	1332.67
7/24	CHECK	100.00	1232.67
7/26	DEPOSIT	792.60	2025.27
7/28	CHECK	200.00	1825.27
7/30	CLOSING BALANCE		1825.27

APPLYING THE CONCEPTS

16.   When a check is written, the amount is _____ from the balance.

17.   When a deposit is made, the amount is _____ to the balance.

18.   In checking the bank statement, _____ to the checkbook balance all checks that have been written but not processed.

19.   In checking the bank balance, _____ any service charge and any deposits not yet recorded.

20.   Define the words *credit* and *debit* as they apply to checkbooks.

Focus on Problem Solving

Counterexamples

An example that is given to show that a statement is not true is called a **counterexample.** For instance, suppose someone makes the statement "All colors are red." A counterexample to that statement would be to show someone the color blue or some other color.

If a statement is *always* true, there are no counterexamples. The statement "All even numbers are divisible by 2" is always true. It is not possible to give an example of an even number that is not divisible by 2.

In mathematics, statements that are always true are called *theorems,* and mathematicians are always searching for theorems. Sometimes a conjecture by a mathematician appears to be a theorem, that is, the statement appears to be always true, but later on someone finds a counterexample.

One example of this occurred when the French mathematician Pierre de Fermat (1601–1665) conjectured that $2^{(2^n)} + 1$ was always a prime number for any natural number n. For instance, when $n = 3$, we have $2^{(2^3)} + 1 = 2^8 + 1 = 257$ and 257 is a prime number. However, in 1732 Leonard Euler (1707–1783) showed that when $n = 5$, $2^{(2^5)} + 1 = 4{,}294{,}967{,}297$ and that $4{,}294{,}967{,}297 = 641 \cdot 6{,}700{,}417$—without a calculator! Since $4{,}294{,}967{,}297$ was the product of two numbers (other than itself and 1), it was not a prime number. This counterexample showed that Fermat's conjecture was not a theorem.

For Exercises 1 and 5, find at least one counterexample.

1. All numbers are positive.

2. All prime numbers are odd numbers.

3. The square of any number is always bigger than the number.

4. The reciprocal of a number is always less than 1.

5. A number ending in 9 is always larger than a number ending in 3.

When a problem is posed, it may not be known whether the problem statement is true or false. For instance, Christian Goldbach (1690–1764) stated that every even number greater than 2 can be written as the sum of two prime numbers. For example,

$$12 = 5 + 7 \qquad 32 = 3 + 29$$

Although this problem is approximately 250 years old, mathematicians have not been able to prove it is a theorem, nor have they been able to find a counterexample.

For Exercises 6 to 9, answer true if the statement is always true. If there is an instance when the statement is false, give a counterexample.

6. The sum of two positive numbers is always larger than either of the two numbers.

7. The product of two positive numbers is always larger than either of the two numbers.

8. Percents always represent a number less than or equal to 1.

9. It is never possible to divide by zero.

Projects and Group Activities

Annuities   Suppose you buy a car and finance $4300 for three years at an annual interest rate of 9%. To completely repay the loan, you must make 36 monthly payments of $136.74. A series of equal payments ($136.74) at regular intervals (monthly in this case) is called an **annuity**.

The mortgage payments and car payments that you calculated in this chapter are examples of annuities. These calculations were accomplished by using the tables in the Appendix. It is possible, however, to use a formula to calculate the amount of a monthly payment.

$$\text{Payment} = B \times \left[\frac{i}{1 - \dfrac{1}{(1+i)^n}} \right]$$

where B is the amount borrowed, $i = \dfrac{\text{annual interest rate as a decimal}}{\text{number of payments per year}}$, and n is the total number of payments.

Although the formula looks quite complicated, it can be evaluated by using a scientific calculator. Here are some keystrokes that will work on most scientific calculators.

$B \;\boxed{\times}\; i \;\boxed{\div}\; \boxed{(}\; 1 \;\boxed{-}\; 1 \;\boxed{\div}\; \boxed{(}\; 1 \;\boxed{+}\; i \;\boxed{)}\; \boxed{y^x}\; n \;\boxed{)}\; \boxed{=}$ Payment

For the car payment above, calculate $i = \dfrac{0.09}{12} = 0.0075$ and $n = 3 \times 12 = 36$. Then the payment is calculated as

$4300 \;\boxed{\times}\; .0075 \;\boxed{\div}\; \boxed{(}\; 1 \;\boxed{-}\; 1 \;\boxed{\div}\; \boxed{(}\; 1 \;\boxed{+}\; .0075 \;\boxed{)}\; \boxed{y^x}\; 36 \;\boxed{)}\; \boxed{=}$

The result will be 136.73885, which is $136.74 rounded to the nearest cent. The advantage of having a formula is that you can calculate the result for any interest rate and time period. You are not restricted to the values in a table.

If you were to trade in your car for a new one before the end of the loan, you would have to *pay off* the remaining amount owed. A modification of the formula above can be used to calculate the payoff.

$$\text{Payoff} = \text{Payment} \div \left[\frac{i}{1 - \dfrac{1}{(1+i)^n}} \right]$$

In this case, n is the number of *remaining* payments and i is as before.

For instance, suppose you have kept the car for 2 years (24 months) and decide to trade it in for a new car. Then you have 12 payments remaining ($12 = 36 - 24$). The payoff is calculated as

$136.74 \;\boxed{\div}\; \boxed{(}\; .0075 \;\boxed{\div}\; \boxed{(}\; 1 \;\boxed{-}\; 1 \;\boxed{\div}\; \boxed{(}\; 1 \;\boxed{+}\; .0075 \;\boxed{)}\; \boxed{y^x}\; 12 \;\boxed{)}\; \boxed{)}\; \boxed{=}$

The result is $1563.61. This is the amount still owed on the car loan.

1. Ford Motor Company offered a loan rate of 1.9% for 3 years on selected models of a new car. Find the monthly payment for a loan of $12,000 on one of these cars.

2. The median price of a home in the United States is approximately $118,000. If you purchase a home at the median price and make a 20% down payment, what are the monthly payments for a 30-year loan for which the annual interest rate is 7.5%?

3. Suppose you keep the home in Exercise 2 for 10 years and then decide to sell it.

    a. How much do you still owe on the house?

    b. How much of the loan have you repaid?

    c. You have owned the home for $\frac{1}{3}$ of the 30-year period. Have you repaid $\frac{1}{3}$ of the loan?

    d. By experimenting with various values of n, try to determine how many months you must own the home before you have repaid $\frac{1}{2}$ of the loan amount.

Credit Card Finance Charges

All credit card companies charge a *fee* (finance charge) when a credit card balance is not paid within a certain number of days of the *billing date* (the date the credit card bill is sent). There may also be an annual fee. The table below shows the charges for five banks in June of 1997. (If you have Internet access, you can get current rates by visiting http://www.bankrate.com.)

Bank	Annual Interest Rate	Annual Fee
Pulaski Bank and Trust	7.99%	$35
Oak Brook Bank	8.45%	$85
Huntington National Bank	8.50%	$70
AFBA Industrial Bank	11.40%	$0
Pullman Bank & Trust	12.50%	$0

The amount of the monthly finance charge is based (usually) on the average daily balance. For instance, using Huntington National Bank, if your average daily balance for one month was $275.89, then the monthly finance charge is calculated using the simple interest formula.

Interest = principal × annual interest rate × time

$$= 275.89 \times 0.085 \times \frac{1}{12} \qquad \bullet \; 1 \text{ month} = \frac{1}{12} \text{ year.}$$

$$\approx 1.95$$

The finance charge is $1.95.

1. Suppose you have an average daily balance of $321.65 for 12 months. With which bank will your *annual* finance charge (including the annual fee) be least?

2. Suppose you have an average daily balance of $725.91 for 12 months. With which bank will your *annual* finance charge (including the annual fee) be least?

Chapter Summary

Key Words

The *unit cost* is the cost of one item.

Percent increase is used to show how much a quantity has increased over its original value.

Cost is the price a business pays for a product.

Selling price is the price at which a business sells a product to a customer.

Markup is the difference between selling price and cost.

Markup rate is the markup expressed as a percent of a product's cost.

Percent decrease is used to show how much a quantity has decreased from its original value.

Sale price is the price that has been reduced from the regular price.

Discount is the difference between the regular price and the sale price.

Discount rate is the discount as a percent of a product's regular price.

Interest is the amount of money paid for the privilege of using someone else's money.

Principal is the amount of money originally deposited or borrowed.

The percent used to determine the amount of interest is the *interest rate*.

Interest computed on the original amount is called *simple interest*.

Compound interest is computed not only on the original principal but also on interest already earned.

The *mortgage* is the amount that is borrowed to buy real estate.

The loan origination fee is usually a percent of the mortgage and is expressed in *points*.

Commissions are usually paid to salespersons and are calculated as a percent of total sales.

An employee who receives an *hourly wage* is paid a certain amount for each hour worked.

An employee who is paid a *salary* receives payment based on a weekly, biweekly, monthly, or annual time schedule.

Essential Rules	***To Find Unit Cost***	To find the unit cost, divide the total cost by the number of units.
	To Find Total Cost	To find the total cost, multiply the unit cost by the number of units.
	Basic Markup Equations	Selling price = cost + markup
		Markup = markup rate × cost
	Basic Discount Equations	Sale price = regular price − discount
		Discount = discount rate × regular price
	Annual Simple Interest Equation	$\text{Principal} \times \dfrac{\text{annual}}{\text{interest rate}} \times \dfrac{\text{time}}{\text{in years}} = \text{interest}$

Chapter Review

1. A 20-ounce box of cereal costs $2.90. Find the unit cost.

2. An account executive had car expenses of $1025.58 for insurance, $605.82 for gas, $37.92 for oil, and $188.27 for maintenance during a year in which 15,320 miles were driven. Find the cost per mile for these four items. Round to the nearest tenth of a cent.

3. An oil stock was bought for 42\frac{3}{8}$ per share. Six months later, the stock was selling for 55\frac{1}{4}$ per share. Find the percent increase in the price of the stock for the 6 months. Round to the nearest tenth of a percent.

4. A sporting goods store uses a markup rate of 40%. What is the markup on a ski suit that costs the store $180?

5. A contractor borrowed $100,000 from a credit union for 9 months at an annual interest rate of 9%. What is the simple interest due on the loan?

6. A computer programmer invested $25,000 in a retirement account that pays 6% interest, compounded daily. What is the value of the investment in 10 years? Use the table in the Appendix. Round to the nearest cent.

7. Last year an oil company had earnings of $4.12 per share. This year the earnings are $4.73 per share. What is the percent increase in earnings per share? Round to the nearest percent.

8. The monthly mortgage payment for a condominum is $523.67. The owner must pay an annual property tax of $658.32. Find the total monthly payment for the mortgage and property tax.

9. A pickup truck with a slide-in camper is purchased for $14,450. A down payment of 8% is made, and the remaining cost is financed for 4 years at an annual interest rate of 9%. Find the monthly payment. Use the monthly schedule in the Appendix. Round to the nearest cent.

10. A fast-food restaurant invested $50,000 in an account that pays 7% annual interest compounded quarterly. What is the value of the investment in 1 year? Use the table in the Appendix.

11. Paula Mason purchased a home for $125,000. The lender requires a down payment of 15%. Find the amount of the down payment.

12. A plumber bought a truck for $13,500. A state license of $315 and a sales tax of 6.25% of the purchase price are required. Find the total cost of the sales tax and the license fee.

13. Techno-Center uses a markup rate of 35% on all computer systems. Find the selling price of a computer system that costs the store $1540.

14. Mien pays a monthly car payment of $122.78. During a month in which $25.45 is principal, how much of the payment is interest?

15. The manager of the retail store at a ski resort receives a commission of 3% on all sales at the alpine shop. Find the total commission received during a month in which the shop had $108,000 in sales.

16. A suit that regularly costs $235 is on sale for 40% off the regular price. Find the sale price.

17. Luke had a checking account balance of $1568.45 before writing checks for $123.76, $756.45, and $88.77. He then deposited a check for $344.21. Find Luke's current checkbook balance.

18. Pros' Sporting Goods borrowed $30,000 at an annual interest rate of 8% for 6 months. Find the simple interest due on the loan.

19. A credit union requires a borrower to pay $2\frac{1}{2}$ points for a loan. Find the origination fee for a loan of $75,000.

20. Twenty-four ounces of a mouthwash cost $3.49. A 60-ounce container of the same kind of mouthwash costs $8.40. Which is the better buy?

21. The Sweeneys bought a home for $156,000. The family made a 10% down payment and financed the remainder with a 30-year loan with an annual interest rate of 7%. Find the monthly mortgage payment. Use the monthly payment table in the Appendix. Round to the nearest cent.

22. Richard Valdez receives $12.60 per hour for working 40 hours a week and time and a half for working over 40 hours. Find his total income during a week in which he worked 48 hours.

23. The business checking account of a donut shop showed a balance of $9567.44 before checks of $1023.55, $345.44, and $23.67 were written and checks of $555.89 and $135.91 were deposited. Find the current checkbook balance.

24. A professional baseball player received a salary of $1 million last year. This year the player signed a contract paying $12 million over 4 years. Find the yearly percent increase in the player's salary.

Chapter Test

1. Twenty feet of lumber cost $138.40. What is the cost per foot?

2. Find the more economical purchase: 5 pounds of tomatoes for $1.65, or 8 pounds for $2.72.

3. Red snapper costs $4.15 per pound. Find the cost of $3\frac{1}{2}$ pounds. Round to the nearest cent.

4. An exercise bicycle increased in price from $415 to $498. Find the percent increase in the cost of the exercise bicycle.

5. Fifteen years ago a painting was priced at $6000. Today the same painting has a value of $15,000. Find the percent increase in the price of the painting during the 15 years.

6. A department store uses a 40% markup rate. Find the selling price of a compact disk player that the store purchased for $215.

7. A bookstore bought a paperback book for $5 and used a markup rate of 25%. Find the selling price of the book.

8. The price of gold dropped from $390 per ounce to $360 per ounce. What percent decrease does this amount represent? Round to the nearest tenth of a percent.

9. The price of a video camera dropped from $1120 to $896. What percent decrease does this price drop represent?

10. A corner hutch with a regular price of $299 is on sale for 30% off the regular price. Find the sale price.

11. A box of stationery that regularly sells for $4.50 is on sale for $2.70. Find the discount rate.

12. A construction company borrowed $75,000 for 4 months at an annual interest rate of 8%. Find the simple interest due on the loan.

13. Jorge, who is self-employed, placed $30,000 in an account that pays 6% annual interest compounded quarterly. How much interest was earned in 10 years? Use the table in the Appendix.

14. A savings and loan institution is giving mortgage loans that have a loan origination fee of $2\frac{1}{2}$ points. Find the loan origination fee on a home purchased with a loan of $134,000.

15. A new housing development offers homes with a mortgage of $222,000 for 25 years at an annual interest rate of 8%. Find the monthly mortgage payment. Use the table in the Appendix.

16. A Chevrolet Blazer was purchased for $23,750, and a 20% down payment was made. Find the amount financed.

17. A rancher purchased a GMC Jimmy for $23,714 and made a down payment of 15% of the cost. The balance was financed for 4 years at an annual interest rate of 7%. Find the monthly truck payment. Use the table in the Appendix.

18. Shaney receives an hourly wage of $13.40 an hour as an emergency room nurse. When called in at night, she receives time and a half. How much does Shaney earn in a week when she works 30 hours at normal rates and 15 hours during the night?

19. The business checking account for a pottery store had a balance of $7349.44 before checks for $1349.67 and $344.12 were written. The store manager then made a deposit of $956.60. Find the current checkbook balance.

20. Balance the checkbook shown.

RECORD ALL CHARGES OR CREDITS THAT AFFECT YOUR ACCOUNT

NUMBER	DATE	DESCRIPTION OF TRANSACTION	PAYMENT/DEBIT (-)	√T	FEE (IF ANY) (-)	DEPOSIT/CREDIT (+)	BALANCE $ 422 13
	8/1	House Payment	$ 213 72	$		$	208 41
	8/4	Deposit				552 60	761 01
	8/5	Plane Tickets	162 40				598 61
	8/6	Groceries	66 44				532 17
	8/10	Car Payment	122 37				409 80
	8/15	Deposit				552 60	962 40
	8/16	Credit Card	213 45				748 95
	8/18	Doctor	92 14				656 81
	8/22	Utilities	72 30				584 51
	8/28	T. V. Repair	78 20				506 31

CHECKING ACCOUNT Monthly Statement		Account Number: 122-345-1	
Date	Transaction	Amount	Balance
8/1	OPENING BALANCE		422.13
8/3	CHECK	213.72	208.41
8/4	DEPOSIT	552.60	761.01
8/8	CHECK	66.44	694.57
8/8	CHECK	162.40	532.17
8/15	DEPOSIT	552.60	1084.77
8/23	CHECK	72.30	1012.47
8/24	CHECK	92.14	920.33
9/1	CLOSING BALANCE		920.33

Cumulative Review

1. Simplify $12 - (10 - 8)^2 \div 2 + 3$.

2. Add: $3\frac{1}{3} + 4\frac{1}{8} + 1\frac{1}{12}$

3. Find the difference between $12\frac{3}{16}$ and $9\frac{5}{12}$.

4. Find the product of $5\frac{5}{8}$ and $1\frac{9}{15}$.

5. Divide: $3\frac{1}{2} \div 1\frac{3}{4}$

6. Simplify $\left(\frac{3}{4}\right)^2 \div \left(\frac{3}{8} - \frac{1}{4}\right) + \frac{1}{2}$.

7. Divide: $0.059\overline{)3.0792}$
Round to the nearest tenth.

8. Convert $\frac{17}{12}$ to a decimal. Round to the nearest thousandth.

9. Write "$410 in 8 hours" as a unit rate.

10. Solve the proportion $\frac{5}{n} = \frac{16}{35}$.
Round to the nearest hundredth.

11. Write $\frac{5}{8}$ as a percent.

12. Find 6.5% of 420.

13. Write 18.2% as a decimal.

14. What percent of 20 is 8.4?

15. 30 is 12% of what?

16. 65 is 42% of what? Round to the nearest hundredth.

17. A series of late summer storms produced rainfall of $3\frac{3}{4}$, $8\frac{1}{2}$, and $1\frac{2}{3}$ inches during a 3-week period. Find the total rainfall during the 3 weeks.

18. The Homer family pays $\frac{1}{5}$ of its total monthly income for taxes. The family has a total monthly income of $2850. Find the amount of the monthly income that the Homers pay in taxes.

19. In 5 years, the cost of a scientific calculator went from $75 to $30. What is the ratio of the decrease in price to the original price?

20. A compact car was driven 417.5 miles on 12.5 gallons of gasoline. Find the number of miles driven per gallon of gasoline.

21. A 14-pound turkey costs $12.96. Find the unit cost. Round to the nearest cent.

22. Eighty shares of a stock paid a dividend of $112. At the same rate, find the dividend on 200 shares of the stock.

23. A video camera that regularly sells for $900 is on sale for 20% off the regular price. What is the sale price?

24. A department store bought a portable disk player for $85 and used a markup rate of 40%. Find the selling price of the disk player.

25. Sook Kim, an elementary school teacher, received an increase in salary from $2800 per month to $3024 per month. Find the percent increase in her salary.

26. A contractor borrowed $120,000 for 6 months at an annual interest rate of 10%. How much simple interest is due on the loan?

27. A red Ford Mustang was purchased for $26,900, and a down payment of $2000 was made. The balance is financed for 3 years at an annual interest rate of 9%. Find the monthly payment. Use the table in the Appendix. Round to the nearest cent.

28. A family had a checking account balance of $1846.78. A check of $568.30 was deposited into the account, and checks of $123.98 and $47.33 were written. Find the new checking account balance.

29. During one year, Anna Gonzalez spent $840 on gasoline and oil, $520 on insurance, $185 on tires, and $432 on repairs. Find the cost per mile to drive the car 10,000 miles during the year. Round to the nearest cent.

30. A house has a mortgage of $72,000 for 20 years at an annual interest rate of 11%. Find the monthly mortgage payment. Use the table in the Appendix. Round to the nearest cent.

Appendix

Compound Interest Table

Compounded Annually

	4%	5%	6%	7%	8%	9%	10%
1 year	1.04000	1.05000	1.06000	1.07000	1.08000	1.09000	1.10000
5 years	1.21665	1.27628	1.33823	1.40255	1.46933	1.53862	1.61051
10 years	1.48024	1.62890	1.79085	1.96715	2.15893	2.36736	2.59374
15 years	1.80094	2.07893	2.39656	2.75903	3.17217	3.64248	4.17725
20 years	2.19112	2.65330	3.20714	3.86968	4.66095	5.60441	6.72750

Compounded Semiannually

	4%	5%	6%	7%	8%	9%	10%
1 year	1.04040	1.05062	1.06090	1.07123	1.08160	1.09203	1.10250
5 years	1.21899	1.28008	1.34392	1.41060	1.48024	1.55297	1.62890
10 years	1.48595	1.63862	1.80611	1.98979	2.19112	2.41171	2.65330
15 years	1.81136	2.09757	2.42726	2.80679	3.24340	3.74531	4.32194
20 years	2.20804	2.68506	3.26204	3.95926	4.80102	5.81634	7.03999

Compounded Quarterly

	4%	5%	6%	7%	8%	9%	10%
1 year	1.04060	1.05094	1.06136	1.07186	1.08243	1.09308	1.10381
5 years	1.22019	1.28204	1.34686	1.41478	1.48595	1.56051	1.63862
10 years	1.48886	1.64362	1.81402	2.00160	2.20804	2.43519	2.68506
15 years	1.81670	2.10718	2.44322	2.83182	3.28103	3.80013	4.39979
20 years	2.21672	2.70148	3.29066	4.00639	4.87544	5.93015	7.20957

Compounded Daily

	4%	5%	6%	7%	8%	9%	10%
1 year	1.04080	1.05127	1.06183	1.07250	1.08328	1.09416	1.10516
5 years	1.22139	1.28400	1.34983	1.41902	1.49176	1.56823	1.64861
10 years	1.49179	1.64866	1.82203	2.01362	2.22535	2.45933	2.71791
15 years	1.82206	2.11689	2.45942	2.85736	3.31968	3.85678	4.48077
20 years	2.22544	2.71810	3.31979	4.05466	4.95217	6.04830	7.38703

To use this table:
1. Locate the section which gives the desired compounding period.
2. Locate the interest rate in the top row of that section.
3. Locate the number of years in the left-hand column of that section.
4. Locate the number where the interest-rate column and the number-of-years row meet. This is the compound interest factor.

Example An investment yields an annual interest rate of 10% compounded quarterly for 5 years.
The compounding period is "compounded quarterly."
The interest rate is 10%.
The number of years is 5.
The number where the row and column meet is 1.63862. This is the compound interest factor.

Compound Interest Table

			Compounded Annually				
	11%	*12%*	*13%*	*14%*	*15%*	*16%*	*17%*
1 year	1.11000	1.12000	1.13000	1.14000	1.15000	1.16000	1.17000
5 years	1.68506	1.76234	1.84244	1.92542	2.01136	2.10034	2.19245
10 years	2.83942	3.10585	3.39457	3.70722	4.04556	4.41144	4.80683
15 years	4.78459	5.47357	6.25427	7.13794	8.13706	9.26552	10.53872
20 years	8.06239	9.64629	11.52309	13.74349	16.36654	19.46076	23.10560

			Compounded Semiannually				
	11%	*12%*	*13%*	*14%*	*15%*	*16%*	*17%*
1 year	1.11303	1.12360	1.13423	1.14490	1.15563	1.16640	1.17723
5 years	1.70814	1.79085	1.87714	1.96715	2.06103	2.15893	2.26098
10 years	2.91776	3.20714	3.52365	3.86968	4.24785	4.66096	5.11205
15 years	4.98395	5.74349	6.61437	7.61226	8.75496	10.06266	11.55825
20 years	8.51331	10.28572	12.41607	14.97446	18.04424	21.72452	26.13302

			Compounded Quarterly				
	11%	*12%*	*13%*	*14%*	*15%*	*16%*	*17%*
1 year	1.11462	1.12551	1.13648	1.14752	1.15865	1.16986	1.18115
5 years	1.72043	1.80611	1.89584	1.98979	2.08815	2.19112	2.29891
10 years	2.95987	3.26204	3.59420	3.95926	4.36038	4.80102	5.28497
15 years	5.09225	5.89160	6.81402	7.87809	9.10513	10.51963	12.14965
20 years	8.76085	10.64089	12.91828	15.67574	19.01290	23.04980	27.93091

			Compounded Daily				
	11%	*12%*	*13%*	*14%*	*15%*	*16%*	*17%*
1 year	1.11626	1.12747	1.13880	1.15024	1.16180	1.17347	1.18526
5 years	1.73311	1.82194	1.91532	2.01348	2.11667	2.22515	2.33918
10 years	3.00367	3.31946	3.66845	4.05411	4.48031	4.95130	5.47178
15 years	5.20569	6.04786	7.02625	8.16288	9.48335	11.01738	12.79950
20 years	9.02203	11.01883	13.45751	16.43582	20.07316	24.51534	29.94039

Monthly Payment Table

	4%	5%	6%	7%	8%	9%
1 year	0.0851499	0.0856075	0.0860664	0.0865267	0.0869884	0.0874515
2 years	0.0434249	0.0438714	0.0443206	0.0447726	0.0452273	0.0456847
3 years	0.0295240	0.0299709	0.0304219	0.0308771	0.0313364	0.0317997
4 years	0.0225791	0.0230293	0.0234850	0.0239462	0.0244129	0.0248850
5 years	0.0184165	0.0188712	0.0193328	0.0198012	0.0202764	0.0207584
20 years	0.0060598	0.0065996	0.0071643	0.0077530	0.0083644	0.0089973
25 years	0.0052784	0.0058459	0.0064430	0.0070678	0.0077182	0.0083920
30 years	0.0047742	0.0053682	0.0059955	0.0066530	0.0073376	0.0080462

	10%	11%	12%	13%		
1 year	0.0879159	0.0883817	0.0888488	0.0893173		
2 years	0.0461449	0.0466078	0.0470735	0.0475418		
3 years	0.0322672	0.0327387	0.0332143	0.0336940		
4 years	0.0253626	0.0258455	0.0263338	0.0268275		
5 years	0.0212470	0.0217424	0.0222445	0.0227531		
20 years	0.0096502	0.0103219	0.0110109	0.0117158		
25 years	0.0090870	0.0098011	0.0105322	0.0112784		
30 years	0.0087757	0.0095232	0.0102861	0.0110620		

To use this table:
1. Locate the desired interest rate in the top row.
2. Locate the number of years in the left-hand column.
3. Locate the number where the interest-rate column and the number-of-years row meet. This is the monthly payment factor.

Example A home has a 30-year mortgage at an annual interest rate of 12%.
The interest rate is 12%.
The number of years is 30.
The number where the row and column meet is 0.0102861. This is the monthly payment factor.

Solutions to Chapter 1 "You Try It"

SECTION 1.1

You Try It 1

0 1 2 3 4 5 6 7 8 9 10 11 12 13 14

You Try It 2   **a.** $45 > 29$ **b.** $27 > 0$

You Try It 3   Thirty-six million four hundred sixty-two thousand seventy-five

You Try It 4   452,007

You Try It 5 $60,000 + 8000 + 200 + 80 + 1$

You Try It 6 $100,000 + 9000 + 200 + 7$

You Try It 7   370,000

You Try It 8   4000

SECTION 1.2

You Try It 1

$$\begin{array}{r} 347 \\ + 12,453 \\ \hline 12,800 \end{array}$$

347 increased by 12,453 is 12,800.

You Try It 2

$$\begin{array}{r} \overset{2}{95} \\ 88 \\ + 67 \\ \hline 250 \end{array}$$

You Try It 3

$$\begin{array}{r} \overset{1\,1\ \,2\,1}{392} \\ 4,079 \\ 89,035 \\ + 4,992 \\ \hline 98,498 \end{array}$$

You Try It 4

Strategy   To find the total amount budgeted for the three items each month, add the three amounts ($475, $275, and $120).

Solution

$$\begin{array}{r} \$475 \\ 275 \\ + 120 \\ \hline \$870 \end{array}$$

The total amount budgeted for the three items is $870.

SECTION 1.3

You Try It 1

$$\begin{array}{r} 8925 \\ - 6413 \\ \hline 2512 \end{array}$$ *Check:* $\begin{array}{r} 6413 \\ + 2512 \\ \hline 8925 \end{array}$

You Try It 2

$$\begin{array}{r} 17,504 \\ - 9,302 \\ \hline 8,202 \end{array}$$ *Check:* $\begin{array}{r} 9,302 \\ + 8,202 \\ \hline 17,504 \end{array}$

You Try It 3

$$\begin{array}{r} \overset{2\ \ 14\ \ 7\ \ 11}{\cancel{3}\ \cancel{4}\ \cancel{8}\ \cancel{1}} \\ - \quad 8\ 6\ 5 \\ \hline 2\ 6\ 1\ 6 \end{array}$$ *Check:* $\begin{array}{r} 865 \\ + 2616 \\ \hline 3481 \end{array}$

You Try It 4

$$\begin{array}{r} \overset{\quad\ \ 15}{\overset{4\ \ \cancel{5}\ \ 12}{5\ 4,\cancel{3}\ \cancel{6}\ \cancel{2}}} \\ - 1\ 4,4\ 8\ 5 \\ \hline 4\ 0,0\ 7\ 7 \end{array}$$ *Check:* $\begin{array}{r} 14,485 \\ + 40,077 \\ \hline 54,562 \end{array}$

You Try It 5

$$\begin{array}{r} \overset{13\ \ 9\ \ 9}{\overset{5\ \ \cancel{3}\ \ \cancel{10}\ \cancel{10}\ 13}{\cancel{6}\ \cancel{4},\cancel{0}\ \cancel{0}\ \cancel{3}}} \\ - 5\ 4,9\ 3\ 6 \\ \hline 9,0\ 6\ 7 \end{array}$$ *Check:* $\begin{array}{r} 54,936 \\ + 9,067 \\ \hline 64,003 \end{array}$

You Try It 6

Strategy   To find the difference, subtract the number of residents who migrated to Montana in 1996 (5250) from the number who migrated to Colorado in 1996 (30,049).

Solution

$$\begin{array}{r} 30,049 \\ - 5,250 \\ \hline 24,799 \end{array}$$

24,799 more residents migrated to Colorado than to Montana in 1996.

You Try It 7

Strategy   To find your take-home pay:
- Add to find the total of the deductions ($127 + $18 + $35).
- Subtract the total of the deductions from your total salary ($638).

Solution

$$\begin{array}{r} 127 \\ 18 \\ + 35 \\ \hline 180 \end{array} \text{ deductions} \qquad \begin{array}{r} 638 \\ - 180 \\ \hline 458 \end{array}$$

Your take-home pay is $458.

SECTION 1.4

You Try It 1

$$
\begin{array}{r}
{}^{3\,5} \\
648 \\
\times \quad 7 \\
\hline
4536
\end{array}
$$

You Try It 2

$$
\begin{array}{r}
756 \\
\times\ 305 \\
\hline
3780 \\
22680\ \ \\
\hline
230{,}580
\end{array}
$$

You Try It 3

Strategy   To find the number of cars the dealer will receive in 12 months, multiply the number of months (12) by the number of cars received each month (37).

Solution

$$
\begin{array}{r}
37 \\
\times\ 12 \\
\hline
74 \\
37\ \ \\
\hline
444
\end{array}
$$

The dealer will receive 444 cars in 12 months.

You Try It 4

Strategy   To find the total cost of the order:
• Find the cost of the sports jackets by multiplying the number of jackets (25) by the cost for each jacket ($23).
• Add the product to the cost for the suits ($4800).

Solution

$$
\begin{array}{r}
\$23 \\
\times\ 25 \\
\hline
115 \\
46\ \ \\
\hline
\$575
\end{array}
\qquad
\begin{array}{r}
\$4800 \\
+\quad 575 \\
\hline
\$5375
\end{array}
$$

$575   cost for jackets

The total cost of the order is $5375.

SECTION 1.5

You Try It 1

$$
\begin{array}{r}
7 \\
9\overline{)63}
\end{array}
$$

Check: $7 \times 9 = 63$

You Try It 2

$$
\begin{array}{r}
453 \\
9\overline{)\ 4077} \\
-36\ \ \\
\hline
47 \\
-45 \\
\hline
27 \\
-27 \\
\hline
0
\end{array}
$$

Check: $453 \times 9 = 4077$

You Try It 3

$$
\begin{array}{r}
705 \\
9\overline{)\ 6345} \\
-63\ \ \ \\
\hline
04 \\
-\ 0 \\
\hline
45 \\
-45 \\
\hline
0
\end{array}
$$

Check: $705 \times 9 = 6345$

You Try It 4

$$
\begin{array}{r}
870\ \text{r}5 \\
6\overline{)\ 5225} \\
-48\ \ \ \\
\hline
42 \\
-42 \\
\hline
05 \\
-\ 0 \\
\hline
5
\end{array}
$$

Check: $(870 \times 6) + 5 =$
$5220 + 5 = 5225$

You Try It 5

$$
\begin{array}{r}
3{,}058\ \text{r}3 \\
7\overline{)\ 21{,}409} \\
-21\ \ \ \ \ \\
\hline
0\,4 \\
-\ 0 \\
\hline
40 \\
-35 \\
\hline
59 \\
-56 \\
\hline
3
\end{array}
$$

Check: $(3058 \times 7) + 3 =$
$21{,}406 + 3 = 21{,}409$

You Try It 6

$$
\begin{array}{r}
109 \\
42\overline{)\ 4578} \\
-42\ \ \ \\
\hline
37 \\
-\ 0 \\
\hline
378 \\
-378 \\
\hline
0
\end{array}
$$

Check: $109 \times 42 = 4578$

You Try It 7

$$\begin{array}{r} 470 \text{ r}29 \\ 39\overline{)\ 18{,}359} \\ -15\ 6 \\ \hline 2\ 75 \\ -2\ 73 \\ \hline 29 \\ -\ 0 \\ \hline 29 \end{array}$$

Check: $(470 \times 39) + 29 =$
$18{,}330 + 29 = 18{,}359$

You Try It 8

$$\begin{array}{r} 62 \text{ r}111 \\ 534\overline{)\ 33{,}219} \\ -32\ 04 \\ \hline 1\ 179 \\ -1\ 068 \\ \hline 111 \end{array}$$

Check: $(62 \times 534) + 111 =$
$33{,}108 + 111 = 33{,}219$

You Try It 9

$$\begin{array}{r} 421 \text{ r}33 \\ 515\overline{)\ 216{,}848} \\ -206\ 0 \\ \hline 10\ 84 \\ -10\ 30 \\ \hline 548 \\ -515 \\ \hline 33 \end{array}$$

Check: $(421 \times 515) + 33 =$
$216{,}815 + 33 = 216{,}848$

You Try It 10

Strategy To find the number of tires that can be stored on each shelf, divide the number of tires (270) by the number of shelves (15).

Solution

$$\begin{array}{r} 18 \\ 15\overline{)\ 270} \\ -15 \\ \hline 120 \\ -120 \\ \hline 0 \end{array}$$

Each shelf can store 18 tires.

You Try It 11

Strategy To find the number of cases produced in 8 hours:
- Find the number of cases produced in one hour by dividing the number of cans produced (12,600)

by the number of cans to a case (24).
- Multiply the number of cases produced in one hour by 8.

Solution

$$\begin{array}{r} 525 \quad \text{cases produced} \\ 24\overline{)\ 12{,}600} \quad \text{in one hour} \\ -12\ 0 \\ \hline 60 \\ -48 \\ \hline 120 \\ -120 \\ \hline 0 \end{array}$$

$$\begin{array}{r} 525 \\ \times\quad 8 \\ \hline 4200 \end{array}$$

In 8 hours, 4200 cases are produced.

SECTION 1.6

You Try It 1 $2^4 \cdot 3^3$

You Try It 2 10^7

You Try It 3 $2^3 \cdot 5^2 = (2 \cdot 2 \cdot 2) \cdot (5 \cdot 5) = 8 \cdot 25$
$= 200$

You Try It 4 $5 \cdot (8 - 4)^2 \div 4 - 2$
$= 5 \cdot 4^2 \div 4 - 2$
$= 5 \cdot 16 \div 4 - 2$
$= 80 \div 4 - 2$
$= 20 - 2$
$= 18$

SECTION 1.7

You Try It 1 1, 2, 4, 5, 8, 10, 20, and 40 are factors of 40.

You Try It 2 $44 = 2 \cdot 2 \cdot 11$

You Try It 3 $177 = 3 \cdot 59$

Solutions to Chapter 2 "You Try It"

SECTION 2.1

You Try It 1

	2	3	5	7
50 =	2		5 · 5	
84 =	2 · 2	3		7
135 =		3 · 3 · 3	5	

The LCM = $2 \cdot 2 \cdot 3 \cdot 3 \cdot 3 \cdot 5 \cdot 5 \cdot 7$
$= 18,900$

You Try It 2

	2	3	5
36 =	2 · 2	3 · 3	
60 =	2 · 2	3	5
72 =	2 · 2 · 2	3 · 3	

The GCF = $2 \cdot 2 \cdot 3 = 12$.

You Try It 3

	2	3	5	11
11 =				11
24 =	2 · 2 · 2	3		
30 =	2	3	5	

Since no numbers are circled, the GCF = 1.

SECTION 2.2

You Try It 1 $4\frac{1}{4}$

You Try It 2 $\frac{17}{4}$

You Try It 3

$$\begin{array}{r} 4 \\ 5\overline{)22} \\ -20 \\ \hline 2 \end{array} \qquad \frac{22}{5} = 4\frac{2}{5}$$

You Try It 4

$$\begin{array}{r} 4 \\ 7\overline{)28} \\ -28 \\ \hline 0 \end{array} \qquad \frac{28}{7} = 4$$

You Try It 5 $14\frac{5}{8} = \frac{112 + 5}{8} = \frac{117}{8}$

SECTION 2.3

You Try It 1 $45 \div 5 = 9 \qquad \frac{3}{5} = \frac{3 \cdot 9}{5 \cdot 9} = \frac{27}{45}$

$\frac{27}{45}$ is equivalent to $\frac{3}{5}$.

You Try It 2 Write 6 as $\frac{6}{1}$.

$18 \div 1 = 18 \qquad 6 = \frac{6 \cdot 18}{1 \cdot 18} = \frac{108}{18}$

$\frac{108}{18}$ is equivalent to 6.

You Try It 3 $\frac{16}{24} = \frac{\overset{1}{\cancel{2}} \cdot \overset{1}{\cancel{2}} \cdot \overset{1}{\cancel{2}} \cdot 2}{\underset{1}{\cancel{2}} \cdot \underset{1}{\cancel{2}} \cdot \underset{1}{\cancel{2}} \cdot 3} = \frac{2}{3}$

You Try It 4 $\frac{8}{56} = \frac{\overset{1}{\cancel{2}} \cdot \overset{1}{\cancel{2}} \cdot \overset{1}{\cancel{2}}}{\underset{1}{\cancel{2}} \cdot \underset{1}{\cancel{2}} \cdot \underset{1}{\cancel{2}} \cdot 7} = \frac{1}{7}$

You Try It 5 $\frac{15}{32} = \frac{3 \cdot 5}{2 \cdot 2 \cdot 2 \cdot 2 \cdot 2} = \frac{15}{32}$

You Try It 6 $\frac{48}{36} = \frac{\overset{1}{\cancel{2}} \cdot \overset{1}{\cancel{2}} \cdot 2 \cdot 2 \cdot \overset{1}{\cancel{3}}}{\underset{1}{\cancel{2}} \cdot \underset{1}{\cancel{2}} \cdot \underset{1}{\cancel{3}} \cdot 3} = \frac{4}{3} = 1\frac{1}{3}$

SECTION 2.4

You Try It 1

$$\begin{array}{r} \frac{3}{8} \\ +\frac{7}{8} \\ \hline \frac{10}{8} = \frac{5}{4} = 1\frac{1}{4} \end{array}$$

You Try It 2

$$\begin{array}{r} \frac{5}{12} = \frac{20}{48} \\ +\frac{9}{16} = \frac{27}{48} \\ \hline \frac{47}{48} \end{array}$$

You Try It 3

$$\begin{array}{r} \frac{7}{8} = \frac{105}{120} \\ +\frac{11}{15} = \frac{88}{120} \\ \hline \frac{193}{120} = 1\frac{73}{120} \end{array}$$

You Try It 4

$$\frac{3}{4} = \frac{30}{40}$$

$$\frac{4}{5} = \frac{32}{40}$$

$$+\frac{5}{8} = \frac{25}{40}$$

$$\overline{\frac{87}{40} = 2\frac{7}{40}}$$

You Try It 5 $7 + \frac{6}{11} = 7\frac{6}{11}$

You Try It 6

$$29$$

$$+ 17\frac{5}{12}$$

$$\overline{46\frac{5}{12}}$$

You Try It 7

$$7\frac{4}{5} = 7\frac{24}{30}$$

$$6\frac{7}{10} = 6\frac{21}{30}$$

$$+ 13\frac{11}{15} = 13\frac{22}{30}$$

$$\overline{26\frac{67}{30} = 28\frac{7}{30}}$$

You Try It 8

$$9\frac{3}{8} = 9\frac{45}{120}$$

$$17\frac{7}{12} = 17\frac{70}{120}$$

$$+ 10\frac{14}{15} = 10\frac{112}{120}$$

$$\overline{36\frac{227}{120} = 37\frac{107}{120}}$$

You Try It 9

Strategy   To find the total time spent on the activities, add the three times $\left(4\frac{1}{2}, 3\frac{3}{4}, 1\frac{1}{3}\right)$.

Solution

$$4\frac{1}{2} = 4\frac{6}{12}$$

$$3\frac{3}{4} = 3\frac{9}{12}$$

$$+ 1\frac{1}{3} = 1\frac{4}{12}$$

$$\overline{8\frac{19}{12} = 9\frac{7}{12}}$$

The total time spent on the three activities was $9\frac{7}{12}$ hours.

You Try It 10

Strategy   To find the overtime pay:
- Find the total number of overtime hours $\left(1\frac{2}{3} + 3\frac{1}{3} + 2\right)$.
- Multiply the total number of hours by the overtime hourly wage ($24).

Solution

$$1\frac{2}{3}$$

$$3\frac{1}{3}$$

$$+ 2$$

$$\overline{6\frac{3}{3} = 7 \text{ hours}}$$

$$\begin{array}{r} \$24 \\ \times \quad 7 \\ \hline \$168 \end{array}$$

Jeff earned $168 in overtime pay.

SECTION 2.5

You Try It 1

$$\frac{16}{27}$$

$$- \frac{7}{27}$$

$$\overline{\frac{9}{27} = \frac{1}{3}}$$

You Try It 2

$$\frac{13}{18} = \frac{52}{72}$$

$$- \frac{7}{24} = \frac{21}{72}$$

$$\overline{\frac{31}{72}}$$

You Try It 3

$$17\frac{5}{9} = 17\frac{20}{36}$$

$$- 11\frac{5}{12} = 11\frac{15}{36}$$

$$\overline{6\frac{5}{36}}$$

You Try It 4

$$8 = 7\frac{13}{13}$$

$$- 2\frac{4}{13} = 2\frac{4}{13}$$

$$\overline{5\frac{9}{13}}$$

You Try It 5

$$21\frac{7}{9} = 21\frac{28}{36} = 20\frac{64}{36}$$

$$- 7\frac{11}{12} = 7\frac{33}{36} = 7\frac{33}{36}$$

$$\overline{13\frac{31}{36}}$$

You Try It 6

Strategy To find the time remaining before the plane lands, subtract the number of hours already in the air $\left(2\frac{3}{4}\right)$ from the total time of the trip $\left(5\frac{1}{2}\right)$.

Solution

$$5\frac{1}{2} = 5\frac{2}{4} = 4\frac{6}{4}$$
$$-2\frac{3}{4} = 2\frac{3}{4} = 2\frac{3}{4}$$
$$\overline{\qquad\qquad 2\frac{3}{4} \text{ hours}}$$

The plane will land in $2\frac{3}{4}$ hours.

You Try It 7

Strategy To find the amount of weight to be lost during the third month:
- Find the total weight loss during the first two months $\left(7\frac{1}{2} + 5\frac{3}{4}\right)$.
- Subtract the total weight loss from the goal (24 pounds).

Solution

$$7\frac{1}{2} = \ 7\frac{2}{4}$$
$$+5\frac{3}{4} = \ 5\frac{3}{4}$$
$$\overline{\qquad\quad 12\frac{5}{4} = 13\frac{1}{4} \text{ pounds lost}}$$

$$24 \ = 23\frac{4}{4}$$
$$-13\frac{1}{4} = 13\frac{1}{4}$$
$$\overline{\qquad\quad 10\frac{3}{4} \text{ pounds}}$$

The patient must lose $10\frac{3}{4}$ pounds to achieve the goal.

SECTION 2.6

You Try It 1

$$\frac{4}{21} \times \frac{7}{44} = \frac{4 \cdot 7}{21 \cdot 44}$$

$$= \frac{\overset{1}{\cancel{2}} \cdot \overset{1}{\cancel{2}} \cdot \overset{1}{\cancel{7}}}{3 \cdot \cancel{7} \cdot \cancel{2} \cdot \cancel{2} \cdot 11} = \frac{1}{33}$$

You Try It 2

$$\frac{2}{21} \times \frac{10}{33} = \frac{2 \cdot 10}{21 \cdot 33}$$

$$= \frac{2 \cdot 2 \cdot 5}{3 \cdot 7 \cdot 3 \cdot 11} = \frac{20}{693}$$

You Try It 3

$$\frac{16}{5} \times \frac{15}{24} = \frac{16 \cdot 15}{5 \cdot 24}$$

$$= \frac{\overset{1}{\cancel{2}} \cdot \overset{1}{\cancel{2}} \cdot \overset{1}{\cancel{2}} \cdot 2 \cdot \overset{1}{\cancel{3}} \cdot \overset{1}{\cancel{5}}}{\cancel{5} \cdot \cancel{2} \cdot \cancel{2} \cdot \cancel{2} \cdot \cancel{3}} = 2$$

You Try It 4

$$5\frac{2}{5} \times \frac{5}{9} = \frac{27}{5} \times \frac{5}{9} = \frac{27 \cdot 5}{5 \cdot 9}$$

$$= \frac{\overset{1}{\cancel{3}} \cdot \overset{1}{\cancel{3}} \cdot 3 \cdot \overset{1}{\cancel{5}}}{\cancel{5} \cdot \cancel{3} \cdot \cancel{3}} = 3$$

You Try It 5

$$3\frac{2}{5} \times 6\frac{1}{4} = \frac{17}{5} \times \frac{25}{4} = \frac{17 \cdot 25}{5 \cdot 4}$$

$$= \frac{17 \cdot \overset{1}{\cancel{5}} \cdot 5}{\cancel{5} \cdot 2 \cdot 2} = \frac{85}{4} = 21\frac{1}{4}$$

You Try It 6

$$3\frac{2}{7} \times 6 = \frac{23}{7} \times \frac{6}{1} = \frac{23 \cdot 6}{7 \cdot 1}$$

$$= \frac{23 \cdot 3 \cdot 2}{7 \cdot 1} = \frac{138}{7} = 19\frac{5}{7}$$

You Try It 7

Strategy To find the value of the house today, multiply the old value of the house ($30,000) by $3\frac{1}{2}$.

Solution

$$30,000 \times 3\frac{1}{2} = \frac{30,000}{1} \times \frac{7}{2}$$
$$= \frac{30,000 \cdot 7}{1 \cdot 2}$$
$$= 105,000$$

The value of the house today is $105,000.

You Try It 8

Strategy To find the cost of the air compressor:
- Multiply to find the value of the drying chamber $\left(\frac{4}{5} \times \$60,000\right)$.
- Subtract the value of the drying chamber from the total value of the two items ($60,000).

Solution

$$\frac{4}{5} \times \frac{\$60,000}{1} = \frac{\$240,000}{5}$$
$$= \$48,000$$

$$\$60,000$$
$$-\ 48,000$$
$$\overline{\quad \$12,000}$$

The cost of the air compressor was $12,000.

SECTION 2.7

You Try It 1 $\dfrac{3}{7} \div \dfrac{2}{3} = \dfrac{3}{7} \times \dfrac{3}{2} = \dfrac{3 \cdot 3}{7 \cdot 2} = \dfrac{9}{14}$

You Try It 2 $\dfrac{3}{4} \div \dfrac{9}{10} = \dfrac{3}{4} \times \dfrac{10}{9}$

$$= \dfrac{3 \cdot 10}{4 \cdot 9} = \dfrac{\overset{1}{\cancel{3}} \cdot \overset{1}{\cancel{2}} \cdot 5}{\underset{1}{\cancel{2}} \cdot 2 \cdot \underset{1}{\cancel{3}} \cdot 3} = \dfrac{5}{6}$$

You Try It 3 $\dfrac{5}{7} \div 6 = \dfrac{5}{7} \div \dfrac{6}{1}$

$$= \dfrac{5}{7} \times \dfrac{1}{6} = \dfrac{5 \cdot 1}{7 \cdot 6}$$

$$= \dfrac{5}{7 \cdot 2 \cdot 3} = \dfrac{5}{42}$$

You Try It 4 $12\dfrac{3}{5} \div 7 = \dfrac{63}{5} \div \dfrac{7}{1} = \dfrac{63}{5} \times \dfrac{1}{7}$

$$= \dfrac{63 \cdot 1}{5 \cdot 7} = \dfrac{3 \cdot 3 \cdot \overset{1}{\cancel{7}}}{5 \cdot \underset{1}{\cancel{7}}} = \dfrac{9}{5} = 1\dfrac{4}{5}$$

You Try It 5 $3\dfrac{2}{3} \div 2\dfrac{2}{5} = \dfrac{11}{3} \div \dfrac{12}{5}$

$$= \dfrac{11}{3} \times \dfrac{5}{12} = \dfrac{11 \cdot 5}{3 \cdot 12}$$

$$= \dfrac{11 \cdot 5}{3 \cdot 2 \cdot 2 \cdot 3} = \dfrac{55}{36} = 1\dfrac{19}{36}$$

You Try it 6 $2\dfrac{5}{6} \div 8\dfrac{1}{2} = \dfrac{17}{6} \div \dfrac{17}{2}$

$$= \dfrac{17}{6} \times \dfrac{2}{17} = \dfrac{17 \cdot 2}{6 \cdot 17}$$

$$= \dfrac{\overset{1}{\cancel{17}} \cdot \overset{1}{\cancel{2}}}{\underset{1}{\cancel{2}} \cdot 3 \cdot \underset{1}{\cancel{17}}} = \dfrac{1}{3}$$

You Try It 7 $6\dfrac{2}{5} \div 4 = \dfrac{32}{5} \div \dfrac{4}{1}$

$$= \dfrac{32}{5} \times \dfrac{1}{4} = \dfrac{32 \cdot 1}{5 \cdot 4}$$

$$= \dfrac{2 \cdot 2 \cdot 2 \cdot \overset{1}{\cancel{2}} \cdot \overset{1}{\cancel{2}}}{5 \cdot \underset{1}{\cancel{2}} \cdot \underset{1}{\cancel{2}}} = \dfrac{8}{5} = 1\dfrac{3}{5}$$

You Try It 8

Strategy To find the price of one ounce of gold, divide the total price of the coin ($195) by the number of ounces $\left(\dfrac{1}{2}\right)$.

Solution $195 \div \dfrac{1}{2} = \dfrac{195}{1} \div \dfrac{1}{2}$

$$= \dfrac{195}{1} \times \dfrac{2}{1} = \dfrac{195 \cdot 2}{1 \cdot 1} = 390$$

The price of one ounce of gold is $390.

You Try It 9

Strategy To find the length of the remaining piece:
- Divide the total length of the board (16 feet) by the length of each shelf $\left(3\dfrac{1}{3} \text{ feet}\right)$.
- Multiply the fraction left over by the length of one shelf to determine the length of the remaining piece.

Solution $16 \div 3\dfrac{1}{3} = 16 \div \dfrac{10}{3}$

$$= \dfrac{16}{1} \times \dfrac{3}{10} = \dfrac{16 \cdot 3}{1 \cdot 10}$$

$$= \dfrac{\overset{1}{\cancel{2}} \cdot 2 \cdot 2 \cdot 2 \cdot 3}{\underset{1}{\cancel{2}} \cdot 5} = \dfrac{24}{5}$$

$$= 4\dfrac{4}{5}$$

$\dfrac{4}{5} \times 3\dfrac{1}{3} = \dfrac{4}{5} \times \dfrac{10}{3}$

$$= \dfrac{4 \cdot 10}{5 \cdot 3} = \dfrac{8}{3} = 2\dfrac{2}{3}$$

The length of the piece remaining is $2\dfrac{2}{3}$ feet.

SECTION 2.8

You Try It 1 $\dfrac{9}{14} = \dfrac{27}{42}$ $\quad \dfrac{13}{21} = \dfrac{26}{42}$ $\quad \dfrac{9}{14} > \dfrac{13}{21}$

You Try It 2 $\left(\dfrac{7}{11}\right)^2 \cdot \left(\dfrac{2}{7}\right) = \left(\dfrac{7}{11} \cdot \dfrac{7}{11}\right) \cdot \left(\dfrac{2}{7}\right)$

$$= \dfrac{\overset{1}{\cancel{7}} \cdot 7 \cdot 2}{11 \cdot 11 \cdot \underset{1}{\cancel{7}}} = \dfrac{14}{121}$$

You Try It 3

$$\left(\frac{1}{13}\right)^2 \cdot \left(\frac{1}{4} + \frac{1}{6}\right) \div \frac{5}{13}$$

$$\left(\frac{1}{13}\right)^2 \cdot \left(\frac{5}{12}\right) \div \frac{5}{13}$$

$$\left(\frac{1}{169}\right) \cdot \left(\frac{5}{12}\right) \div \frac{5}{13}$$

$$\left(\frac{1 \cdot 5}{13 \cdot 13 \cdot 12}\right) \div \frac{5}{13}$$

$$\left(\frac{1 \cdot 5}{13 \cdot 13 \cdot 12}\right) \times \frac{13}{5}$$

$$\frac{1 \cdot \overset{1}{\cancel{5}} \cdot \overset{1}{\cancel{13}}}{\underset{1}{\cancel{13}} \cdot 13 \cdot 12 \cdot \underset{1}{\cancel{5}}}$$

$$\frac{1}{156}$$

Solutions to Chapter 3 "You Try It"

SECTION 3.1

You Try It 1   Two hundred nine and five thousand eight hundred thirty-eight hundred-thousandths

You Try It 2   42,000.000207

You Try It 3   4.35

You Try It 4   3.29053

SECTION 3.2

You Try It 1

$$
\begin{array}{r}
\overset{1\ \ 2}{}\\
4.62\\
27.9\\
+\ \ 0.62054\\
\hline
33.14054
\end{array}
$$

You Try It 2

$$
\begin{array}{r}
\overset{1}{}\\
6.05\\
12.\\
+\ \ 0.374\\
\hline
18.424
\end{array}
$$

You Try It 3

Strategy   To find the total, add the cost of each kind and number of tulip. ($16.40 + $5.80 + $16.40).

Solution

$$
\begin{array}{r}
\$16.40\\
5.80\\
+\ \ 16.40\\
\hline
\$38.60
\end{array}
$$

The cost of the tulips is $38.60.

You Try It 4

Strategy   To find the total income, add the four commissions ($485.60, $599.46,

$326.75, and $725.42) to the salary ($425.00).

Solution

$$
\begin{array}{r}
\$485.60\\
599.46\\
326.75\\
725.42\\
+\ \ 425.00\\
\hline
\$2562.23
\end{array}
$$

Anita's total income was $2562.23.

SECTION 3.3

You Try It 1

$$
\begin{array}{r}
\overset{11\ 9}{\underset{6\ \cancel{7}\ \cancel{10}\ 13}{}}\\
7\,2.0\,3\,9\\
-\ \ \ \ 8.4\,7\\
\hline
6\,3.5\,6\,9
\end{array}
$$

Check:

$$
\begin{array}{r}
\overset{1\ \ 1}{}\\
8.47\\
+\ 63.569\\
\hline
72.039
\end{array}
$$

You Try It 2

$$
\begin{array}{r}
\overset{14\ 9}{\underset{2\ \cancel{4}\ \cancel{10}\ 10}{}}\\
3\,5.0\,0\\
-\ \ \ 9.6\,7\\
\hline
2\,5.3\,3
\end{array}
$$

Check:

$$
\begin{array}{r}
\overset{1\ \ 1}{}\\
9.67\\
+\ 25.33\\
\hline
35.00
\end{array}
$$

You Try It 3

$$
\begin{array}{r}
\overset{16\ 9\ 9}{\underset{2\ \cancel{6}\ \cancel{10}\ \cancel{10}\ 10}{}}\\
3.7\,0\,0\,0\\
-\ 1.9\,7\,1\,5\\
\hline
1.7\,2\,8\,5
\end{array}
$$

Check:

$$
\begin{array}{r}
\overset{1\ 1\ 1\ 1}{}\\
1.9715\\
+\ 1.7285\\
\hline
3.7000
\end{array}
$$

You Try It 4

Strategy   To find the amount of change, subtract the amount paid ($3.85) from $5.00.

Solution

$$
\begin{array}{r}
\$5.00\\
-\ 3.85\\
\hline
\$1.15
\end{array}
$$

Your change was $1.15.

You Try It 5

Strategy   To find the new balance:
- Add to find the total of the three checks ($1025.60 + $79.85 + $162.47).
- Subtract the total from the previous balance ($2472.69).

Solution

$1025.60 $2472.69
    79.85   − 1267.92
+ 162.47 $1204.77
$1267.92

The new balance is $1204.77.

SECTION 3.4

You Try It 1

$$\begin{array}{r} 870 \\ \times\ 4.6 \\ \hline 522\ 0 \\ 3480 \\ \hline 4002.0 \end{array}$$

You Try It 2

$$\begin{array}{r} 0.000086 \\ \times\ 0.057 \\ \hline 602 \\ 430 \\ \hline 0.000004902 \end{array}$$

You Try It 3

$$\begin{array}{r} 4.68 \\ \times\ 6.03 \\ \hline 1404 \\ 28\ 080 \\ \hline 28.2204 \end{array}$$

You Try It 4 $6.9 \times 1000 = 6900$

You Try It 5 $4.0273 \times 10^2 = 402.73$

You Try It 6

Strategy   To find the total bill:
- Find the number of gallons of water used by multiplying the number of gallons used per day (5000) by the number of days (62).
- Find the cost of water by multiplying the cost per 1000 gallons ($1.39) by the number of 1000-gallon units used.
- Add the cost of the water to the meter fee ($133.70).

Solution
    Number of gallons = 5000(62) = 310,000

$$\text{Cost of water} = \frac{310{,}000}{1000} \times 1.39 = 430.90$$

Total cost = 133.70 + 430.90 = 564.60

The total bill is $564.60.

You Try It 7

Strategy   To find the cost of running the freezer for 210 hours, multiply the hourly cost ($.035) by the number of hours the freezer has run (210).

Solution

$$\begin{array}{r} \$.035 \\ \times\ 210 \\ \hline \$7.35 \end{array}$$

The cost of running the freezer for 210 hours is $7.35.

You Try It 8

Strategy   To find the total cost of the stereo:
- Multiply the monthly payment ($37.18) by the number of months (18).
- Add the total to the down payment ($175.00).

Solution

$37.18 $669.24
×   18   + 175.00
$669.24 $844.24

The total cost of the stereo is $844.24.

SECTION 3.5

You Try It 1

$$\begin{array}{r} 2.7 \\ 0.052.\overline{)0.140.4} \\ -104 \\ \hline 36\ 4 \\ -36\ 4 \\ \hline 0 \end{array}$$

You Try It 2

$$\begin{array}{r} 0.4873 \approx 0.487 \\ 76\overline{)37.0420} \\ -30\ 4 \\ \hline 6\ 64 \\ -6\ 08 \\ \hline 562 \\ -532 \\ \hline 300 \\ -228 \end{array}$$

You Try It 3

$$\begin{array}{r} 72.73 \approx 72.7 \\ 5.09.\overline{)370.20.00} \\ \underline{-356\ 3} \\ 13\ 90 \\ \underline{10\ 18} \\ 3\ 720 \\ \underline{-3\ 563} \\ 1570 \\ \underline{-1527} \end{array}$$

You Try It 4 $309.21 \div 10{,}000 = 0.030921$

You Try It 5 $42.93 \div 10^4 = 0.004293$

You Try It 6

Strategy   To find the amount she paid in gasoline taxes:
- Find the total number of gallons of gas used by dividing the total number of miles driven (9675) by the number of miles driven per gallon of gas (22.5).
- Multiply the state tax ($.875) by the total number of gallons of gas used.

Solution $9675 \div 22.5 = 430$
$0.875 \times 430 = 376.25$

Susan paid $376.25 in gasoline taxes.

You Try It 7

Strategy   To find the average number of people watching TV:
- Add the number of people watching each day of the week.
- Divide the total number of people watching by 7.

Solution $91.9 + 89.8 + 90.6 + 93.9 + 78.0 + 77.1 + 87.7 = 609$

$$\frac{609}{7} = 87$$

An average of 87 million people watch television per day.

SECTION 3.6

You Try It 1

$$\begin{array}{r} 0.56 \approx 0.6 \\ 16\overline{)9.00} \end{array}$$

You Try It 2

$$4\frac{1}{6} = \frac{25}{6}$$

$$\begin{array}{r} 4.166 \approx 4.17 \\ 6\overline{)25.00} \end{array}$$

You Try It 3

$$0.56 = \frac{56}{100} = \frac{14}{25}$$

$$5.35 = 5\frac{35}{100} = 5\frac{7}{20}$$

You Try It 4

$$0.12\frac{7}{8} = \frac{12\frac{7}{8}}{100} = 12\frac{7}{8} \div 100$$

$$= \frac{103}{8} \times \frac{1}{100} = \frac{103}{800}$$

You Try It 5

$$\frac{5}{8} = 0.625$$

$$0.63 > 0.625$$

$$0.63 > \frac{5}{8}$$

Solutions to Chapter 4 "You Try It"

SECTION 4.1

You Try It 1

$$\frac{20\ \text{pounds}}{24\ \text{pounds}} = \frac{20}{24} = \frac{5}{6}$$

$20\ \text{pounds}:24\ \text{pounds} = 20:24 = 5:6$

$20\ \text{pounds to}\ 24\ \text{pounds} = 20\ \text{to}\ 24$
$\qquad\qquad\qquad\qquad = 5\ \text{to}\ 6$

You Try It 2

$$\frac{64\ \text{miles}}{8\ \text{miles}} = \frac{64}{8} = \frac{8}{1}$$

$64\ \text{miles}:8\ \text{miles} = 64:8 = 8:1$

$64\ \text{miles to}\ 8\ \text{miles} = 64\ \text{to}\ 8 = 8\ \text{to}\ 1$

You Try It 3

Strategy   To find the ratio, write the ratio of board feet of cedar (12,000) to board feet of ash (18,000) in simplest form.

Solution $$\frac{12{,}000}{18{,}000} = \frac{2}{3}$$

The ratio is $\frac{2}{3}$.

You Try It 4

Strategy   To find the ratio, write the ratio of the amount spent on radio advertising ($15,000) to the amount spent on radio and television advertising ($15,000 + $20,000) in simplest form.

Solution $\dfrac{\$15,000}{\$15,000 + \$20,000} = \dfrac{\$15,000}{\$35,000} = \dfrac{3}{7}$

The ratio is $\dfrac{3}{7}$.

SECTION 4.2

You Try It 1 $\dfrac{15 \text{ pounds}}{12 \text{ trees}} = \dfrac{5 \text{ pounds}}{4 \text{ trees}}$

You Try It 2 $\dfrac{260 \text{ miles}}{8 \text{ hours}}$

$$8)\overline{260.0} \quad 32.5$$

32.5 miles/hour

You Try It 3

Strategy To find Erik's profit per ounce:
• Find the total profit by subtracting the cost ($1625) from the selling price ($1720).
• Divide the total profit by the number of ounces (5).

Solution
$$\begin{array}{r} 1720 \\ -\ 1625 \\ \hline 95 \text{ total profit} \end{array} \qquad 5)\overline{95} \quad 19$$

The profit is $19 per ounce.

SECTION 4.3

You Try It 1 $\dfrac{6}{10} \diagup\!\!\!\!\diagdown \dfrac{9}{15}$ ➤ $10 \times 9 = 90$
$6 \times 15 = 90$

The proportion is true.

You Try It 2 $\dfrac{32}{6} \diagup\!\!\!\!\diagdown \dfrac{90}{8}$ ➤ $6 \times 90 = 540$
$32 \times 8 = 256$

The proportion is not true.

You Try It 3 $\dfrac{n}{14} = \dfrac{3}{7}$

$n \times 7 = 14 \times 3$
$n \times 7 = 42$
$n = 42 \div 7$
$n = 6$

Check: $\dfrac{6}{14} \diagup\!\!\!\!\diagdown \dfrac{3}{7}$ ➤ $14 \times 3 = 42$
$6 \times 7 = 42$

You Try It 4 $5 \times 20 = 8 \times n$
$100 = 8 \times n$
$100 \div 8 = n$
$12.5 = n$

You Try It 5 $15 \times n = 20 \times 12$
$15 \times n = 240$
$n = 240 \div 15$
$n = 16$

Check: $\dfrac{15}{20} \diagup\!\!\!\!\diagdown \dfrac{12}{16}$ ➤ $20 \times 12 = 240$
$15 \times 16 = 240$

You Try It 6 $12 \times 4 = 7 \times n$
$48 = 7 \times n$
$48 \div 7 = n$
$6.86 \approx n$

You Try It 7 $n \times 1 = 12 \times 4$
$n \times 1 = 48$
$n = 48 \div 1$
$n = 48$

Check: $\dfrac{48}{12} \diagup\!\!\!\!\diagdown \dfrac{4}{1}$ ➤ $12 \times 4 = 48$
$48 \times 1 = 48$

You Try It 8 $3 \times n = 12 \times 8$
$3 \times n = 96$
$n = 96 \div 3$
$n = 32$

Check: $\dfrac{3}{8} \diagup\!\!\!\!\diagdown \dfrac{12}{32}$ ➤ $8 \times 12 = 96$
$3 \times 32 = 96$

You Try It 9

Strategy To find the number of jars that can be packed in 15 boxes, write and solve a proportion, using n to represent the number of jars.

Solution $\dfrac{24 \text{ jars}}{6 \text{ boxes}} = \dfrac{n \text{ jars}}{15 \text{ boxes}}$

$24 \times 15 = 6 \times n$
$360 = 6 \times n$
$360 \div 6 = n$
$60 = n$

60 jars can be packed in 15 boxes.

You Try It 10

Strategy   To find the number of tablespoons of fertilizer needed, write and solve a proportion, using n to represent the number of tablespoons of fertilizer.

Solution $\dfrac{3 \text{ tablespoons}}{4 \text{ gallons}} = \dfrac{n \text{ tablespoons}}{10 \text{ gallons}}$

$$3 \times 10 = 4 \times n$$
$$30 = 4 \times n$$
$$30 \div 4 = n$$
$$7.5 = n$$

For 10 gallons of water, 7.5 tablespoons of fertilizer are required.

Solutions to Chapter 5 "You Try It"

SECTION 5.1

You Try It 1 $125\% = 125 \times \dfrac{1}{100} = \dfrac{125}{100} = 1\dfrac{1}{4}$

$125\% = 125 \times 0.01 = 1.25$

You Try It 2 $33\dfrac{1}{3}\% = 33\dfrac{1}{3} \times \dfrac{1}{100}$

$= \dfrac{100}{3} \times \dfrac{1}{100}$

$= \dfrac{100}{300} = \dfrac{1}{3}$

You Try It 3 $0.25\% = 0.25 \times 0.01 = 0.0025$

You Try It 4 $0.048 = 0.048 \times 100\% = 4.8\%$

You Try It 5 $3.67 = 3.67 \times 100\% = 367\%$

You Try It 6 $0.62\dfrac{1}{2} = 0.62\dfrac{1}{2} \times 100\%$

$= 62\dfrac{1}{2}\%$

You Try It 7 $\dfrac{5}{6} = \dfrac{5}{6} \times 100\% = \dfrac{500\%}{6} = 83\dfrac{1}{3}\%$

You Try It 8 $1\dfrac{4}{9} = \dfrac{13}{9} = \dfrac{13}{9} \times 100\%$

$= \dfrac{1300\%}{9} \approx 144.4\%$

SECTION 5.2

You Try It 1 $n = 0.063 \times 150$
$n = 9.45$

You Try It 2 $n = \dfrac{1}{6} \times 66$

$n = 11$

You Try It 3

Strategy   To find the new hourly wage:
- Find the amount of the raise. Write and solve a basic percent equation, using n to represent the amount of the raise (amount). The percent is 8%. The base is $13.50.
- Add the amount of the raise to the old wage.

Solution

$$8\% \times \$13.50 = n \qquad \$13.50$$
$$0.08 \times \$13.50 = n \qquad \underline{+\ \ 1.08}$$
$$\$1.08 = n \qquad \$14.58$$

The new hourly wage is $14.58.

SECTION 5.3

You Try It 1 $n \times 32 = 16$
$n = 16 \div 32$
$n = 0.50$
$n = 50\%$

You Try It 2 $n \times 15 = 48$
$n = 48 \div 15$
$n = 3.20$
$n = 320\%$

You Try It 3 $30 = n \times 45$
$30 \div 45 = n$
$\dfrac{2}{3} = n$
$66\dfrac{2}{3}\% = n$

You Try It 4

Strategy   To find what percent of the income the income tax is, write and solve a basic percent equation, using n to represent the percent. The base is $33,500 and the amount is $5025.

Solution
$$n \times 33{,}500 = \$5025$$
$$n = \$5025 \div \$33{,}500$$
$$n = 0.15 = 15\%$$

The income tax is 15% of the income.

You Try It 5

Strategy To find the percent that are under the age of 30:
- Subtract to find the number of people that are under the age of 30 $(1{,}300{,}000 - 416{,}000)$.
- Write and solve a basic percent equation, using n to represent the percent. The base is 1,300,000 and the amount is the number of people who are under the age of 30.

Solution $1{,}300{,}000 - 416{,}000 = 884{,}000$

$$n \times 1{,}300{,}000 = 884{,}000$$
$$n = 884{,}000 \div 1{,}300{,}000$$
$$n = 0.68 = 68\%$$

68% of the people are under the age of 30.

SECTION 5.4

You Try It 1
$$0.86 \times n = 215$$
$$n = 215 \div 0.86$$
$$n = 250$$

You Try It 2
$$0.025 \times n = 15$$
$$n = 15 \div 0.025$$
$$n = 600$$

You Try It 3
$$\frac{1}{6} \times n = 5$$
$$n = 5 \div \frac{1}{6}$$
$$n = 30$$

You Try It 4

Strategy To find the original value of the car, write and solve a basic percent equation, using n to represent the original value (base). The percent is 42%. The amount is $5229.

Solution
$$42\% \times n = \$5229$$
$$0.42 \times n = \$5229$$
$$n = \$5229 \div 0.42$$
$$n = 12{,}450$$

The original value of the car was $12,450.

You Try It 5

Strategy To find the difference between the original price and the sale price:
- Find the original price. Write and solve a basic percent equation, using n to represent the original price (base). The percent is 80%. The amount is $44.80.
- Subtract the sale price ($44.80) from the original price.

Solution
$$80\% \times n = \$44.80$$
$$0.80 \times n = \$44.80$$
$$n = \$44.80 \div 0.80$$
$$n = \$56.00 \text{ original price}$$

$$\$56.00 - \$44.80 = \$11.20$$

The difference between the original price and the sale price is $11.20.

SECTION 5.5

You Try It 1
$$\frac{26}{100} = \frac{22}{n}$$
$$26 \times n = 100 \times 22$$
$$26 \times n = 2200$$
$$n = 2200 \div 26$$
$$n \approx 84.62$$

You Try It 2
$$\frac{16}{100} = \frac{n}{132}$$
$$16 \times 132 = 100 \times n$$
$$2112 = 100 \times n$$
$$2112 \div 100 = n$$
$$21.12 = n$$

You Try It 3

Strategy To find the number of days it snowed, write and solve a proportion, using n to represent the number of days (amount). The percent is 64%. The base is 150.

Solution
$$\frac{64}{100} = \frac{n}{150}$$
$$64 \times 150 = 100 \times n$$
$$9600 = 100 \times n$$
$$9600 \div 100 = n$$
$$96 = n$$

It snowed 96 days.

You Try It 4

Strategy To find the percent of pens that were not defective:

- Subtract to find the number of pens that were not defective $(200 - 5)$.
- Write and solve a proportion, using n to represent the percent of pens that were not defective. The base is 200, and the amount is the number of pens not defective.

Solution $200 - 5 = 195$ number of pens not defective

$$\frac{n}{100} = \frac{195}{200}$$

$200 \times n = 195 \times 100$
$200 \times n = 19{,}500$
$ n = 19{,}500 \div 200$
$ n = 97.5\%$

97.5% of the pens were not defective.

Solutions to Chapter 6 "You Try It"

SECTION 6.1

You Try It 1

Strategy   To find the unit cost, divide the total cost by the number of units.

Solution   **a.** $5.59 \div 8 = 0.69875$
 \$.699 for each battery
**b.** $1.89 \div 15 = 0.126$
 \$.126 per ounce

You Try It 2

Strategy   To find the more economical purchase, compare the unit costs.

Solution $2.52 \div 6 = 0.42$
 $1.66 \div 4 = 0.415$
 $\$.415 < \$.42$

The more economical purchase is 4 cans for \$1.66.

You Try It 3

Strategy   To find the total cost, multiply the unit cost (\$4.96) by the number of units (7).

Solution $4.96 \times 7 = 34.72$

The total cost is \$34.72.

SECTION 6.2

You Try It 1

Strategy   To find the percent increase:
- Find the amount of the increase.
- Solve the basic percent equation for *percent*.

Solution
$$\begin{array}{ll} 150 & n \times 80 = 70 \\ -\ 80 & n = 70 \div 80 \\ \hline 70 & n = 0.875 = 87.5\% \end{array}$$

The percent increase was 87.5%.

You Try It 2

Strategy   To find the new hourly wage:
- Solve the basic percent equation for *amount*.
- Add the amount of the increase to the original wage.

Solution $0.14 \times 8.50 = n$
$ 1.19 = n$
$8.50 + 1.19 = 9.69$

The new hourly wage is \$9.69.

You Try It 3

Strategy   To find the markup, solve the basic percent equation for *amount*.

Solution $0.20 \times 8 = n$
$ 1.60 = n$

The markup is \$1.60.

You Try It 4

Strategy   To find the selling price:
- Find the markup by solving the basic percent equation for *amount*.
- Add the markup to the cost.

Solution $0.55 \times 72 = n$
$ 39.60 = n$
$72 + 39.60 = 111.60$

The selling price is \$111.60.

You Try It 5

Strategy   To find the percent decrease:
- Find the amount of the decrease.
- Solve the basic percent equation for *percent*.

Solution $6670 - 3100 = 3570$
$ n \times 6670 = 3570$
$ n = 3570 \div 6670$
$ n \approx 0.535$

The percent decrease is 54%.

You Try It 6

Strategy   To find the visibility:
- Find the amount of decrease by solving the basic percent equation for *amount*.
- Subtract the amount of decrease from the original visibility.

Solution $0.40 \times 5 = n$
$\qquad\qquad 2 = n$

$5 - 2 = 3$

The visibility was 3 miles.

You Try It 7

Strategy   To find the discount rate:
- Find the discount.
- Solve the basic percent equation for *percent*.

Solution $4.25 - 3.75 = 0.50$
$\qquad n \times 4.25 = 0.50$
$\qquad\qquad\quad n = 0.50 \div 4.25$
$\qquad\qquad\quad n \approx 0.1176$

The discount rate is 11.8%.

You Try It 8

Strategy   To find the sale price:
- Find the discount by solving the basic percent equation for *amount*.
- Subtract the find the sale price.

Solution $0.15 \times 110 = n$
$\qquad\qquad 16.5 = n$

$110 - 16.5 = 93.5$

The sale price is $93.50.

SECTION 6.3

You Try It 1

Strategy   To find the simple interest, multiply the principal by the annual interest rate by the time (in years).

Solution $15,000 \times 0.08 \times 1.5 = 1800$

The interest due is $1800.

You Try It 2

Strategy   To find the interest due, multiply the principal by the monthly interest rate by the time (in months).

Solution $400 \times 0.012 \times 2 = 9.60$

The interest charge is $9.60.

You Try It 3

Strategy   To find the interest earned:
- Find the new principal by multiplying the original principal by the factor (3.29066) found in the compound interest table.
- Subtract the original principal from the new principal.

Solution $1000 \times 3.29066 = 3290.66$

The new principal is $3290.66.

$3290.66 - 1000 = 2290.66$

The interest earned is $2290.66.

SECTION 6.4

You Try It 1

Strategy   To find the mortgage:
- Find the down payment by solving the basic percent equation for *amount*.
- Subtract the down payment from the purchase price.

Solution $0.25 \times 216,000 = n$
$\qquad\qquad\quad 54,000 = n$

The down payment is $54,000.

$216,000 - 54,000 = 162,000$

The mortgage is $162,000.

You Try It 2

Strategy   To find the loan origination fee, solve the basic percent equation for *amount*.

Solution $0.045 \times 80,000 = n$
$\qquad\qquad\quad 3600 = n$

The loan origination fee was $3600.

You Try It 3

Strategy   To find the monthly mortgage payment:
- Subtract the down payment from the purchase price to find the mortgage.
- Multiply the mortgage by the factor found in the monthly payment table.

Solution $75,000 - 15,000 = 60,000$
The mortgage is $60,000.
$60,000 \times 0.0089973 = 539.838$

The monthly mortgage payment is $539.84.

You Try It 4

Strategy   To find the interest:
- Multiply the mortgage by the factor found in the monthly payment table to find the monthly mortgage payment.
- Subtract the principal from the monthly mortgage payment.

Solution $125,000 \times 0.0083920 = 1049$

The monthly mortgage payment is $1049.

$1049 - 492.65 = 556.35$

The interest on the mortgage is $556.35.

You Try It 5

Strategy   To find the monthly payment:
- Divide the annual property tax by 12 to find the monthly property tax.
- Add the monthly property tax to the monthly mortgage payment.

Solution $744 \div 12 = 62$

The monthly property tax is $62.

$415.20 + 62 = 477.20$

The total monthly payment is $477.20.

SECTION 6.5

You Try It 1

Strategy   To find the amount financed:
- Find the down payment by solving the basic percent equation for *amount*.
- Subtract the down payment from the purchase price.

Solution $0.20 \times 9200 = n$
$\qquad\qquad 1840 = n$

The down payment is $1840.

$9200 - 1840 = 7360$

The amount financed is $7360.

You Try It 2

Strategy   To find the license fee, solve the basic percent equation for *amount*.

Solution $0.015 \times 7350 = n$
$\qquad\qquad 110.25 = n$

The license fee is $110.25.

You Try It 3

Strategy   To find the cost of operating the car, multiply the cost per mile by the number of miles driven.

Solution $23,000 \times 0.22 = 5060$

The cost of operating the car is $5060.

You Try It 4

Strategy   To find the cost per mile for the car insurance, divide the cost for insurance by the number of miles driven.

Solution $360 \div 15,000 = 0.024$

The cost per mile for insurance is $.024.

You Try It 5

Strategy   To find the monthly payment:
- Subtract the down payment from the purchase price to find the amount financed.
- Multiply the amount financed by the factor found in the monthly payment table.

Solution $15,900 - 3975 = 11,925$

The amount financed is $11,925.

$11,925 \times 0.0244129 \approx 291.123$

The monthly payment is $291.12.

SECTION 6.6

You Try It 1

Strategy   To find the worker's earnings:
- Find the worker's overtime wage by multiplying the hourly wage by 2.
- Multiply the number of overtime hours worked by the overtime wage.

Solution $8.50 \times 2 = 17$

The hourly wage for overtime is $17.

$17 \times 8 = 136$

The construction worker earns $136.

You Try It 2

Strategy To find the salary per month, divide the annual salary by the number of months in a year (12).

Solution 28,224 ÷ 12 = 2352

The contractor's monthly salary is $2352.

You Try It 3

Strategy To find the total earnings:
- Find the commission earned by multiplying the commission rate by the sales over $50,000.
- Add the commission to the annual salary.

Solution 175,000 − 50,000 = 125,000

Sales over $50,000 totaled $125,000.

125,000 × 0.095 = 11,875

Earnings from commissions totaled $11,875.

12,000 + 11,875 = 23,875

The insurance agent earned $23,875.

SECTION 6.7

You Try It 1

Strategy To find the current balance:
- Subtract the amount of the check from the old balance.
- Add the amount of each deposit.

Solution
$$
\begin{array}{ll}
302.46 & \\
-\ \ 20.59 & \text{check} \\
\hline
281.87 & \\
176.86 & \text{first deposit} \\
+\ \ 94.73 & \text{second deposit} \\
\hline
553.46 & \\
\end{array}
$$

The current checking account balance is $553.46.

You Try It 2

Current checkbook balance:	653.41
Check: 237	+ 48.73
	702.14
Interest:	+ 2.11
	704.25
Deposit:	−523.84
	180.41

Closing bank balance from bank statement: $180.41.

Checkbook balance: $180.41.

The bank statement and checkbook balance.

Answers to Chapter 1 Odd-Numbered Exercises

SECTION 1.1

1. **3.** **5.** $37 < 49$ **7.** $101 > 87$

9. $245 > 158$ **11.** $0 < 45$ **13.** $815 < 928$ **15.** Three thousand seven hundred ninety

17. Fifty-eight thousand four hundred seventy three **19.** Four hundred ninety-eight thousand five hundred twelve

21. Six million eight hundred forty-two thousand seven hundred fifteen **23.** 357 **25.** 63,780 **27.** 7,024,709

29. $6000 + 200 + 90 + 5$ **31.** $400,000 + 50,000 + 3000 + 900 + 20 + 1$ **33.** $300,000 + 1000 + 800 + 9$

35. $3,000,000 + 600 + 40 + 2$ **37.** 850 **39.** 4000 **41.** 53,000 **43.** 250,000 **45.** 999; 10,000

47. No. Round 3846 to the nearest hundred.

SECTION 1.2

1. 28 **3.** 125 **5.** 102 **7.** 154 **9.** 1489 **11.** 828 **13.** 1584 **15.** 1219 **17.** 102,317 **19.** 79,326

21. 1804 **23.** 1579 **25.** 19,740 **27.** 7420 **29.** 120,570 **31.** 207,453 **33.** 24,218 **35.** 11,974

37. 9323 **39.** 77,139 **41.** 14,383 **43.** 9473 **45.** 33,247 **47.** 5058 **49.** 1992 **51.** 68,263

53. Est.: 17,700 **55.** Est.: 2900 **57.** Est.: 101,000 **59.** Est.: 158,000 **61.** Est.: 260,000 **63.** Est.: 940,000
Cal.: 17,754   Cal.: 2872   Cal.: 101,712   Cal.: 158,763   Cal.: 261,595   Cal.: 946,718

65. Est.: 33,000,000 **67.** Est.: 34,000,000 **69.** The total amount of trade with the United States was $603 billion.
Cal.: 32,691,621   Cal.: 34,420,922

71. The total number of yards gained by passing was 307. **73.** The total income from the five Disney productions was $900,600,000 **75.** Yes, the total income from the two productions with the lowest box-office incomes does exceed the income from *Aladdin*. **77a.** 1285 miles will be driven. **b.** The odometer will read 69,977 miles. **79.** The total average amount for Americans ages 16 to 34 in checking accounts, savings accounts, and U.S. Savings Bonds is $1796. **81.** The sum of the average amounts invested in home equity and retirement is greater for all Americans than that same sum for Americans between the ages of 16 and 34. **83.** 11 different sums **85.** No. $0 + 0 = 0$ **87.** 10 numbers

SECTION 1.3

1. 4 **3.** 4 **5.** 10 **7.** 4 **9.** 9 **11.** 22 **13.** 60 **15.** 66 **17.** 31 **19.** 901 **21.** 791 **23.** 1125

25. 3131 **27.** 47 **29.** 925 **31.** 4561 **33.** 3205 **35.** 1222 **37.** 3021 **39.** 3022 **41.** 3040

43. 212 **45.** 60,245 **47.** 65 **49.** 17 **51.** 23 **53.** 456 **55.** 57 **57.** 375 **59.** 3139 **61.** 3621

63. 738 **65.** 3545 **67.** 749 **69.** 5343 **71.** 66,463 **73.** 16,590 **75.** 52,404 **77.** 38,777 **79.** 4638

81. 3612 **83.** 2913 **85.** 2583 **87.** 5268 **89.** 71,767 **91.** 11,239 **93.** 8482 **95.** 625 **97.** 76,725

99. Est.: 30,000 **101.** Est.: 40,000 **103.** Est.: 100,000
Cal.: 29,837   Cal.: 36,668   Cal.: 101,998

105. Sam saved $2479 off the sticker price. **107.** The amount that remains to be paid is $4775. **109.** The national debt increased $220,000,000,000 in one year. **111.** Florida has lost 9,286,713 acres of wetlands over the past 200 years. **113.** The Chevrolet with all the options will cost more. **115.** It will take 27 more months to withdraw principal and interest in 2012 than it did in 1992. **117.** Answers will vary.

SECTION 1.4

1. 12 **3.** 35 **5.** 25 **7.** 0 **9.** 72 **11.** 198 **13.** 335 **15.** 2492 **17.** 5463 **19.** 4200 **21.** 6327

23. 1896 **25.** 5056 **27.** 1685 **29.** 46,963 **31.** 59,976 **33.** 19,120 **35.** 19,790 **37.** 108

39. 3664 **41.** 20,036 **43.** 71,656 **45.** 432 **47.** 1944 **49.** 41,832 **51.** 43,620 **53.** 335,195

55. 594,625 **57.** 321,696 **59.** 342,171 **61.** 279,220 **63.** 191,800 **65.** 463,712 **67.** 180,621

69. 478,800 **71.** 158,422 **73.** 4,696,714 **75.** 5,542,452 **77.** 51,443 **79.** 18,834 **81.** 260,178

83. 315,109,895 **85.** Est.: 450,000 **87.** Est.: 4,200,000 **89.** Est.: 6,300,000 **91.** Est.: 18,000,000
Cal.: 440,076   Cal.: 4,315,403   Cal.: 6,491,166   Cal.: 18,728,744

93. Est.: 54,000,000 **95.** Est.: 56,000,000 **97.** The plane used 5190 gallons of fuel on a 6-hour flight.
Cal.: 57,691,192   Cal.: 56,324,340

99. The machine can fill and cap 168,000 bottles in 40 hours.   **101.** The estimated attendance for the 11 home games is 300,000. The total attendance for the 15 games is 397,159.   **103.** The lighting designer can save $84.   **105.** The total wages of the four plumbers are $1380.   **107a.** Always true   **b.** Always true   **c.** Sometimes true; c is a true statement except in the cases described in parts a and b.   **109.** 17,112 acres are deforested each day. 513,360 acres are deforested each month. 6,245,880 acres are deforested each year.   **111.** Answers will vary.

SECTION 1.5

1. 2   **3.** 6   **5.** 7   **7.** 16   **9.** 210   **11.** 44   **13.** 703   **15.** 910   **17.** 21,560   **19.** 3580   **21.** 482   **23.** 1075   **25.** 2 rl   **27.** 5 r2   **29.** 13 rl   **31.** 10 r3   **33.** 90 r2   **35.** 230 rl   **37.** 204 r3   **39.** 1347 r3   **41.** 1720 r2   **43.** 409 r2   **45.** 6214 r2   **47.** 8708 r2   **49.** 1080 r2   **51.** 4200   **53.** 19,600   **55.** 1 r38   **57.** 1 r26   **59.** 21 r21   **61.** 30 r22   **63.** 5 r40   **65.** 9 r17   **67.** 200 r21   **69.** 303 r1   **71.** 67 r13   **73.** 176 r13   **75.** 1086 r7   **77.** 403   **79.** 12 r456   **81.** 4 r160   **83.** 160 r27   **85.** 1669 r14   **87.** 7950   **89.** Est.: 2000 Cal.: 2225   **91.** Est.: 10,000 Cal.: 11,016   **93.** Est.: 30,000 Cal.: 26,656   **95.** Est.: 500 Cal.: 504   **97.** Est.: 500 Cal.: 541   **99.** Est.: 20,000 Cal.: 20,621   **101.** The average number of miles traveled on each gallon of gas was 27.   **103.** Belle receives $156,250 for each home run.   **105.** The number of bytes of information on 1 disk is 368,640.   **107. a.** The remaining balance to be paid is $8976.   **b.** The monthly payment is $187.   **109.** Lemieux's hourly wage is $5660.   **111.** Belle's average salary per game is $61,728. Aikman's average salary per game is $335,625. Aikman's per game salary is approximately 5 times Belle's per game salary.   **113.** The smallest four-digit palindromic number is 2112.   **115a.** False; division by zero is not allowed.   **b.** False; division by zero is not allowed.   **c.** True

SECTION 1.6

1. 2^3 **3.** $6^3 \cdot 7^4$ **5.** $2^3 \cdot 3^3$ **7.** $5 \cdot 7^5$ **9.** $3^3 \cdot 6^4$ **11.** $3^3 \cdot 5 \cdot 9^3$ **13.** 8   **15.** 400   **17.** 900   **19.** 972   **21.** 120   **23.** 360   **25.** 0   **27.** 90,000   **29.** 540   **31.** 4050   **33.** 11,025   **35.** 25,920   **37.** 4,320,000   **39.** 5   **41.** 10   **43.** 47   **45.** 8   **47.** 5   **49.** 8   **51.** 6   **53.** 53   **55.** 44   **57.** 19   **59.** 67   **61.** 168   **63.** 27   **65.** 14   **67.** 10   **69.** 9   **71.** 12   **73.** 32   **75.** 39   **77.** 1024   **79.** Yes. Answers will vary.

SECTION 1.7

1. 1, 2, 4   **3.** 1, 2, 5, 10   **5.** 1, 7   **7.** 1, 3, 9   **9.** 1, 13   **11.** 1, 2, 3, 6, 9, 18   **13.** 1, 2, 4, 7, 8, 14, 28, 56   **15.** 1, 3, 5, 9, 15, 45   **17.** 1, 29   **19.** 1, 2, 11, 22   **21.** 1, 2, 4, 13, 26, 52   **23.** 1, 2, 41, 82   **25.** 1, 3, 19, 57   **27.** 1, 2, 3, 4, 6, 8, 12, 16, 24, 48   **29.** 1, 5, 19, 95   **31.** 1, 2, 3, 6, 9, 18, 27, 54   **33.** 1, 2, 3, 6, 11, 22, 33, 66   **35.** 1, 2, 4, 5, 8, 10, 16, 20, 40, 80   **37.** 1, 2, 3, 4, 6, 8, 12, 16, 24, 32, 48, 96   **39.** 1, 2, 3, 5, 6, 9, 10, 15, 18, 30, 45, 90   **41.** $2 \cdot 3$ **43.** Prime   **45.** $2 \cdot 2 \cdot 2 \cdot 3$ **47.** $3 \cdot 3 \cdot 3$ **49.** $2 \cdot 2 \cdot 3 \cdot 3$ **51.** Prime   **53.** $2 \cdot 3 \cdot 3 \cdot 5$ **55.** $5 \cdot 23$ **57.** $2 \cdot 3 \cdot 3$ **59.** $2 \cdot 2 \cdot 7$ **61.** Prime   **63.** $2 \cdot 31$ **65.** $2 \cdot 11$ **67.** Prime   **69.** $2 \cdot 3 \cdot 11$ **71.** $2 \cdot 37$ **73.** Prime   **75.** $5 \cdot 11$ **77.** $2 \cdot 2 \cdot 2 \cdot 3 \cdot 5$ **79.** $2 \cdot 2 \cdot 2 \cdot 2 \cdot 2 \cdot 5$ **81.** $2 \cdot 2 \cdot 2 \cdot 3 \cdot 3 \cdot 3$ **83.** $5 \cdot 5 \cdot 5 \cdot 5$ **85.** 3 and 5, 5 and 7, 11 and 13. Other answers are possible.   **87.** Answers will vary.

CHAPTER REVIEW

1. 600   [1.6A]   **2.** 10,000 + 300 + 20 + 7   [1.1C]   **3.** 1, 2, 3, 6, 9, 18   [1.7A]   **4.** 12,493   [1.2A]   **5.** 1749   [1.3B]   **6.** 2135   [1.5A]   **7.** 101 > 87   [1.1A]   **8.** $5^2 \cdot 7^5$ [1.6A]   **9.** 619,833   [1.4B]   **10.** 5409   [1.3B]   **11.** 1081   [1.2A]   **12.** 2   [1.6B]   **13.** $2 \cdot 3 \cdot 7$ [1.7B]   **14.** Two hundred seventy-six thousand fifty-seven   [1.1B]   **15.** 1306 r59   [1.5C]   **16.** 2,011,044   [1.1B]   **17.** 1, 2, 3, 5, 6, 10, 15, 30   [1.7A]   **18.** 17   [1.6B]   **19.** 32   [1.6B]   **20.** $2 \cdot 2 \cdot 2 \cdot 3 \cdot 3$ [1.7B]   **21.** $2^4 \cdot 5^3$ [1.6A]   **22.** 22,761   [1.4B]   **23.** Vincent's total pay for the last week was $384.   [1.4C]   **24.** He drove 27 miles per gallon of gasoline.   [1.5D]   **25.** The monthly car payment is $155.   [1.5D]   **26.** The total income from commissions was $2567.   [1.2B]   **27.** The total amount deposited was $301. The new checking account balance is $817.   [1.2B]   **28.** The total of the car payments is $1476.   [1.4C]   **29.** A U.S. company spends $1985 per person more than an Italian company.   [1.3C]   **30.** A U.S. company spends $563 more for business travel and entertainment.   [1.3C]   **31.** No   [1.3C]

CHAPTER TEST

1. 432 [1.6A]   **2.** Two hundred seven thousand sixty-eight [1.1B]   **3.** 9333 [1.3A]   **4.** 1, 2, 4, 5, 10, 20 [1.7A]
5. 6,854,144 [1.4B]   **6.** 9 [1.6B]   **7.** 900,000 + 6000 + 300 + 70 + 8 [1.1C]   **8.** 75,000 [1.1D]
9. 1121 r27 [1.5C]   **10.** $3^3 \cdot 7^2$ [1.6A]   **11.** 135,915 [1.2A]   **12.** $2 \cdot 2 \cdot 3 \cdot 7$ [1.7B]   **13.** 4 [1.6B]
14. 726,104 [1.4A]   **15.** 1,204,006 [1.1B]   **16.** 8710 r2 [1.5B]   **17.** 21 > 19 [1.1A]   **18.** 703 [1.5A]
19. 96,798 [1.2A]   **20.** 19,922 [1.3B]   **21.** The salary difference between mechanical engineers and industrial
engineers is $2120. [1.3C]   **22.** The average salary of the degree candidates is $39,788. [1.5D]   **23.** The farmer
needed 3000 boxes. [1.5D]   **24.** The investor will receive $2844 over 12 months. [1.4C]   **25a.** They drove
855 miles. [1.2B]   **b.** The odometer reading was 48,481. [1.2B]

Answers to Chapter 2 Odd-Numbered Exercises

SECTION 2.1

1. 40   **3.** 24   **5.** 30   **7.** 12   **9.** 24   **11.** 60   **13.** 56   **15.** 9   **17.** 32   **19.** 36   **21.** 660
23. 9384   **25.** 24   **27.** 30   **29.** 24   **31.** 576   **33.** 1680   **35.** 1   **37.** 3   **39.** 5   **41.** 25   **43.** 1
45. 4   **47.** 4   **49.** 6   **51.** 4   **53.** 1   **55.** 7   **57.** 5   **59.** 8   **61.** 1   **63.** 25   **65.** 7   **67.** 8
69. Numbers with no common factors except the factor 1. 4, 5; 8, 9; 16, 21 (Answers will vary.)   **71.** The LCM of 2 and 3
is 6, of 5 and 7 is 35, of 11 and 19 is 209. The LCM of two prime numbers is the product of the two numbers. The LCM of
three prime numbers is the product of the three numbers.   **73.** Yes. Each number contains all the factors of the GCF.
75. 44 solar years; 73 ritual years

SECTION 2.2

1. $\frac{3}{4}$ **3.** $\frac{7}{8}$ **5.** $1\frac{1}{2}$ **7.** $2\frac{5}{8}$ **9.** $3\frac{3}{5}$ **11.** $\frac{5}{4}$ **13.** $\frac{8}{3}$ **15.** $\frac{28}{8}$ **17.**   **19.**

21.   **23.** $2\frac{3}{4}$ **25.** 5   **27.** $1\frac{1}{8}$ **29.** $2\frac{3}{10}$ **31.** 3   **33.** $1\frac{1}{7}$ **35.** $2\frac{1}{3}$ **37.** 16   **39.** $2\frac{1}{8}$

41. $2\frac{2}{5}$ **43.** 1   **45.** 9   **47.** $\frac{7}{3}$ **49.** $\frac{13}{2}$ **51.** $\frac{41}{6}$ **53.** $\frac{37}{4}$ **55.** $\frac{21}{2}$ **57.** $\frac{73}{9}$ **59.** $\frac{58}{11}$ **61.** $\frac{21}{8}$

63. $\frac{13}{8}$ **65.** $\frac{100}{9}$ **67.** $\frac{27}{8}$ **69.** $\frac{85}{13}$ **71.** $\frac{8}{50}$; $\frac{8}{50}$ **73.** Answers will vary.

SECTION 2.3

1. $\frac{5}{10}$ **3.** $\frac{9}{48}$ **5.** $\frac{12}{32}$ **7.** $\frac{9}{51}$ **9.** $\frac{12}{16}$ **11.** $\frac{27}{9}$ **13.** $\frac{20}{60}$ **15.** $\frac{44}{60}$ **17.** $\frac{12}{18}$ **19.** $\frac{35}{49}$ **21.** $\frac{10}{18}$

23. $\frac{21}{3}$ **25.** $\frac{35}{45}$ **27.** $\frac{60}{64}$ **29.** $\frac{21}{98}$ **31.** $\frac{30}{48}$ **33.** $\frac{15}{42}$ **35.** $\frac{102}{144}$ **37.** $\frac{153}{408}$ **39.** $\frac{340}{800}$ **41.** $\frac{1}{3}$ **43.** $\frac{1}{2}$

45. $\frac{1}{6}$ **47.** $1\frac{1}{9}$ **49.** 0   **51.** $\frac{9}{22}$ **53.** 3   **55.** $\frac{4}{21}$ **57.** $\frac{12}{35}$ **59.** $\frac{7}{11}$ **61.** $1\frac{1}{3}$ **63.** $\frac{3}{5}$ **65.** $\frac{1}{11}$

67. 4   **69.** $\frac{1}{3}$ **71.** $\frac{3}{5}$ **73.** $2\frac{1}{4}$ **75.** $\frac{1}{5}$ **77.** Answers will vary. Examples are $\frac{12}{4}$, $\frac{15}{5}$, $\frac{21}{7}$, $\frac{24}{8}$, and $\frac{30}{10}$.
79. Answers will vary.

SECTION 2.4

1. $\frac{3}{7}$ **3.** 1   **5.** $1\frac{4}{11}$ **7.** $3\frac{2}{5}$ **9.** $2\frac{4}{5}$ **11.** $2\frac{1}{4}$ **13.** $1\frac{3}{8}$ **15.** $1\frac{7}{15}$ **17.** $\frac{15}{16}$ **19.** $1\frac{4}{11}$ **21.** 1

23. $1\frac{7}{8}$ **25.** $1\frac{1}{6}$ **27.** $\frac{13}{14}$ **29.** $\frac{53}{60}$ **31.** $1\frac{1}{56}$ **33.** $\frac{23}{60}$ **35.** $\frac{56}{57}$ **37.** $1\frac{17}{18}$ **39.** $1\frac{11}{48}$ **41.** $1\frac{9}{20}$

43. $1\frac{109}{180}$ **45.** $1\frac{91}{144}$ **47.** $2\frac{5}{72}$ **49.** $\frac{39}{40}$ **51.** $1\frac{19}{24}$ **53.** $1\frac{65}{72}$ **55.** $3\frac{2}{3}$ **57.** $10\frac{1}{12}$ **59.** $9\frac{2}{7}$ **61.** $6\frac{7}{40}$

63. $9\frac{47}{48}$ **65.** $8\frac{22}{39}$ **67.** $16\frac{29}{120}$ **69.** $24\frac{29}{40}$ **71.** $33\frac{7}{24}$ **73.** $10\frac{5}{36}$ **75.** $10\frac{5}{12}$ **77.** $14\frac{73}{90}$ **79.** $10\frac{13}{48}$

81. $42\frac{13}{54}$ **83.** $8\frac{1}{36}$ **85.** $14\frac{1}{12}$ **87.** $9\frac{37}{72}$ **89.** The fractional amount of income spent on these three items is $\frac{17}{24}$.

91. The value of the stock after the increase is $\$84\frac{3}{4}$.   **93.** The length of the shaft is $8\frac{9}{16}$ inches.   **95a.** The total number of overtime hours worked was 12.   **b.** The overtime pay is $264.   **97.** The value of the stock at the end of the three months is $\$30\frac{5}{8}$.   **99.** Answers will vary.   **101.** No. These activities account for only 22 hours.

SECTION 2.5

1. $\frac{2}{17}$ **3.** $\frac{1}{3}$ **5.** $\frac{1}{10}$ **7.** $\frac{5}{13}$ **9.** $\frac{1}{3}$ **11.** $\frac{4}{7}$ **13.** $\frac{1}{4}$ **15.** $\frac{9}{23}$ **17.** $\frac{1}{4}$ **19.** $\frac{1}{2}$ **21.** $\frac{19}{56}$ **23.** $\frac{1}{2}$

25. $\frac{11}{60}$ **27.** $\frac{15}{56}$ **29.** $\frac{37}{51}$ **31.** $\frac{17}{70}$ **33.** $\frac{49}{120}$ **35.** $\frac{8}{45}$ **37.** $\frac{11}{21}$ **39.** $\frac{23}{60}$ **41.** $\frac{1}{18}$ **43.** $5\frac{1}{5}$ **45.** $10\frac{9}{17}$

47. $4\frac{7}{8}$ **49.** $\frac{16}{21}$ **51.** $5\frac{1}{2}$ **53.** $5\frac{4}{7}$ **55.** $7\frac{5}{24}$ **57.** $48\frac{31}{70}$ **59.** $85\frac{2}{9}$ **61.** $9\frac{5}{13}$ **63.** $15\frac{11}{20}$ **65.** $4\frac{23}{24}$

67. $2\frac{11}{15}$ **69.** $9\frac{1}{2}$ inches.   **71.** Meyfarth jumped $9\frac{3}{8}$ inches farther than Coachman. Kostadinova jumped $5\frac{1}{4}$ inches farther than Meyfarth.   **73a.** It is $7\frac{17}{24}$ miles to the second checkpoint.   **b.** It is $4\frac{7}{24}$ miles from the second checkpoint to the finish line.   **75a.** Yes.   **b.** The wrestler must lose $3\frac{1}{4}$ pounds in the third week.   **77.** The difference in price is $\$45\frac{5}{8}$.   **79.** $2\frac{5}{6}$ **81.**

$\frac{3}{8}$	$\frac{3}{4}$	$\frac{3}{4}$
1	$\frac{5}{8}$	$\frac{1}{4}$
$\frac{1}{2}$	$\frac{1}{2}$	$\frac{7}{8}$

SECTION 2.6

1. $\frac{7}{12}$ **3.** $\frac{7}{48}$ **5.** $\frac{1}{48}$ **7.** $\frac{11}{14}$ **9.** $\frac{1}{7}$ **11.** $\frac{1}{8}$ **13.** 6   **15.** $\frac{5}{12}$ **17.** 6   **19.** $\frac{2}{3}$ **21.** $\frac{3}{16}$ **23.** $\frac{3}{80}$

25. 10   **27.** $\frac{1}{15}$ **29.** $\frac{2}{3}$ **31.** $\frac{7}{26}$ **33.** 4   **35.** $\frac{100}{357}$ **37.** $\frac{5}{24}$ **39.** $\frac{1}{12}$ **41.** $\frac{4}{15}$ **43.** $1\frac{1}{2}$ **45.** 4

47. $\frac{4}{9}$ **49.** $\frac{1}{2}$ **51.** $16\frac{1}{2}$ **53.** 10   **55.** $6\frac{3}{7}$ **57.** $18\frac{1}{3}$ **59.** $1\frac{5}{7}$ **61.** $3\frac{1}{2}$ **63.** $25\frac{5}{8}$ **65.** $1\frac{11}{16}$ **67.** 16

69. 3   **71.** $8\frac{37}{40}$ **73.** $6\frac{19}{28}$ **75.** $7\frac{2}{5}$ **77.** 16   **79.** $32\frac{4}{5}$ **81.** $28\frac{1}{2}$ **83.** 9   **85.** $\frac{5}{8}$ **87.** $3\frac{1}{40}$

89. The salmon cost $22.   **91.** The Honda Civic can travel 361 miles on $9\frac{1}{2}$ gallons of gasoline.   **93.** 535,000 gallons of propellant is used before burnout.   **95a.** 30 students passed the chemistry course.   **b.** 6 of the chemistry students received an A grade.   **97.** $267,000 of the monthly income remained.   **99.** The weight is $54\frac{19}{36}$ pounds.   **101.** $\frac{3}{5}$ of the park is heavily wooded.   **103.** $\frac{1}{2}$ **105.** Yes   **107.** A

SECTION 2.7

1. $\frac{5}{6}$ **3.** 1   **5.** 0   **7.** $\frac{1}{2}$ **9.** $2\frac{73}{256}$ **11.** $1\frac{1}{9}$ **13.** $\frac{1}{6}$ **15.** $\frac{7}{10}$ **17.** 2   **19.** 2   **21.** $\frac{1}{6}$ **23.** 6

25. $\frac{1}{15}$ **27.** 2   **29.** $2\frac{1}{2}$ **31.** 3   **33.** 6   **35.** $\frac{1}{2}$ **37.** $1\frac{1}{6}$ **39.** $3\frac{1}{3}$ **41.** $\frac{1}{30}$ **43.** $1\frac{4}{5}$ **45.** 13

47. $\frac{25}{288}$ **49.** 3 **51.** $\frac{1}{5}$ **53.** $\frac{11}{28}$ **55.** $\frac{5}{86}$ **57.** 120 **59.** $\frac{11}{40}$ **61.** $\frac{33}{40}$ **63.** $4\frac{4}{9}$ **65.** $\frac{13}{32}$ **67.** $10\frac{2}{3}$

69. $9\frac{39}{40}$ **71.** $\frac{12}{53}$ **73.** $4\frac{62}{191}$ **75.** 68 **77.** $8\frac{2}{7}$ **79.** $3\frac{13}{49}$ **81.** 4 **83.** $1\frac{3}{5}$ **85.** $\frac{9}{34}$ **87.** The package

contains twelve $1\frac{1}{3}$-ounce servings. **89.** Each acre cost $24,000. **91.** The nut makes 12 turns to move $1\frac{7}{8}$ inches.

93a. The total weight of the fat and bone is $1\frac{5}{12}$ pounds. **b.** There will be 28 servings. **95a.** The actual length of wall

a is $12\frac{1}{2}$ feet. **b.** The actual length of wall b is 18 feet. **c.** The actual length of wall c is $15\frac{3}{4}$ feet. **97.** No. For

example, $4 \div \frac{1}{2} = 8$. Answers will vary. **99.** Each column is $2\frac{1}{4}$ inches wide. **101a.** $2\frac{1}{2}$ **b.** $\frac{2}{3}$ **c.** $\frac{1}{3}$ **d.** $2\frac{1}{2}$

SECTION 2.8

1. $\frac{11}{40} < \frac{19}{40}$ **3.** $\frac{2}{3} < \frac{5}{7}$ **5.** $\frac{5}{8} > \frac{7}{12}$ **7.** $\frac{7}{9} < \frac{11}{12}$ **9.** $\frac{13}{14} > \frac{19}{21}$ **11.** $\frac{7}{24} < \frac{11}{30}$ **13.** $\frac{9}{64}$ **15.** $\frac{8}{729}$ **17.** $\frac{1}{24}$

19. $\frac{8}{245}$ **21.** $\frac{1}{121}$ **23.** $\frac{81}{625}$ **25.** $\frac{7}{36}$ **27.** $\frac{27}{49}$ **29.** $\frac{16}{75}$ **31.** $\frac{5}{6}$ **33.** $1\frac{5}{12}$ **35.** $1\frac{1}{5}$ **37.** $\frac{7}{48}$ **39.** $\frac{29}{36}$

41. $\frac{55}{72}$ **43.** $\frac{35}{54}$ **45.** 2 **47.** $\frac{9}{19}$ **49.** $\frac{7}{32}$ **51.** $\frac{64}{75}$ **53.** Answers will vary.

CHAPTER REVIEW

1. $\frac{2}{3}$ [2.3B] **2.** $\frac{5}{16}$ [2.8B] **3.** $\frac{13}{4}$ [2.2A] **4.** $1\frac{13}{18}$ [2.4B] **5.** $\frac{11}{18} < \frac{17}{24}$ [2.8A] **6.** $14\frac{19}{42}$ [2.5C]

7. $\frac{5}{36}$ [2.8C] **8.** $9\frac{1}{24}$ [2.6B] **9.** 2 [2.7B] **10.** $\frac{25}{48}$ [2.5B] **11.** $3\frac{1}{3}$ [2.7B] **12.** 4 [2.1B]

13. $\frac{24}{36}$ [2.3A] **14.** $\frac{3}{4}$ [2.7A] **15.** $\frac{32}{44}$ [2.3A] **16.** $16\frac{1}{2}$ [2.6B] **17.** 36 [2.1A] **18.** $\frac{4}{11}$ [2.3B]

19. $1\frac{1}{8}$ [2.4A] **20.** $10\frac{1}{8}$ [2.5C] **21.** $18\frac{13}{54}$ [2.4C] **22.** 5 [2.1B] **23.** $3\frac{2}{5}$ [2.2B] **24.** $\frac{1}{15}$ [2.8C]

25. $5\frac{7}{8}$ [2.4C] **26.** 54 [2.1A] **27.** $\frac{1}{3}$ [2.5A] **28.** $\frac{19}{7}$ [2.2B] **29.** 2 [2.7A] **30.** $\frac{1}{15}$ [2.6A]

31. $\frac{1}{8}$ [2.6A] **32.** $1\frac{7}{8}$ [2.2A] **33.** The total rainfall was $21\frac{7}{24}$ inches. [2.4D] **34.** Each acre cost $36,000. [2.7C]

35. The second checkpoint is $4\frac{3}{4}$ miles from the finish line. [2.5D] **36.** The car can travel 243 miles. [2.6C]

CHAPTER TEST

1. $\frac{4}{9}$ [2.6A] **2.** 8 [2.1B] **3.** $1\frac{3}{7}$ [2.7A] **4.** $\frac{7}{24}$ [2.8C] **5.** $\frac{49}{5}$ [2.2B] **6.** 8 [2.6B] **7.** $\frac{5}{8}$ [2.3B]

8. $\frac{3}{8} < \frac{5}{12}$ [2.8A] **9.** $\frac{5}{6}$ [2.8C] **10.** 120 [2.1A] **11.** $\frac{1}{4}$ [2.5A] **12.** $3\frac{3}{5}$ [2.2B] **13.** $2\frac{2}{19}$ [2.7B]

14. $\frac{45}{72}$ [2.3A] **15.** $1\frac{61}{90}$ [2.4B] **16.** $13\frac{81}{88}$ [2.5C] **17.** $\frac{7}{48}$ [2.5B] **18.** $\frac{1}{6}$ [2.8B] **19.** $1\frac{11}{12}$ [2.4A]

20. $22\frac{4}{15}$ [2.4C] **21.** $\frac{11}{4}$ [2.2A] **22.** The electrician's total earnings are $420. [2.6C] **23.** There were 11 lots

available for sale. [2.7C] **24.** The value of one share is $27\frac{7}{8}$. [2.5D] **25.** The total rainfall was $21\frac{11}{24}$ inches. [2.4D]

CUMULATIVE REVIEW

1. 290,000 [1.1D] **2.** 291,278 [1.3B] **3.** 73,154 [1.4B] **4.** 540 r12 [1.5C] **5.** 1 [1.6B]

6. $2 \cdot 2 \cdot 11$ [1.7B] **7.** 210 [2.1A] **8.** 20 [2.1B] **9.** $\frac{23}{3}$ [2.2B] **10.** $6\frac{1}{4}$ [2.2B] **11.** $\frac{15}{48}$ [2.3A]

12. $\frac{2}{5}$ [2.3B]   **13.** $1\frac{7}{48}$ [2.4B]   **14.** $14\frac{11}{48}$ [2.4C]   **15.** $\frac{13}{24}$ [2.5B]   **16.** $1\frac{7}{9}$ [2.5C]   **17.** $\frac{7}{20}$ [2.6A]

18. $7\frac{1}{2}$ [2.6B]   **19.** $1\frac{1}{20}$ [2.7A]   **20.** $2\frac{5}{8}$ [2.7B]   **21.** $\frac{1}{9}$ [2.8B]   **22.** $5\frac{5}{24}$ [2.8C]

23. There is $862 in the checking account at the end of the week.   [1.3C]   **24.** The total income is $705.   [1.4C]

25. The total weight is $12\frac{1}{24}$ pounds.   [2.4D]   **26.** The length of the remaining piece is $4\frac{17}{24}$ feet.   [2.5D]

27. The car traveled 225 miles.   [2.6C]   **28.** 25 parcels can be sold from the remaining land.   [2.7C]

Answers to Chapter 3 Odd-Numbered Exercises

SECTION 3.1

1. Twenty-seven hundredths   **3.** One and five thousandths   **5.** Thirty-six and four tenths   **7.** Thirty-five hundred-thousandths   **9.** Ten and seven thousandths   **11.** Fifty-two and ninety-five hundred-thousandths   **13.** Two hundred ninety-three ten-thousandths   **15.** Six and three hundred twenty-four thousandths   **17.** Two hundred seventy-six and three thousand two hundred ninety-seven ten-thousandths   **19.** Two hundred sixteen and seven hundred twenty-nine ten-thousandths   **21.** Four thousand six hundred twenty-five and three hundred seventy-nine ten-thousandths   **23.** One and one hundred-thousandth   **25.** 0.762   **27.** 0.000062   **29.** 8.0304   **31.** 304.07   **33.** 362.048   **35.** 3048.2002   **37.** 7.4   **39.** 23.0   **41.** 22.68   **43.** 7.073   **45.** 62.009   **47.** 0.0123   **49.** 2.07924   **51.** 0.100975   **53.** Answers will vary.   **55a.** 5.5 and 6.4   **b.** 10.15 and 10.24

SECTION 3.2

1. 150.1065   **3.** 95.8446   **5.** 69.644   **7.** 92.883   **9.** 113.205   **11.** 0.69   **13.** 16.305   **15.** 110.7666   **17.** 104.4959   **19.** Est.: 234   Cal.: 234.192   **21.** Est.: 782   Cal.: 781.943   **23.** The length of the shaft is 5.65 inches.   **25.** The total amount of gas used is 47.8 gallons.   **27.** The amount in the checking account is $3664.20.   **29.** The total number of television viewers is 366.2 million.   **31.** No, the customer does not have enough money.   **33.** No, the rope cannot be wrapped around the box.

SECTION 3.3

1. 5.627   **3.** 113.6427   **5.** 6.7098   **7.** 215.697   **9.** 53.8776   **11.** 72.7091   **13.** 4.685   **15** 10.0365   **17.** 0.7727   **19.** 3.273   **21.** 791.247   **23.** 547.951   **25.** 403.8557   **27.** 22.479   **29.** Est.: 600   Cal.: 590.25   **31.** Est.: 30   Cal.: 35.194   **33.** Est.: 3   Cal.: 2.74506   **35.** Est.: 7   Cal.: 7.14925   **37.** The missing dimension is 7.55 inches.   **39a.** The total amount of the checks is $607.36.   **b.** Your new balance is $422.38.   **41.** The rainfall was 0.86 inch below normal.   **43.** The difference in average speed was 40.124 mph.   **45.** $167.61 would be saved by buying from Waterhouse Securities.

SECTION 3.4

1. 0.36   **3.** 0.30   **5.** 0.25   **7.** 0.45   **9.** 6.93   **11.** 1.84   **13.** 4.32   **15.** 0.74   **17.** 39.5   **19.** 2.72   **21.** 0.603   **23.** 0.096   **25.** 13.50   **27.** 79.80   **29.** 4.316   **31.** 1.794   **33.** 0.06   **35.** 0.072   **37.** 0.1323   **39.** 0.03568   **41.** 0.0784   **43.** 0.076   **45.** 34.48   **47.** 580.5   **49.** 20.148   **51.** 0.04255   **53.** 0.17686   **55.** 0.19803   **57.** 14.8657   **59.** 0.0006608   **61.** 53.9961   **63.** 0.536335   **65.** 0.429   **67.** 2.116   **69.** 0.476   **71.** 1.022   **73.** 37.96   **75.** 2.318   **77.** 3.2   **79** 6.5   **81.** 6285.6   **83.** 3200   **85.** 35,700   **87.** 6.3   **89.** 3.9   **91.** 49,000   **93.** 6.7   **95.** 0.012075   **97.** 0.0117796   **99.** 0.31004   **101.** 0.082845   **103.** 5.175   **105.** Est.: 90   Cal.: 91.2   **107.** Est.: 0.8   Cal.: 1.0472   **109.** Est.: 4.5   Cal.: 3.897   **111.** Est.: 12   Cal.: 11.2406   **113.** Est.: 0.32   Cal.: 0.371096   **115.** Est.: 30   Cal.: 31.8528

117. Est.: 2000   **119.** Est.: 0.00005   **121.** The cost to rent the car is $188.40.
Cal.: 1941.069459   Cal.: 0.000043512

123. The amount received from recycling is $14.06.   **125.** The broker's fee is $173.25.   **127a.** The amount of the payments is $4590.   **b.** The total cost of the car is $6590.   **129.** The total cost to rent the car is $64.20.

131. Grade 1 of steel costs $21.12, Grade 2 costs $29.84, and Grade 3 costs $201.66.   **133a.** The cost is $52.90. **b.** The cost is $79.60.   **c.** The cost is $61.45.   **135.** The complete solution is in the Solutions Manual.   **137.** The total cost for labor is $1029.88.   **139.** Answers will vary.   **141.** $1.3 \times 2.31 = \frac{13}{10} \times \frac{231}{100} = \frac{3003}{1000} = 3.003$

SECTION 3.5

1. 0.82   **3.** 4.8   **5.** 89   **7.** 60   **9.** 84.3   **11.** 32.3   **13.** 5.06   **15.** 1.3   **17.** 0.11   **19.** 3.8   **21.** 6.3
23. 0.6   **25.** 2.5   **27.** 1.1   **29.** 130.6   **31.** 0.81   **33.** 42.40   **35.** 40.70   **37.** 0.46   **39.** 0.019
41. 0.087   **43.** 0.360   **45.** 0.103   **47.** 0.009   **49.** 1   **51.** 3   **53.** 1   **55.** 57   **57.** 0.407   **59.** 4.267
61. 0.01037   **63.** 0.008295   **65.** 0.82537   **67.** 0.032   **69.** 0.23627   **71.** 0.000053   **73.** 0.0018932
75. 18.42   **77.** 16.07   **79.** 0.0135   **81.** 0.023678   **83.** 0.112   **85.** Est.: 10   **87.** Est.: 1000
Cal.: 11.1632   Cal.: 884.0909

89. Est.: 1.5   **91.** Est.: 50   **93.** Est.: 100   **95.** Est.: 0.0025   **97.** Ramon averaged 6.23 yards per carry.
Cal.: 1.8269   Cal.: 58.8095   Cal.: 72.3053   Cal.: 0.0023

99. The car travels 25.5 miles on 1 gallon of gasoline.   **101.** Three shelves can be cut from the board.   **103.** The amount of the dividend is $1.72.   **105a.** The amount to be paid in monthly payments is $842.58.   **b.** The monthly payment is $46.81.   **107.** The possible answers are 1¢, 3¢, 5¢, 9¢, 15¢, or 45¢.   **109.** Answers will vary.   **111.** ×   **113.** ×
115. ÷   **117.** 2.53

SECTION 3.6

1. 0.625   **3.** 0.667   **5.** 0.167   **7.** 0.417   **9.** 1.750   **11.** 1.500   **13.** 4.000   **15.** 0.003   **17.** 7.080

19. 37.500   **21.** 0.375   **23.** 0.208   **25.** 3.333   **27.** 5.444   **29.** 0.313   **31.** $\frac{4}{5}$ **33.** $\frac{8}{25}$ **35.** $\frac{1}{8}$

37. $1\frac{1}{4}$ **39.** $16\frac{9}{10}$ **41.** $8\frac{2}{5}$ **43.** $8\frac{437}{1000}$ **45.** $2\frac{1}{4}$ **47.** $\frac{23}{150}$ **49.** $\frac{703}{800}$ **51.** $1\frac{17}{25}$ **53.** $\frac{9}{200}$ **55.** $16\frac{18}{25}$

57. $\frac{33}{100}$ **59.** $\frac{1}{3}$ **61.** $0.15 < 0.5$ **63.** $6.65 > 6.56$ **65.** $2.504 > 2.054$ **67.** $\frac{3}{8} > 0.365$ **69.** $\frac{2}{3} > 0.65$

71. $\frac{5}{9} > 0.55$ **73.** $0.62 > \frac{7}{15}$ **75.** $0.161 > \frac{1}{7}$ **77.** $0.86 > 0.855$ **79.** $1.005 > 0.5$ **81a.** False   **b.** False

c. True   **83.** No. 0.0402 rounded to hundredths is 0.04, to thousandths is 0.040.   **85.** Answers will vary.

CHAPTER REVIEW

1. 54.5 [3.5A]   **2.** 833.958 [3.2A]   **3.** $0.055 < 0.1$ [3.6C]   **4.** Twenty-two and ninety-two ten-thousandths [3.1A]

5. 0.05678 [3.1B]   **6.** 2.33 [3.6A]   **7.** $\frac{3}{8}$ [3.6B]   **8.** 36.714 [3.2A]   **9.** 34.025 [3.1A]   **10.** $\frac{5}{8} > 0.62$ [3.6C]

11. 0.778 [3.6A]   **12.** $\frac{2}{3}$ [3.6B]   **13.** 22.8635 [3.3A]   **14.** 7.94 [3.1B]   **15.** 8.932 [3.4A]

16. Three hundred forty-two and thirty-seven hundredths [3.1A]   **17.** 3.06753 [3.1A]   **18.** 25.7446 [3.4A]
19. 6.594 [3.5A]   **20.** 4.8785 [3.3A]   **21.** The new checking account balance is $661.51. [3.3B]
22. The amount of income tax paid was $5600. [3.4B]   **23.** The amount of each monthly payment is $123.45. [3.5B]
24. The checking account balance is $478.02. [3.2B]

CHAPTER TEST

1. $0.66 < 0.666$ [3.6C]   **2.** 4.087 [3.3A]   **3.** Forty-five and three hundred two ten-thousandths [3.1A]

4. 0.692 [3.6A]   **5.** $\frac{33}{40}$ [3.6B]   **6.** 0.0740 [3.1B]   **7.** 1.538 [3.5A]   **8.** 27.76626 [3.3A]   **9.** 7.095 [3.1B]

10. 232 [3.5A]   **11.** 458.581 [3.2A]   **12.** 1.37 inches [3.3B]   **13.** 0.00548 [3.4A]   **14.** 255.957 [3.2A]
15. 209.07086 [3.1A]   **16.** The amount of each monthly payment is $142.85. [3.5B]   **17.** Your total income was
$1543.57. [3.2B]   **18.** The cost of the long-distance call is $4.63. [3.4B]   **19.** The cost of mailing the 55 pieces of mail
was $19.11. [3.4B]   **20.** The cost of mailing the 112 pieces of mail was $50.47. [3.4B]

CUMULATIVE REVIEW

1. 235 r17 [1.5C]   **2.** 128 [1.6A]   **3.** 3 [1.6B]   **4.** 72 [2.1A]   **5.** $4\frac{2}{5}$ [2.2B]   **6.** $\frac{37}{8}$ [2.2B]

7. $\frac{25}{60}$ [2.3A]   **8.** $1\frac{17}{48}$ [2.4B]   **9.** $8\frac{35}{36}$ [2.4C]   **10.** $5\frac{23}{36}$ [2.5C]   **11.** $\frac{1}{12}$ [2.6A]   **12.** $9\frac{1}{8}$ [2.6B]

13. $1\frac{2}{9}$ [2.7A]   **14.** $\frac{19}{20}$ [2.7B]   **15.** $\frac{3}{16}$ [2.8B]   **16.** $2\frac{5}{18}$ [2.8C]   **17.** Sixty-five and three hundred nine

ten-thousandths [3.1A]   **18.** 504.6991 [3.2A]   **19.** 21.0764 [3.3A]   **20.** 55.26066 [3.4A]   **21.** 2.154 [3.5A]

22. 0.733 [3.6A]   **23.** $\frac{1}{6}$ [3.6B]   **24.** $\frac{8}{9} < 0.98$ [3.6C]   **25.** There were 234 passengers on the continuing

flight. [1.3C]   **26.** The value of each share is $32\frac{7}{8}$. [2.5D]   **27.** The checking account balance was $617.38. [3.3B]

28. The resulting thickness of the bushing was 1.395 inches. [3.3B]   **29.** The amount of income tax paid last year was
$6008.80. [3.4B]   **30.** The monthly payment is $23.87. [3.5B]

Answers to Chapter 4 Odd-Numbered Exercises

SECTION 4.1

1. $\frac{1}{5}$ 1:5 1 to 5   **3.** $\frac{2}{1}$ 2:1 2 to 1   **5.** $\frac{3}{8}$ 3:8 3 to 8   **7.** $\frac{37}{24}$ 37:24 37 to 24   **9.** $\frac{1}{1}$ 1:1 1 to 1

11. $\frac{7}{10}$ 7:10 7 to 10   **13.** $\frac{1}{2}$ 1:2 1 to 2   **15.** $\frac{2}{1}$ 2:1 2 to 1   **17.** $\frac{3}{4}$ 3:4 3 to 4   **19.** $\frac{5}{3}$ 5:3 5 to 3

21. $\frac{2}{3}$ 2:3 2 to 3   **23.** $\frac{2}{1}$ 2:1 2 to 1   **25.** The ratio of housing cost to total expenses is $\frac{1}{3}$.   **27.** The ratio of

utilities cost to food cost is $\frac{3}{8}$.   **29.** The ratio of first-year college basketball players to high school senior basketball

players is $\frac{2}{77}$.   **31.** The ratio of the turns in the primary coil to the turns in the secondary coil is $\frac{1}{12}$.   **33a.** The amount

of increase is $20,000.   **b.** The ratio of the increase to the original value is $\frac{2}{9}$.   **35.** The ratio of the increase in price to

the original price is $\frac{5}{16}$.   **37.** No. $\frac{265}{778}$ is greater than $\frac{1}{3}$.   **39.** No. Answers will vary.

SECTION 4.2

1. $\frac{3 \text{ pounds}}{4 \text{ people}}$ **3.** $\frac{\$20}{3 \text{ boards}}$ **5.** $\frac{20 \text{ miles}}{1 \text{ gallon}}$ **7.** $\frac{5 \text{ children}}{2 \text{ families}}$ **9.** $\frac{8 \text{ gallons}}{1 \text{ hour}}$ **11.** 2.5 feet/second   **13.** $325/week
15. 110 trees/acre   **17.** $4.71/hour   **19.** 52.4 miles/hour   **21.** 28 miles/gallon   **23.** $1.65/pound   **25.** The car
was driven 28.4 miles per gallon of gas.   **27.** The rocket uses 213,600 gallons of fuel per minute.   **29a.** The estimated
cost is $40 per share.   **b.** The actual cost is $42.25 per share.   **31a.** There were 4878 compact disks meeting company
standards.   **b.** The cost was $5.44 per disk.   **33.** The profit was $.60 per box of strawberries.   **35.** The pay scale of
$18 per hour gives you the highest yearly income.

SECTION 4.3

1. True   **3.** Not true   **5.** Not true   **7.** True   **9.** True   **11.** Not true   **13.** True   **15.** True   **17.** True
19. Not true   **21.** True   **23.** Not true   **25.** 3   **27.** 6   **29.** 9   **31.** 5.67   **33.** 4   **35.** 4.38   **37.** 88

39. 3.33   **41.** 26.25   **43.** 96   **45.** 9.78   **47.** 3.43   **49.** 1.34   **51.** 50.4   **53.** There are 50 calories in the serving of cereal.   **55.** There were 50 pounds of fertilizer used.   **57.** There were 375 wooden bats produced.   **59.** The distance between the two cities is 16 miles.   **61.** 1.25 ounces of medication are required.   **63.** There would be 160,000 people voting in the election.   **65.** The monthly cost of the life insurance is $176.75.   **67.** You will own 400 shares of stock.   **69.** Students used the computer 300 hours.   **71.** No. $\frac{2}{5} + \frac{3}{4} = \frac{23}{20}$, which is more than the number of voters.   **73.** The length of the door on the car is 3.25 feet.   **75.** Answers will vary.

CHAPTER REVIEW

1. True [4.3A]   **2.** $\frac{2}{5}$ 2:5 2 to 5 [4.1A]   **3.** 62.5 miles/hour [4.2B]   **4.** True [4.3A]   **5.** 68 [4.3B]

6. $7.50/hour [4.2B]   **7.** $1.75/pound [4.2B]   **8.** $\frac{2}{7}$ 2:7 2 to 7 [4.1A]   **9.** 36 [4.3B]   **10.** 19.44 [4.3B]

11. $\frac{2}{5}$ 2:5 2 to 5 [4.1A]   **12.** Not true [4.3A]   **13.** $\frac{\$15}{4\text{ hours}}$ [4.2A]   **14.** 27.2 miles/gallon [4.2B]

15. $\frac{1}{1}$ 1:1 1 to 1 [4.1A]   **16.** True [4.3A]   **17.** 65.45 [4.3B]   **18.** $\frac{100\text{ miles}}{3\text{ hours}}$ [4.2A]   **19.** The ratio of the decrease in price to the original price is $\frac{2}{5}$. [4.1B]   **20.** The property tax is $2400. [4.3C]   **21.** The ratio of the high temperature to the low temperature is 2:1. [4.1B]   **22.** The cost per radio is $37.50. [4.2C]   **23.** It would take 1344 blocks to build the wall. [4.3C]   **24.** The ratio of TV advertising to newspaper advertising is $\frac{5}{2}$. [4.1B]   **25.** The cost is $.68/pound. [4.2C]   **26.** Mahesh drove an average of 56.8 miles/hour. [4.2C]   **27.** The cost of the insurance is $193.50. [4.3C]   **28.** The cost is $44.75 per share. [4.2C]   **29.** There were 22.5 pounds of fertilizer used. [4.3C]

30. The ratio of the increase to the original value is $\frac{1}{2}$. [4.1B]

CHAPTER TEST

1. $1836.40/month [4.2B]   **2.** $\frac{1}{6}$ 1:6 1 to 6 [4.1A]   **3.** $\frac{9\text{ supports}}{4\text{ feet}}$ [4.2A]   **4.** Not true [4.3A]

5. $\frac{3}{2}$ 3:2 3 to 2 [4.1A]   **6.** 144 [4.3B]   **7.** 30.5 miles/gallon [4.2B]   **8.** $\frac{1}{3}$ 1:3 1 to 3 [4.1A]

9. True [4.3A]   **10.** 40.5 [4.3B]   **11.** $\frac{\$27}{4\text{ boards}}$ [4.2A]   **12.** $\frac{3}{5}$ 3:5 3 to 5 [4.1A]   **13.** The dividend would be $625. [4.3C]   **14.** The ratio of the city temperature to the desert temperature is $\frac{43}{56}$. [4.1B]   **15.** The speed is 538 miles/hour. [4.2C]   **16.** The number of pounds of water is 132. [4.3C]   **17.** The per-foot cost is $1.73. [4.2C]   **18.** The amount of medication needed is 0.875 ounce. [4.3C]   **19.** The ratio of the cost of radio advertising to the total cost of advertising is $\frac{8}{13}$. [4.1B]   **20.** The property tax is $1200. [4.3C]

CUMULATIVE REVIEW

1. 9158 [1.3B]   **2.** $2^4 \cdot 3^3$ [1.6A]   **3.** 3 [1.6B]   **4.** $2 \cdot 2 \cdot 2 \cdot 2 \cdot 2 \cdot 5$ [1.7B]   **5.** 36 [2.1A]   **6.** 14 [2.1B]

7. $\frac{5}{8}$ [2.3B]   **8.** $8\frac{3}{10}$ [2.4C]   **9.** $5\frac{11}{18}$ [2.5C]   **10.** $2\frac{5}{6}$ [2.6B]   **11.** $4\frac{2}{3}$ [2.7B]   **12.** $\frac{23}{30}$ [2.8C]

13. Four and seven hundred nine ten-thousandths [3.1A]   **14.** 2.10 [3.1B]   **15.** 1.990 [3.5A]   **16.** $\frac{1}{15}$ [3.6B]

17. $\frac{1}{8}$ [4.1A]   **18.** $\frac{29¢}{2\text{ bars}}$ [4.2A]   **19.** 33.4 miles/gallon [4.2B]   **20.** 4.25 [4.3B]   **21.** 57.2 miles/hour [4.2C]

22. 36 [4.3B]   **23.** The new balance is $744. [1.3C]   **24.** The monthly payment is $370. [1.5D]   **25.** There are 105 pages remaining to be read. [2.6C]   **26.** The cost for each acre is $36,000. [2.7C]   **27.** The amount of change is $19.62. [3.3B]   **28.** The player's batting average is 0.271. [3.5B]   **29.** There will be 25 inches eroded. [4.3C]

30. The person needs 1.6 ounces of medication. [4.3C]

Answers to Chapter 5 Odd-Numbered Exercises

SECTION 5.1

1. $\frac{1}{4}$, 0.25   **3.** $1\frac{3}{10}$, 1.30   **5.** 1, 1.00   **7.** $\frac{73}{100}$, 0.73   **9.** $3\frac{83}{100}$, 3.83   **11.** $\frac{7}{10}$, 0.70   **13.** $\frac{22}{25}$, 0.88

15. $\frac{8}{25}$, 0.32   **17.** $\frac{2}{3}$ **19.** $\frac{5}{6}$ **21.** $\frac{1}{9}$ **23.** $\frac{5}{11}$ **25.** $\frac{3}{70}$ **27.** $\frac{1}{15}$ **29.** 0.065   **31.** 0.0055   **33.** 0.0825

35. 0.0675   **37.** 0.0045   **39.** 0.804   **41.** 16%   **43.** 5%   **45.** 1%   **47.** 70%   **49.** 124%   **51.** 0.4%

53. 0.6%   **55.** 310.6%   **57.** 54%   **59.** 33.3%   **61.** 62.5%   **63.** 16.7%   **65.** 17.5%   **67.** 177.8%

69. 30%   **71.** $23\frac{1}{3}$%   **73.** $237\frac{1}{2}$%   **75.** $216\frac{2}{3}$%   **77a.** False; 4(200%) = 8   **b.** False; $\frac{4}{200\%} = 2$ **c.** True

d. False; 125% = 1.25   **79.** Answers will vary.   **81.** $\frac{1}{2}$ **83.** No; 0.495   **85.** The fraction is less than $\frac{1}{100}$.

SECTION 5.2

1. 8   **3.** 10.8   **5.** 0.075   **7.** 80   **9.** 51.895   **11.** 7.5   **13.** 13   **15.** 3.75   **17.** 20   **19.** 210

21. 5% of 95   **23.** 79% of 16   **25.** 72% of 40   **27.** 25,805.0324   **29.** The invoice cost was $16,750.80.

31. The average number of hours to build a car was 25.2 hours.   **33.** The rebate was $822.50.

35a. The sales tax was $570.   **b.** The total cost was $10,070.   **37.** A total of 671 employees was needed.

39. The actual number of hours spent with family and friends is 48.2 hours.   **41.** The difference is 3.4 hours.

SECTION 5.3

1. 32%   **3.** $16\frac{2}{3}$%   **5.** 200%   **7.** 37.5%   **9.** 18%   **11.** 0.25%   **13.** 20%   **15.** 400%   **17.** 2.5%

19. 37.5%   **21.** 0.25%   **23.** 2.4%   **25.** 9.6%   **27.** Microsoft's revenues were 52.8% of the total revenues.

29. The percent of wasted vegetables was 25.4%.   **31.** Women were 13.7% of the total number of people in the Army.

33a. The speed of the chip was increased by 75 megahertz.   **b.** The speed of the chip was increased by $33\frac{1}{3}$%.

35. 98.5% of the slabs met the safety requirements.   **37.** The percent spent on veterinary care was 26.7%.

39. The cost to refurbish the Statue of Liberty is 3250% of the original cost.   **41.** Answers will vary.

SECTION 5.4

1. 75   **3.** 50   **5.** 100   **7.** 85   **9.** 1200   **11.** 19.2   **13.** 7.5   **15.** 32   **17.** 200   **19.** 80   **21.** 9

23. 504   **25.** 108   **27.** 7122.15   **29.** The average size of a house in 1977 was 1680 square feet.   **31.** The selling price of the car was $16,400.   **33.** The amount spent in 2000 will be $92 billion.   **35a.** The number of computer boards tested was 3000.   **b.** The number of nondefective boards was 2976.   **37.** There is 0.45 milligram of thiamine in one serving.   **39.** No. Answers will vary.

SECTION 5.5

1. 65   **3.** 25%   **5.** 75   **7.** 12.5%   **9.** 400   **11.** 19.5   **13.** 14.8%   **15.** 62.62   **17.** 5   **19.** 45

21. 15   **23.** The club collected $24,500.   **25.** The cost of the calculator 8 years ago was $71.25.

27. Her average golf score per round in 1977 was 73.27.   **29.** The mail order price was 68.3% of the retail price.

31. There were 495 people who worried somewhat about privacy on the Internet.   **33.** The projected percent increase in demand is 3.3%.   **35.** The approximate total land area is 58,750,000 square miles.   **37.** The potency has decreased by 93.75% after 60 hours.

CHAPTER REVIEW

1. 60 [5.2A]   **2.** 20% [5.3A]   **3.** 175% [5.1B]   **4.** 75 [5.4A]   **5.** $\frac{3}{25}$ [5.1A]   **6.** 19.36 [5.2A]

7. 150% [5.3A]   **8.** 504 [5.4A]   **9.** 0.42 [5.1A]   **10.** 5.4 [5.2A]   **11.** 157.5 [5.4A]   **12.** 0.076 [5.1A]

13. 77.5 [5.2A]   **14.** $\frac{1}{6}$ [5.1A]   **15.** 160% [5.5A]   **16.** 75 [5.5A]   **17.** 38% [5.1B]   **18.** 10.9 [5.4A]

19. 7.3% [5.3A]   **20.** 613.3% [5.3A]   **21.** Trent answered 85% of the questions correctly. [5.5B]

22. The company spent $4500 for TV advertising. [5.2B]   **23.** The population increased by 6.25%. [5.3B]

24. The total cost of the camera was $1041.25. [5.2B]   **25.** 78.6% of the women wore sunscreen often. [5.3B]

26. The world's population in 1997 was 5,800,000,000 people. [5.4B]   **27.** The cost of the computer 4 years ago was

$3000. [5.5B]   **28.** The team scored 110 points. [5.4B]

CHAPTER TEST

1. 0.973 [5.1A]   **2.** $\frac{5}{6}$ [5.1A]   **3.** 30% [5.1B]   **4.** 163% [5.1B]   **5.** 150% [5.1B]   **6.** $66\frac{2}{3}$% [5.1B]

7. 50.05 [5.2A]   **8.** 61.36 [5.2A]   **9.** 76% of 13 [5.2A]   **10.** 212% of 12 [5.2A]   **11.** The company spends

$4500 for advertising. [5.2B]   **12.** There were 1170 pounds of vegetables unspoiled. [5.2B]   **13.** One serving provides

14.7% of the daily recommended amount of potassium. [5.3B]   **14.** One serving provides 9.1% of the daily recommended

number of calories. [5.3B]   **15.** The store's temporary employees were 16% of the permanent employees. [5.3B]

16. Conchita answered 91.3% of the questions correctly. [5.3B]   **17.** 80 [5.4A]   **18.** 28.3 [5.4A]   **19.** There were

32,000 transistors tested. [5.4B]   **20.** The increase was 60% of the original price. [5.3B]   **21.** 143.0 [5.5A]

22. 1000% [5.5A]   **23.** The increase in the hourly wage is $1.02. [5.5B]   **24.** The population now is 220% of the

population 10 years ago. [5.5B]   **25.** The value of the car is $6500. [5.5B]

CUMULATIVE REVIEW

1. 4 [1.6B]   **2.** 240 [2.1A]   **3.** $10\frac{11}{24}$ [2.4C]   **4.** $12\frac{41}{48}$ [2.5C]   **5.** $12\frac{4}{7}$ [2.6B]   **6.** $\frac{7}{24}$ [2.7B]

7. $\frac{1}{3}$ [2.8B]   **8.** $\frac{13}{36}$ [2.8C]   **9.** 3.08 [3.1B]   **10.** 1.1196 [3.3A]   **11.** 34.2813 [3.5A]   **12.** 3.625 [3.6A]

13. $1\frac{3}{4}$ [3.6B]   **14.** $\frac{3}{8} < 0.87$ [3.6C]   **15.** 53.3 [4.3B]   **16.** $9.60/hour [4.2B]   **17.** $\frac{11}{60}$ [5.1A]

18. $83\frac{1}{3}$% [5.1B]   **19.** 19.56 [5.2A]   **20.** $133\frac{1}{3}$% [5.3A]   **21.** 9.92 [5.4A]   **22.** 342.9% [5.5A]

23. Sergio's take-home pay is $592. [2.6C]   **24.** The monthly payment is $92.25. [3.5B]   **25.** There were 420 gallons

of gasoline used during the month. [3.5B]   **26.** The real estate tax is $3000. [4.3C]   **27.** The sales tax was 6% of the

purchase price. [5.3B]   **28.** 45% of the people surveyed did not favor the candidate. [5.3B]   **29.** The 1990 value is

250% of the 1985 value. [5.5B]   **30.** 18% of the children tested had levels of lead that exceeded federal standards. [5.5B]

Answers to Chapter 6 Odd-Numbered Exercises

SECTION 6.1

1. The unit cost is $.073 per ounce.   **3.** The unit cost is $.374 per ounce.   **5.** The unit cost is $.040 per tablet.

7. The unit cost is $4.975 per clamp.   **9.** The unit cost is $.153 per ounce.   **11.** The unit cost is $.119 per screw.

13. The more economical purchase is Sutter Home.   **15.** The more economical purchase is La Victoria.   **17.** The more

economical purchase is 400 tablets.   **19.** The more economical purchase is Land to Lake.   **21.** The more economical

purchase is Maxwell House.   **23.** The more economical purchase is Friskies Chef's Blend.   **25.** The total cost is $7.77.

27. The total cost is $1.84.   **29.** The total cost is $6.37.   **31.** The total cost is $1.24.   **33.** The total cost is $5.96.

35. Answers will vary.

SECTION 6.2

1. The percent increase is 20%.   **3.** The percent increase is 62.0%.   **5.** The percent increase is 80%.   **7.** The percent

increase is 87.0%.   **9.** The percent increase is 5852%.   **11.** The markup is $71.25.   **13.** The markup is $24.

15. The markup rate is 30%.   **17a.** The markup is $77.76.   **b.** The selling price is $239.76.   **19a.** The markup is

$25.60.   **b.** The selling price is $57.60.   **21.** The selling price is $227.20.   **23.** The percent decrease is 40%.
25. The percent decrease is 40%.   **27.** The car loses $3360 in value.   **29a.** The amount of the decrease is 12 cameras.
b. The percent decrease is 60%.   **31a.** The amount of the decrease is $15.20.   **b.** The average monthly gasoline bill is
now $60.80.   **33.** The percent decrease is 20%.   **35.** The discount rate is $33\frac{1}{3}$%.   **37.** The discount is $67.50.
39. The discount rate is 25%.   **41a.** The discount is $.17 per pound.   **b.** The sale price is $.68 per pound.   **43a.** The
discount is $4 per gallon.   **b.** The discount rate is 25%.   **45.** Yes.   **47.** No. 50% off the regular price.

SECTION 6.3

1. The simple interest due is $6750.   **3.** The simple interest due is $16.   **5a.** The interest on the loan is $72,000.
b. The monthly payment is $6187.50.   **7a.** The simple interest is $1080.   **b.** The monthly payment is $545.   **9.** The
monthly payment is $1332.50.   **11.** The value will be $1414.78.   **13.** The value will be $12,380.43.   **15.** The value will
be $28,212.   **17a.** The value of the investment will be $6040.86.   **b.** The interest earned on the investment will be
$3040.86.   **19.** Answers will vary.   **21a.** The value after the deposit will be $200.50.   **b.** The value on March 1 will be
$301.50.

SECTION 6.4

1. The mortgage is $82,450.   **3.** The down payment is $7500.   **5.** The down payment was $212,500.   **7.** The loan
origination fee is $3750.   **9a.** The down payment is $7500.   **b.** The mortgage is $142,500.   **11.** The mortgage is
$189,000.   **13.** The monthly mortgage payment is $1157.73.   **15.** No, the couple cannot afford to buy the home.
17. The monthly property tax is $112.35.   **19a.** The monthly mortgage payment is $1678.40.   **b.** The interest is $736.68
during that month.   **21.** The total monthly payment (mortgage and property tax) is $982.70.   **23.** The monthly
mortgage payment is $1430.83.   **25.** $63,408 of interest can be saved.

SECTION 6.5

1. Amanda does not have enough money for the down payment.   **3.** The sales tax was $742.50.   **5.** The license fee is $250.
7a. The sales tax is $420.   **b.** The total cost of the sales tax and the license fee is $595.   **9a.** The down payment is $550.
b. The amount financed is $1650.   **11.** The amount financed is $28,000.   **13.** The monthly truck payment is $348.39.
15. The cost to operate the car is $5120.   **17.** The cost per mile is $.11.   **19.** The interest is $74.75.   **21a.** The
amount financed is $76,600.   **b.** The monthly truck payment is $1590.09.   **23.** The monthly car payment is $709.88.
25. The 10% loan with the application fee has a lower loan cost.

SECTION 6.6

1. Lewis earns $300.   **3.** The commission earned is $3930.   **5.** The commission received was $84.   **7.** Keisha
receives $3244 per month.   **9.** The electrician's overtime hourly wage is $31.60.   **11.** The commission earned is $112.50.
13. The typist earns $393.75.   **15.** The consultant's hourly wage is $85.   **17a.** Mark's hourly wage is $19.35.
b. Mark earns $154.80 on Saturday.   **19a.** The increase in pay is $1.65 per hour.   **b.** The nurse's hourly pay is $18.15.
21. Nicole earned $375.   **23.** The starting salary was $36,281.   **25.** Liberal arts majors received the smaller
amount of increase in starting salary.

SECTION 6.7

1. The new balance is $486.32.   **3.** The balance in the account is $1222.47.   **5.** The current balance is $825.27.
7. The current balance is $3000.82.   **9.** Yes, there is enough money to purchase a refrigerator.   **11.** Yes, there is enough
money to make both purchases.   **13.** The bank statement and checkbook balance.   **15.** The bank statement and
checkbook balance.   **17.** Added   **19.** Subtract

CHAPTER REVIEW

1. The unit cost is 14.5¢ per ounce. [6.1A]   **2.** The cost per mile was 12.1¢. [6.5B]   **3.** The percent increase was
30.4%. [6.2A]   **4.** The markup is $72. [6.2B]   **5.** The simple interest due on the loan is $6750. [6.3A]   **6.** The
value of the investment will be $45,550.75. [6.3B]   **7.** The percent increase was 15%. [6.2A]   **8.** The monthly

payment for mortgage and property tax is $578.53. [6.4B]   **9.** The monthly payment is $330.82. [6.5B]
10. The value of the investment will be $53,593. [6.3B]   **11.** The down payment is $18,750. [6.4A]   **12.** The total of the sales tax and license fee is $1158.75. [6.5A]   **13.** The selling price is $2079. [6.2B]   **14.** The amount of interest is $97.33. [6.5B]   **15.** The commission is $3240. [6.6A]   **16.** The sale price was $141. [6.2D]   **17.** The current balance is $943.68. [6.7A]   **18.** The simple interest due is $1200. [6.3A]   **19.** The loan origination fee is $1875. [6.4A]   **20.** The more economical purchase is 60 ounces for $8.40. [6.1B]   **21.** The monthly mortgage payment is $934.08. [6.4B]   **22.** The total income was $655.20. [6.6A]   **23.** The current checkbook balance is $8866.58. [6.7A]   **24.** The percent increase was 200%. [6.2A]

CHAPTER TEST

1. The cost per foot is $6.92. [6.1A]   **2.** The more economical purchase is 5 pounds for $1.65. [6.1B]
3. The cost is $14.53. [6.1C]   **4.** The percent increase is 20%. [6.2A]   **5.** The percent increase was 150%. [6.2A]
6. The selling price is $301. [6.2B]   **7.** The selling price is $6.25. [6.2B]   **8.** The percent decrease was 7.7%. [6.2C]
9. The percent decrease was 20%. [6.2C]   **10.** The sale price is $209.30. [6.2D]   **11.** The discount rate is 40%. [6.2D]
12. The simple interest due is $2000. [6.3A]   **13.** The interest earned was $24,420.60. [6.3B]   **14.** The loan origination fee is $3350. [6.4A]   **15.** The monthly mortgage payment is $1713.44. [6.4B]   **16.** The amount financed is $19,000. [6.5A]   **17.** The monthly truck payment is $482.68. [6.5B]   **18.** The nurse earns $703.50. [6.6A]
19. The current balance is $6612.25. [6.7A]   **20.** The bank statement and checkbook balance. [6.7B]

CUMULATIVE REVIEW

1. 13 [1.6B]   **2.** $8\frac{13}{24}$ [2.4C]   **3.** $2\frac{37}{48}$ [2.5C]   **4.** 9 [2.6B]   **5.** 2 [2.7B]   **6.** 5 [2.8C]   **7.** 52.2 [3.5A]
8. 1.417 [3.6A]   **9.** $51.25/hour [4.2B]   **10.** 10.94 [4.3B]   **11.** 62.5% [5.1B]   **12.** 27.3 [5.2A]
13. 0.182 [5.1A]   **14.** 42% [5.3A]   **15.** 250 [5.4A]   **16.** 154.76 [5.5A]   **17.** The total rainfall was $13\frac{11}{12}$ inches. [2.4D]   **18.** The amount the family pays in taxes is $570. [2.6C]   **19.** The ratio of decrease in price to original price is $\frac{3}{5}$. [4.1B]   **20.** The car was driven 33.4 miles per gallon. [4.2C]   **21.** The unit cost is $.93/lb. [4.2C]
22. The dividend is $280. [4.3C]   **23.** The sale price is $720. [6.2D]   **24.** The selling price is $119. [6.2B]
25. The percent increase is 8%. [6.2A]   **26.** The simple interest due is $6000. [6.3A]   **27.** The monthly payment is $791.81. [6.5B]   **28.** The new balance is $2243.77. [6.7A]   **29.** The cost per mile is $.20. [6.5B]
30. The monthly mortgage payment is $743.18. [6.4B]

Glossary

addend In addition, one of the numbers added. (Sec. 1.2)

addition The process of finding the total of two numbers. (Sec. 1.2)

Addition Property of Zero Zero added to a number does not change the number. (Sec. 1.2)

Associative Property of Addition Numbers to be added can be grouped (with parentheses, for example) in any order; the sum will be the same. (Sec. 1.2)

Associative Property of Multiplication Numbers to be multiplied can be grouped (with parentheses, for example) in any order; the product will be the same. (Sec. 1.4)

balancing a checkbook Determining if the checking account balance is accurate. (Sec. 6.7)

basic percent equation Percent times base equals amount. (Sec. 5.2)

commission That part of the pay earned by a salesperson that is calculated as a percent of the salesperson's sales. (Sec. 6.6)

common factor A number that is a factor of two or more numbers is a common factor of those numbers. (Sec. 2.1)

common multiple A number that is a multiple of two or more numbers is a common multiple of those numbers. (Sec. 2.1)

Commutative Property of Addition Two numbers can be added in either order; the sum will be the same. (Sec. 1.2)

Commutative Property of Multiplication Two numbers can be multiplied in either order; the product will be the same. (Sec. 1.4)

composite number A number that has whole-number factors besides 1 and itself. For instance, 18 is a composite number. (Sec. 1.7)

compound interest Interest computed not only on the original principal but also on interest already earned. (Sec. 6.3)

cost The price that a business pays for a product. (Sec. 6.2)

decimal A number written in decimal notation. (Sec. 3.1)

decimal notation Notation in which a number consists of a whole-number part, a decimal point, and a decimal part. (Sec. 3.1)

decimal part In decimal notation, that part of the number that appears to the right of the decimal point. (Sec. 3.1)

decimal point In decimal notation, the point that separates the whole-number part from the decimal part. (Sec. 3.1)

denominator The part of a fraction that appears below the fraction bar. (Sec. 2.2)

difference In subtraction, the result of subtracting two numbers. (Sec. 1.3)

discount The difference between the regular price and the sale price. (Sec. 6.2)

discount rate The percent of a product's regular price that is represented by the discount. (Sec. 6.2)

dividend In division, the number into which the divisor is divided to yield the quotient. (Sec. 1.5)

division The process of finding the quotient of two numbers. (Sec. 1.5)

divisor In division, the number that is divided into the dividend to yield the quotient. (Sec. 1.5)

equivalent fractions Equal fractions with different denominators. (Sec. 2.3)

expanded form The number 46,208 can be written in expanded form as $40,000 + 6000 + 200 + 0 + 8$. (Sec. 1.1)

exponent In exponential notation, the raised number that indicates how many times the number to which it is attached is taken as a factor. (Sec. 1.6)

exponential notation The expression of a number to some power, indicated by an exponent. (Sec. 1.6)

factors In multiplication, the numbers that are multiplied. (Sec. 1.4)

fraction The notation used to represent the number of equal parts of a whole. (Sec. 2.2)

fraction bar The bar that separates the numerator of a fraction from the denominator. (Sec. 2.2)

graph of a whole number A heavy dot placed directly above that number on the number line. (Sec. 1.1)

greater than A number that appears to the right of a given number on the number line is greater than that given number. (Sec. 1.1)

greatest common factor The largest common factor of two or more numbers. (Sec. 2.1)

hourly wage Pay calculated on the basis of a certain amount for each hour worked. (Sec. 6.6)

improper fraction A fraction greater than or equal to 1. (Sec. 2.2)

interest The amount of money paid for the privilege of using someone else's money. (Sec. 6.3)

interest rate The percent used to determine the amount of interest. (Sec. 6.3)

inverting a fraction Interchanging the numerator and denominator. (Sec. 2.7)

least common denominator The least common multiple of denominators. (Sec. 2.4)

least common multiple The smallest common multiple of two or more numbers. (Sec. 2.1)

less than A number that appears to the left of a given number on the number line is less than that given number. (Sec. 1.1)

markup The difference between selling price and cost. (Sec. 6.2)

markup rate The percent of a product's selling price that is represented by the markup. (Sec. 6.2)

minuend In subtraction, the number from which another number (the subtrahend) is subtracted. (Sec. 1.3)

mixed number A number greater than 1 that has a whole-number part and a fractional part. (Sec. 2.2)

mortgage The amount borrowed to buy real estate. (Sec. 6.4)

multiples of a number The products of that number and the numbers 1, 2, 3, …. (Sec. 2.1)

multiplication The process of finding the product of two numbers. (Sec. 1.4)

Multiplication Property of One The product of a number and one is the number. (Sec. 1.4)

Multiplication Property of Zero The product of a number and zero is zero. (Sec. 1.4)

number line A line on which a number can be graphed. (Sec. 1.1)

numerator The part of a fraction that appears above the fraction bar. (Sec. 2.2)

Order of Operations Agreement A set of rules that tell us in what order to perform the operations that occur in a numerical expression. (Sec. 1.6)

percent Parts per hundred. (Sec. 5.1)

percent decrease A decrease of a quantity expressed as a percent of its original value. (Sec. 6.2)

percent increase An increase of a quantity expressed as a percent of its original value. (Sec. 6.2)

period In a number written in standard form, each group of digits separated by a comma. (Sec. 1.1)

place value The position of each digit in a number written in standard form determines that digit's place value. (Sec. 1.1)

points A term banks use to mean percent of a mortgage; used to express the loan origination fee. (Sec. 6.4)

prime factorization The expression of a number as the product of its prime factors. (Sec. 1.7)

prime number A number whose only whole-number factors are 1 and itself. For instance, 13 is a prime number. (Sec. 1.7)

principal The amount of money originally deposited or borrowed. (Sec. 6.3)

product In multiplication, the result of multiplying two numbers. (Sec. 1.4)

proper fraction A fraction less than 1. (Sec. 2.2)

proportion An expression of the equality of two ratios or rates. (Sec. 4.3)

quotient In division, the result of dividing the divisor into the dividend. (Sec. 1.5)

rate A comparison of two quantities that have different units. (Sec. 4.2)

rate in simplest form A rate in which the numbers that form the rate have no common factors. (Sec. 4.2)

ratio A comparison of two quantities that have the same units. (Sec. 4.1)

ratio in simplest form A ratio in which the two numbers that form the ratio do not have a common factor. (Sec. 4.1)

reciprocal of a fraction The fraction with the numerator and denominator interchanged. (Sec. 2.7)

remainder In division, the quantity left over when it is not possible to separate objects or numbers into a whole number of equal groups. (Sec. 1.5)

rounding Giving an approximate value of an exact number. (Sec. 1.1)

salary Pay based on a weekly, biweekly, monthly, or annual time schedule. (Sec. 6.6)

sale price The reduced price. (Sec. 6.2)

selling price The price for which a business sells a product to a customer. (Sec. 6.2)

simple interest Interest computed on the original principal. (Sec. 6.3)

simplest form of a fraction A fraction is in simplest form when there are no common factors in the numerator and the denominator. (Sec. 2.3)

standard form A whole number is in standard form when it is written using the digits 0, 1, 2, …, 9. An example is 46,208. (Sec. 1.1)

subtraction The process of finding the difference between two numbers. (Sec. 1.3)

subtrahend In subtraction, the number that is subtracted from another number (the minuend). (Sec. 1.3)

sum In addition, the total of the numbers added. (Sec. 1.2)

true proportion A proportion in which the fractions are equal. (Sec. 4.3)

unit cost The cost of one item. (Sec. 6.1)

unit rate A rate in which the number in the denominator is 1. (Sec. 4.2)

units In the quantity 3 feet, feet are the units in which the measurement is made. (Sec. 4.1)

whole numbers The whole numbers are 0, 1, 2, 3, …. (Sec. 1.1)

whole-number part In decimal notation, that part of the number that appears to the left of the decimal point. (Sec. 3.1)

Index